**Microhydrodynamics, Brownian Motion, and Complex Fluids**

This text is an introduction to the dynamics of fluids at small scales, the physical and mathematical underpinnings of Brownian motion, and the application of these subjects to the dynamics and flow of complex fluids such as colloidal suspensions and polymer solutions. It brings together continuum mechanics, statistical mechanics, polymer and colloid science, and various branches of applied mathematics, in a self-contained and integrated treatment that provides a foundation for understanding complex fluids, with a strong emphasis on fluid dynamics.

Students and researchers will find that this book is extensively cross-referenced to illustrate connections between different aspects of the field. Its focus on fundamental principles and theoretical approaches provides the necessary groundwork for research in the dynamics of flowing complex fluids.

MICHAEL D. GRAHAM is the Vilas Distinguished Achievement Professor and Harvey D. Spangler Professor of Chemical and Biological Engineering at the University of Wisconsin–Madison. His research focuses on theoretical and computational studies of the fluid dynamics of complex fluids. Among his recognitions are a CAREER Award from the National Science Foundation (NSF), the François Frenkiel and Stanley Corrsin Awards from the American Physical Society Division of Fluid Dynamics, and the Kellett Mid-Career Award at UW–Madison. He has served as associate editor of the *Journal of Fluid Mechanics* and editor-in-chief of the *Journal of Non-Newtonian Fluid Mechanics*. He is coauthor of the textbook *Modeling and Analysis Principles for Chemical and Biological Engineers*.

Cambridge Texts in Applied Mathematics

All titles listed below can be obtained from good booksellers or from Cambridge University Press. For a complete series listing, visit www.cambridge.org/mathematics.

*Nonlinear Dispersive Waves*
MARK J. ABLOWITZ

*Flow, Deformation and Fracture*
G. I. BARENBLATT

*Geometric and Topological Inference*
JEAN-DANIEL BOISSONNAT, FRÉDÉRIC CHAZAL & MARIETTE YVINEC

*The Mathematics of Signal Processing*
STEVEN B. DAMELIN & WILLARD MILLER, JR

*Introduction to Magnetohydrodynamics (2nd Edition)*
P. A. DAVIDSON

*An Introduction to Stochastic Dynamics*
JINQIAO DUAN

*Singularities: Formation, Structure and Propagation*
J. EGGERS & M. A. FONTELOS

*A Physical Introduction to Suspension Dynamics*
ÉLISABETH GUAZZELLI & JEFFREY F. MORRIS

*Discrete Systems and Integrability*
J. HIETARINTA, N. JOSHI & F. W. NIJHOFF

*An Introduction to Polynomial and Semi-Algebraic Optimization*
JEAN BERNARD LASSERRE

*An Introduction to Computational Stochastic PDEs*
GABRIEL J. LORD, CATHERINE E. POWELL & TONY SHARDLOW

*Numerical Linear Algebra*
HOLGER WENDLAND

# Microhydrodynamics, Brownian Motion, and Complex Fluids

MICHAEL D. GRAHAM
*University of Wisconsin–Madison*

CAMBRIDGE
UNIVERSITY PRESS

CAMBRIDGE
UNIVERSITY PRESS

Shaftesbury Road, Cambridge CB2 8EA, United Kingdom

One Liberty Plaza, 20th Floor, New York, NY 10006, USA

477 Williamstown Road, Port Melbourne, VIC 3207, Australia

314–321, 3rd Floor, Plot 3, Splendor Forum, Jasola District Centre, New Delhi – 110025, India

103 Penang Road, #05–06/07, Visioncrest Commercial, Singapore 238467

Cambridge University Press is part of Cambridge University Press & Assessment, a department of the University of Cambridge.

We share the University's mission to contribute to society through the pursuit of education, learning and research at the highest international levels of excellence.

www.cambridge.org
Information on this title: www.cambridge.org/9781107024649

DOI: 10.1017/9781139175876



First published 2018

*A catalogue record for this publication is available from the British Library*

ISBN 978-1-107-02464-9 Hardback
ISBN 978-1-107-69593-1 Paperback

*To Mary Ellen, Nathaniel and Jacob*

# Contents

# Preface

This book is an introduction to the dynamics of fluids at small scales, the physical and mathematical underpinnings of the phenomenon of Brownian motion, and the application of these two subjects to gaining a predictive understanding of the dynamics and flow behavior of complex fluids such as colloidal suspensions and polymer solutions. Even the limited class of complex fluids that we are able to consider here (we do not touch on liquid crystals, polymer melts, self-assembling fluids, or many other interesting soft matter systems) arises in an enormous range of settings from the dynamics of the cytoskeleton to the rheology of suspensions of swimming bacteria to the response of polymer solutions to turbulent flow. My aim is not to address these applications, but to focus on common principles and theoretical approaches that will help students and researchers to fundamentally understand complex fluids in a variety of contexts.

Understanding flowing complex fluids requires knowledge from a number of areas, including fluid dynamics, continuum mechanics, statistical mechanics, polymer and colloid physics, and various branches of applied mathematics. A primary goal of this book is to bring these topics together in a reasonably self-contained and integrated treatment that allows the reader to see these topics in the broader context of the field. To illustrate connections between different aspects of the field, the book is extensively cross-referenced. Many results are not presented in full generality – formalism can obscure key issues for the beginning student of the subject. Many other works document detailed formalisms, so I have tried to stick to essentials. The reader is strongly encouraged to peruse the exercises at the end of each chapter. These contain many illustrative examples as well as elaborations and extensions of material covered in the main text.

In writing this book, I envisioned its audience as (1) graduate students in engineering, applied mathematics, physics, or physical chemistry setting out to do thesis research involving microhydrodynamics, complex fluids, and/or their applications, and (2) researchers and practitioners interested in strengthening their understanding of some of the fundamental underpinnings of the field. I have assumed that the reader has had a semester of upper level or graduate fluid dynamics, transport phenomena, or applied partial differential equations as well as some exposure to the basic principles of equilibrium statistical mechanics. Basic knowledge of linear algebra, Fourier and Laplace transform methods, asymptotic analysis, and familiarity with index notation for vectors and tensors is also assumed. The book is structured to form the basis of a one-semester course for advanced graduate students (though not all of the book can be adequately treated in one semester). When I teach this course, students are expected to work homework problems as well as completing a project that includes development and simulation of a mathematical model of a process or phenomenon involving elements of the course material. Topics of projects have included locomotion in porous media, polymer dynamics in flow, chaotic dynamics in sedimentation, thermal fluctuations of membranes, and simulation of the fluctuating Navier–Stokes equations.

The book begins with a brief overview of basic continuum mechanics principles, including kinematics of deforming continua and mass and momentum balances. Then the Stokes flow regime, in which viscous effects are dominant, is described in detail. Fundamental solutions to the Stokes equations are introduced and their connections to the viscosity of a suspension and flow in a porous medium are described. The concepts of hydrodynamic interactions between particles in a fluid and hydrodynamic screening are introduced. Flows resulting from motions of spheres in a fluid are described, including a model of a self-propelled particle such as a microorganism. Dynamics of nonspherical particles are also described, and an important computational method for particles in Stokes flow, Stokesian Dynamics, is introduced. Given that complex fluids are often found in confined geometries both in nature and technology, confinement effects, including wall-induced migration and hydrodynamic screening phenomena, are introduced, and the fundamental equations of the boundary integral method are derived. Then the first effects of inertia on viscous flows are presented, including inertial migration and transient drag forces on a moving sphere.

With the aforementioned material as background, we then introduce, from several points of view, the phenomena of thermal fluctuations and Brownian motion. The classical Langevin equation for a Brownian particle is developed, as is the description of how the Navier–Stokes equations themselves are modified upon consideration of thermal fluctuations in a fluid. With these results in hand, Brownian diffusion is described, as well as an elementary computational method for simulating trajectories of Brownian particles. Turning from a stochastic to a statistical mechanical point of view, we introduce the concept of entropic driving forces. With the physical basis of Brownian motion established, the next topic is an introduction to the mathematical foundations of stochastic processes, emphasizing the Wiener process and stochastic differential equations. This section establishes the mathematical basis for understanding simulation methods for random walk processes.

The equilibrium conformation of a flexible linear polymer is a random walk parametrized not by time but by position along the backbone of the molecule. This observation begins a discussion of coarse-grained models of polymers in solution and their dynamic properties, including diffusion, relaxation, and stress. Evolution equations for stress are developed in some detail for the important special cases of bead-spring dumbbell and rigid rod models of polymers. These models provide context for a development of the principles of linear viscoelasticity; a relatively recent application of this theory, particle-tracking microrheology, is then introduced. This methodology allows the dynamic properties of a viscoelastic material to be inferred from observations of a particle undergoing Brownian motion within it. Then the basic features of nonlinear flow of viscoelastic fluids are described; of particular note are discussions of viscoelasticity-driven flow instabilities as well as scaling arguments for the rheology of polymer solutions at high shear or extension rates. Although the focus of the book is on micromechanical models of complex fluids, the phenomenological concept of material frame indifference is important and is described with regard to its relation to such models.

In an appendix, some mathematical results are reviewed, beginning with definitions and theorems from vector calculus and Fourier and Laplace transforms. The reader is assumed to have some familiarity with these subjects, so the sections covering them are brief. These are followed with more substantial treatments of the delta function and of probability theory and time-correlation functions, including the important central limit and Wiener–Khinchin theorems.

## Acknowledgments

This text began as a set of notes for a short course that Ole Hassager invited Juan de Pablo and myself to teach at the Danish Technical University (DTU) in summer 2006. I am grateful to Ole, Juan, and the DTU for this opportunity. I am also grateful to all of the students and postdocs who have worked with me in this area; I learned much of the material in this book during the course of working with them. Special thanks goes to Juan Hernández-Ortiz, who cotaught a short course on some of this material with me at the 2009 Society of Rheology Annual Meeting. My interest in this field was first piqued while I was a graduate student at Cornell, and I am forever indebted to my teachers there, especially Paul Steen and Don Koch, for providing an intellectual foundation that has served me well over many years.

I have also learned a great deal from many colleagues. Regarding the topics in this book, I have particularly benefited from discussions with Patrick Anderson, Bob Bird, John Brady, Yeng-Long Chen, Juan de Pablo, Aleks Donev, John Hinch, Christel Hohenegger, Bud Homsy, Martien Hulsen, Eric Keaveny, Bamin Khomami, Sangtae Kim, Dan Klingenberg, Ron Larson, Gary Leal, Gareth McKinley, Jeff Morris, Susan Muller, Rob Poole, Michael Renardy, Jay Schieber, Charles Schroeder, Eric Shaqfeh, Saverio Spagnolie, and Patrick Underhill. Several of the aforementioned were kind enough to review parts of the book; I owe them my particular thanks.

For aid in preparation of the figures, I am grateful to Alec Linot, Frank Nguyen, Ashwin Shekar, Yijiang Yu, and Xiao Zhang.

Sometimes a change of academic scenery is conducive to the writing process. The Kavli Institute for Theoretical Physics at the University of California–Santa Barbara provided opportunities for me to visit and work on this project in 2012 and 2017. Much of the final writing and rewriting was done in spring 2017 at the Department of Applied Mathematics and Theoretical Physics at Cambridge University and the Materials Technology group at the Technical University of Eindhoven. I am thankful to Grae Worster and Patrick Anderson for making these productive and memorable visits possible and for their generous hospitality during them.

Finally, it has been my great privilege to reside in a department and university that for many decades has had a culture that values book writing. In particular, Bob Bird and his coauthors over the years have set an inspiring standard for scholarly writing in transport phenomena and related topics, and Jim Rawlings (my coauthor on a separate book project) and Cheryl Rawlings (my publisher on that project) have given me great training in the art of authorship. The department and university have also provided financial resources that have given me breathing room from the usual academic pressures to pursue this project. I am grateful for support from the Hougen and Spangler endowments to the Department of Chemical and Biological Engineering and the Vilas and Kellett endowments to the University of Wisconsin–Madison.

MDG
Madison, Wisconsin

# 1 Kinematics, Balance Equations, and Principles of Stokes Flow

The focus of this book is the dynamics of fluids at small scales and of small objects (e.g., particles, cells, macromolecules) suspended in fluids. As we will see, such suspensions or solutions can have nontrivial dynamical and rheological behavior: i.e., in this regard they are *complex fluids*. Small is relative, of course, and we exclusively consider systems that are not so small that atomistic details of the fluid or objects are important; in particular, we treat the fluid as a continuum to which we can assign properties at every spatial position $\boldsymbol{x}$. For liquids, the continuum approximation is broadly valid for scales of about 1 nanometer (nm) and larger (a water molecule has a size of about 0.2 nm). One way that we know this is through molecular simulations (Schmidt & Skinner 2003, 2004), which show agreement with, for example, the continuum prediction for the drag force on a moving sphere even when the sphere is only several solvent atoms across. In the first several chapters of this book, we will only concern ourselves with the behavior of fluids, and particles within fluids, in the absence of thermal fluctuations. This behavior is governed by the classical equations of continuum mechanics, which are the starting point of the chapter. After reviewing these here, the governing equations for Newtonian fluids are introduced. Our ultimate focus is the Stokes equation, which governs fluid motions when the inertia of the fluid is negligible compared to viscous stresses.

## 1.1 Kinematics of Continua

### 1.1.1 Velocity Fields and the Velocity Gradient

Under the continuum approximation, a material can take on a velocity $\boldsymbol{v}$ (which may vary with time $t$) at every position $\boldsymbol{x}$: i.e., we have a velocity field $\boldsymbol{v}(\boldsymbol{x}, t)$. We further

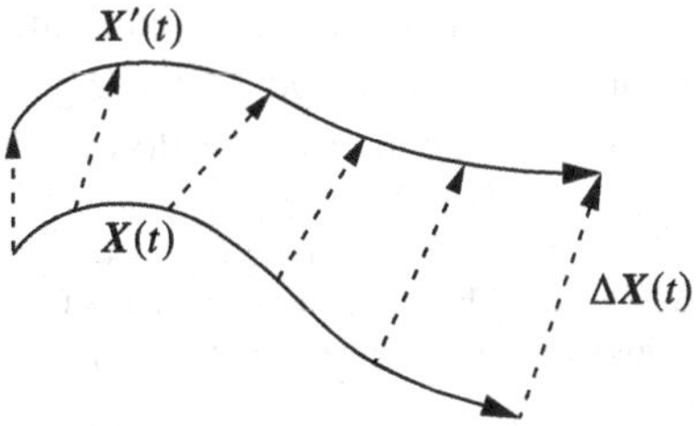


**Figure 1.1** Motions of material points $\boldsymbol{X}(t)$ and $\boldsymbol{X}'(t)$, with the material line $\Delta\boldsymbol{X}(t) = \boldsymbol{X}'(t) - \boldsymbol{X}(t)$ (dashed) shown at several time instants.

assume that this field is differentiable except possibly at interfaces between different materials or phases. Now, consider two neighboring *material points* $\boldsymbol{X}$ and $\boldsymbol{X}'$, which by definition move with the instantaneous velocity of the material at their respective positions, as illustrated in Figure 1.1. Thus

$$\frac{d\boldsymbol{X}}{dt} = \boldsymbol{v}(\boldsymbol{X}, t), \quad \frac{d\boldsymbol{X}'}{dt} = \boldsymbol{v}(\boldsymbol{X}', t).$$

To understand how the material deforms as it flows with velocity $\boldsymbol{v}(\boldsymbol{x}, t)$, we consider the time evolution of a "material line" $\Delta\boldsymbol{X} = \boldsymbol{X}' - \boldsymbol{X}$ connecting points $\boldsymbol{X}$ and $\boldsymbol{X}'$. We can then write

$$\frac{d\Delta\boldsymbol{X}}{dt} = \boldsymbol{v}(\boldsymbol{X}', t) - \boldsymbol{v}(\boldsymbol{X}, t).$$

Now we take the distance between $\boldsymbol{X}$ and $\boldsymbol{X}'$ to be small, so that

$$\boldsymbol{v}(\boldsymbol{X}', t) = \boldsymbol{v}(\boldsymbol{X}, t) + \Delta\boldsymbol{X} \cdot \boldsymbol{\nabla}\boldsymbol{v}(\boldsymbol{X}, t) + O(|\Delta\boldsymbol{X}|^2),$$

where $\boldsymbol{\nabla}\boldsymbol{v}(\boldsymbol{X}, t)$ is the *velocity gradient* evaluated at the material point $\boldsymbol{X}$. Combining these two equations yields that

$$\frac{d\Delta\boldsymbol{X}}{dt} = \Delta\boldsymbol{X} \cdot \boldsymbol{\nabla}\boldsymbol{v}, \tag{1.1}$$

or equivalently

$$\frac{d\Delta\boldsymbol{X}}{dt} = \mathbf{L} \cdot \Delta\boldsymbol{X}, \tag{1.2}$$

where $\mathbf{L} = \boldsymbol{\nabla}\boldsymbol{v}^{\mathrm{T}}$. Therefore, all information about the deformation of an infinitesimal line connecting two neighboring points in the fluid, is contained in the velocity gradient tensor.[1] Note that we use the convention[2] that in Cartesian coordinates

$$(\boldsymbol{\nabla}\boldsymbol{v})_{ij} = \frac{\partial}{\partial x_i} v_j.$$

Flows in which the velocity gradient is independent of position are called *linear flows*, because the velocity is a linear function of position:

$$\boldsymbol{v}(\boldsymbol{x}) - \boldsymbol{v}_0 = \mathbf{L} \cdot (\boldsymbol{x} - \boldsymbol{x}_0), \tag{1.3}$$

where $\boldsymbol{v}_0$ is a constant uniform velocity and $\boldsymbol{x}_0$ is a constant position. By appropriate choice of reference frame, we can always take $\boldsymbol{v}_0$ and $\boldsymbol{x}_0$ to be zero. If in addition $\boldsymbol{\nabla}\boldsymbol{v}$ is independent of position and time, then (1.1) is a simple linear constant coefficient equation and the evolution of $\Delta\boldsymbol{X}$ is completely determined by the eigenvalues and eigenvectors of $\boldsymbol{\nabla}\boldsymbol{v}$. We will exclusively consider incompressible flows, which satisfy

[1] See Section A.1 for a brief summary of vector and tensor notation as used in this book.

[2] This convention is common in the fluid mechanics literature but is not universal. In much of the continuum mechanics literature, e.g. Malvern (1969), Gonzalez & Stuart (2008), $(\boldsymbol{\nabla}\boldsymbol{v})_{ij}$ is defined as $\frac{\partial v_i}{\partial x_j}$. This is the transpose of what we use.

$\nabla \cdot \boldsymbol{v} = 0$ due to mass conservation (see Section 1.2.1), so the eigenvalues $\lambda_i$ of $\nabla\boldsymbol{v}$ must sum to zero. Equivalently,

$$\mathrm{tr}\,\nabla\boldsymbol{v} = \nabla \cdot \boldsymbol{v} = \sum_{i=1}^{3} \lambda_i = 0.$$

We will often make use of the decomposition of the velocity gradient tensor into the symmetric *strain rate* or *deformation rate* tensor $\mathbf{E}$ and antisymmetric *vorticity* tensor[3] $\mathbf{W}$:

$$\nabla\boldsymbol{v} = \mathbf{E} + \mathbf{W},$$

where

$$\mathbf{E} = \frac{1}{2}\left(\nabla\boldsymbol{v} + \nabla\boldsymbol{v}^{\mathrm{T}}\right)$$

and

$$\mathbf{W} = \frac{1}{2}\left(\nabla\boldsymbol{v} - \nabla\boldsymbol{v}^{\mathrm{T}}\right).$$

The vorticity *vector* $\boldsymbol{w}$ is given by

$$\boldsymbol{w} = \nabla \times \boldsymbol{v} \tag{1.4}$$

and is related to the local angular velocity of the fluid, $\omega$, by the simple expression

$$\omega = \frac{1}{2}\boldsymbol{w}. \tag{1.5}$$

These quantities are related to the vorticity tensor as follows:

$$\mathbf{W} = \boldsymbol{\epsilon} \cdot \omega = \frac{1}{2}\boldsymbol{\epsilon} \cdot \boldsymbol{w}, \tag{1.6}$$

where $\boldsymbol{\epsilon}$ is the Levi–Civita symbol, whose properties are summarized in Appendix A.1.2.

If $\mathbf{E} = \mathbf{0}$ at some point in the fluid, then an infinitesimal volume of material at that point, a *fluid element*, is undergoing rigid rotation: there is no stretching of material lines within that volume. On the other hand, if $\mathbf{W} = \mathbf{0}$, then the fluid element is undergoing stretching without any rotation. Now $\nabla\boldsymbol{v}$ is symmetric, so its eigenvectors are orthogonal, forming a coordinate system in which $\nabla\boldsymbol{v}$ can be written

$$\nabla\boldsymbol{v} = \begin{bmatrix} \lambda_1 & 0 & 0 \\ 0 & \lambda_2 & 0 \\ 0 & 0 & \lambda_3 \end{bmatrix}.$$

Defining the extension rate as $\dot{\epsilon}$ ($> 0$), important special cases include the following:

- Uniaxial extension: $\lambda_1 = \dot{\epsilon}, \lambda_2 = \lambda_3 = -\dot{\epsilon}/2$,
- Biaxial extension: $\lambda_1 = \lambda_2 = \dot{\epsilon}, \lambda_3 = -2\dot{\epsilon}$,
- Planar extension: $\lambda_1 = -\lambda_2 = \dot{\epsilon}, \lambda_3 = 0$.

[3] Again, the convention in the fluid mechanics literature is different than in the continuum mechanics literature, where $\mathbf{W} = (\mathbf{L} - \mathbf{L}^{\mathrm{T}})/2$ and is often called the *spin* tensor (Malvern, 1969).

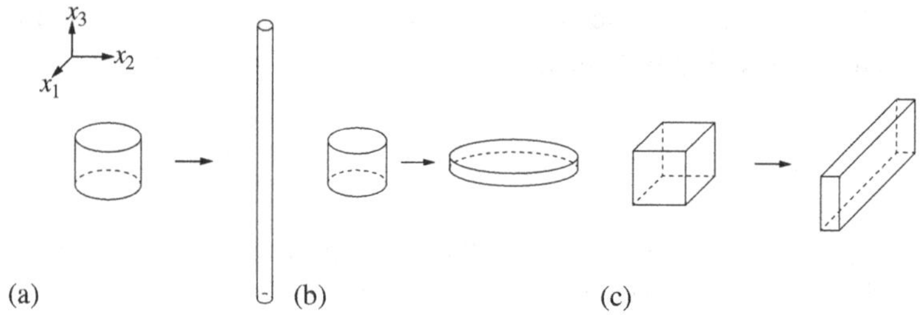


**Figure 1.2** Deformation of a fluid element in (a) uniaxial extension, (b) biaxial extension, and (c) planar extension.

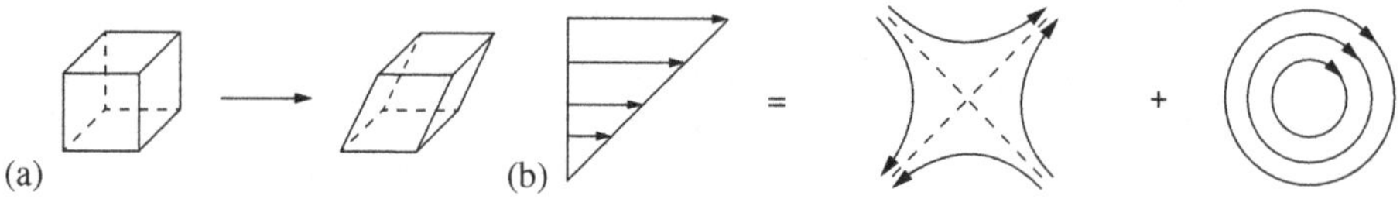


**Figure 1.3** (a) Deformation of a volume of fluid in simple shear. (b) Decomposition of the velocity field of simple shear into equal parts of planar extension and rigid rotation.

In these cases, material lines will either shrink or grow exponentially in time, depending on their initial orientation relative to the eigenvectors of $\nabla \boldsymbol{v}$. Figure 1.2 illustrates how fluid elements evolve in these flows.

Simple shear flow, where $\boldsymbol{v} = \dot{\gamma} y \boldsymbol{e}_x$, is another important special case that deserves particular attention. In this case, illustrated in Figure 1.3(a), elements simply move in the $x$-direction in a straight line, the eigenvalues of $\nabla \boldsymbol{v}$ are all zero, and an arbitrarily oriented material line stretches linearly with time. In Cartesian coordinates,

$$\nabla \boldsymbol{v} = \begin{bmatrix} 0 & 0 & 0 \\ \dot{\gamma} & 0 & 0 \\ 0 & 0 & 0 \end{bmatrix}$$

and thus

$$\mathbf{E} = \frac{1}{2} \begin{bmatrix} 0 & \dot{\gamma} & 0 \\ \dot{\gamma} & 0 & 0 \\ 0 & 0 & 0 \end{bmatrix}$$

and

$$\mathbf{W} = \frac{1}{2} \begin{bmatrix} 0 & -\dot{\gamma} & 0 \\ \dot{\gamma} & 0 & 0 \\ 0 & 0 & 0 \end{bmatrix}.$$

From (1.4) and (1.5),

$$\omega = \begin{bmatrix} 0 \\ 0 \\ -\frac{1}{2}\dot{\gamma} \end{bmatrix}.$$

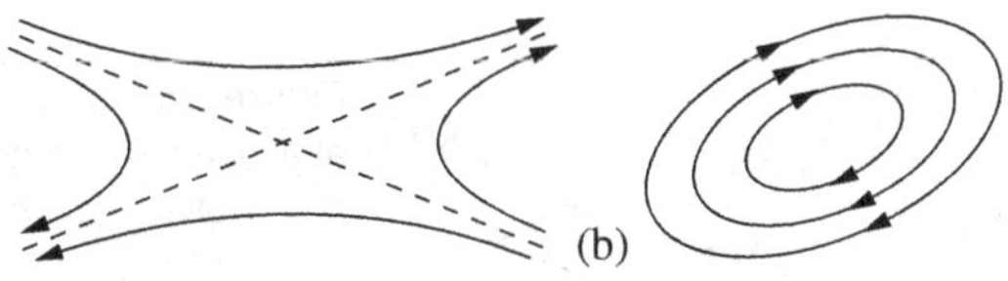


**Figure 1.4** Paths of fluid elements for perturbations of simple shear with (a) $\alpha > 0$ and (b) $\alpha < 0$. The case $\alpha = 0$ is simple shear as shown in Figure 1.3.

Letting $||\cdot||$ denote the Frobenius norm when applied to a second-order tensor,[4] observe that in this case $||\mathbf{E}|| = ||\mathbf{W}||$. If the velocity gradient were composed entirely of $\mathbf{E}$, then fluid elements would undergo planar extension, compressing along the line $x = -y$, the "compressional axis" and stretching along the line $x = y$, the "extensional axis." On the other hand, if the velocity gradient were composed entirely of $\mathbf{W}$, then fluid elements would just rotate clockwise. The overall deformation that occurs during simple shear is an equal superposition of these deformations, as illustrated in Figure 1.3(b), in which the stretching due to $\mathbf{E}$ is just balanced by the rotation due to $\mathbf{W}$, and fluid elements stretch linearly in time, tilting toward the flow direction but without "tumbling." This result is a simple consequence of the fact that the particle paths are all straight lines.

Consider now a small change in the velocity gradient from the simple shear case (Fuller & Leal 1981):

$$\nabla \boldsymbol{v} = \dot{\gamma} \begin{bmatrix} 0 & \alpha & 0 \\ 1 & 0 & 0 \\ 0 & 0 & 0 \end{bmatrix}. \tag{1.7}$$

Now

$$\mathbf{E} = \frac{1}{2}\dot{\gamma} \begin{bmatrix} 0 & 1+\alpha & 0 \\ 1+\alpha & 0 & 0 \\ 0 & 0 & 0 \end{bmatrix}$$

and

$$\mathbf{W} = \frac{1}{2}\dot{\gamma} \begin{bmatrix} 0 & \alpha-1 & 0 \\ 1-\alpha & 0 & 0 \\ 0 & 0 & 0 \end{bmatrix}.$$

Holding $\dot{\gamma}$ constant and restricting $\alpha$ to the range $-1 \leq \alpha \leq 1$, there are three cases as shown in Figure 1.4. If $\alpha = 0$, then the flow is simple shear. If $\alpha > 0$, then strain dominates over vorticity, $||\mathbf{E}|| > ||\mathbf{W}||$, the eigenvalues of $\nabla \boldsymbol{v}$ are real, particle paths are hyperbolas, and a material line stretches exponentially fast – in the limiting case $\alpha = 1$, the vorticity vanishes, and flow is pure planar extension, with compressional axis along the line $x = -y$ and extensional axis along $x = y$. On the other hand, if $\alpha < 0$, vorticity dominates, $||\mathbf{E}|| < ||\mathbf{W}||$, and particle paths are ellipses – an individual fluid element will oscillate in length. Accordingly, $\nabla \boldsymbol{v}$ has a pair of purely imaginary eigenvalues. (There is always one zero eigenvalue, independent of $\alpha$, because there is no motion in the $z$ direction.) When $\alpha = -1$, the strain rate vanishes and the flow is rigid rotation – particle paths are circles and material lines rotate without any change in length. Thus

[4] See Appendix A.1.

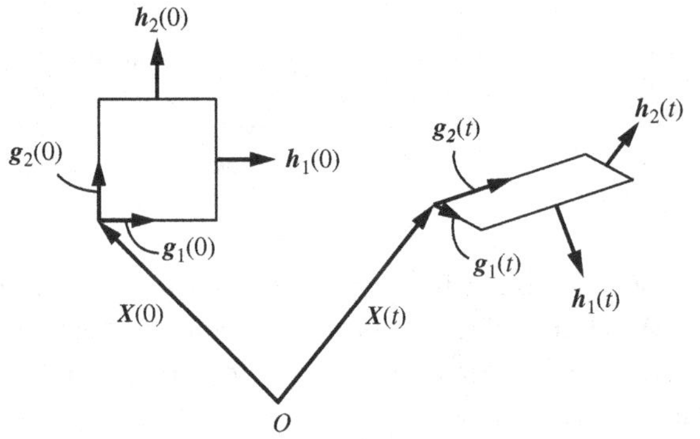


**Figure 1.5** Basis vectors $\boldsymbol{g}_i(t)$ attached to a material point $\boldsymbol{X}(t)$ and evolving as material lines. At $t = 0$ (left) these vectors define the edges of a cube, only two dimensions of which are shown, that is deformed at time $t$ to a parallelepiped. The reciprocal basis vectors $\boldsymbol{h}_i(t)$ define the faces of the parallelepiped.

we see that simple shear is a delicate special case, a point that should be remembered because it is so often used as a model flow.

### 1.1.2 Deformation Tensors

For time-independent linear flows, the eigenvalues and eigenvectors of $\nabla \boldsymbol{v}$ contain all the information needed to determine how material lines stretch and thus how the material deforms. For more complex flow fields, however, we require a more general formalism to characterize deformations (Malvern 1969, Bird, Armstrong, & Hassager 1987, Segel 1987, Gonzalez & Stuart 2008, Morozov & Spagnolie 2015). Consider a set of basis vectors $\boldsymbol{g}_i$ that are attached to a material element $\boldsymbol{X}(t)$. At $t = 0$, the basis vectors correspond to the Cartesian basis vectors: $\boldsymbol{g}_i(\boldsymbol{X}(0), 0) = \boldsymbol{e}_i$, where $\boldsymbol{e}_i$ is the $i$th Cartesian unit basis vector. Thus the $\boldsymbol{g}_i$ are the edges of the parallelepiped that begins as cubic material volume attached to $\boldsymbol{X}(t)$. These basis vectors will be taken to evolve as material lines, as shown in Figure 1.5: i.e.,

$$\frac{d\boldsymbol{g}_i(\boldsymbol{X}(t), t)}{dt} = \mathbf{L}(\boldsymbol{X}(t), t) \cdot \boldsymbol{g}_i(\boldsymbol{X}(t), t). \tag{1.8}$$

Note that $\mathbf{L}$ is time dependent in general. These vectors form a basis that is *codeforming* with the material. Recall that at $t = 0$, the $\boldsymbol{g}_i$ are tangent to the Cartesian coordinate lines. If we take these coordinate lines to be embedded in the material, moving and deforming with it, then because the $\boldsymbol{g}_i$ evolve as material lines, they will each be tangent to the corresponding coordinate line at time $t$; basis vectors that are parallel to coordinate lines are said to be *contravariant* (Aris 1989).

The solution to (1.8) can be written as

$$\boldsymbol{g}_i(\boldsymbol{X}(t), t) = \mathbf{F}(\boldsymbol{X}(t), t) \cdot \boldsymbol{g}_i(\boldsymbol{X}(0), 0), \tag{1.9}$$

where $\mathbf{F}(\boldsymbol{X}(t), t)$ is simply the time-dependent mapping between $\boldsymbol{g}_i(0)$ and $\boldsymbol{g}_i(t)$ for material point $\boldsymbol{X}$. It is called the *deformation gradient tensor*. (For brevity of notation, we occasionally drop the dependence of the $\boldsymbol{g}_i$ on $\boldsymbol{X}$.) Since $\boldsymbol{g}_i(0) = \boldsymbol{e}_i$, we have that

$$\boldsymbol{g}_i(t) = \mathbf{F}(\boldsymbol{X}(t), t) \cdot \boldsymbol{e}_i \tag{1.10}$$

or equivalently, the $i$th column of $\mathbf{F}(\boldsymbol{X}(t),t)$ is the basis vector $\boldsymbol{g}_i(t)$: $F_{ji}(\boldsymbol{X}(t),t) = g_{ji}(\boldsymbol{X}(t),t)$. Inserting (1.10) into (1.8) and factoring out $\boldsymbol{e}_i$ yields that

$$\frac{d\mathbf{F}}{dt} = \mathbf{L}\cdot\mathbf{F} \tag{1.11}$$

with initial condition $\mathbf{F}(\boldsymbol{X}(0),0) = \boldsymbol{\delta}$. In the special case of a steady linear flow, where $\mathbf{L}$ is constant, $\mathbf{F} = e^{\mathbf{L}t}$ and the $\boldsymbol{g}_i$ are independent of position.

Because the $\boldsymbol{g}_i$ evolve as material lines, Equation (1.9) also holds with $\boldsymbol{g}_i$ replaced with $\Delta\boldsymbol{X}$:

$$\Delta\boldsymbol{X}(t) = \mathbf{F}(\boldsymbol{X}(t),t)\cdot\Delta\boldsymbol{X}(0). \tag{1.12}$$

Thus the deformation gradient tensor $\mathbf{F}$ contains all information needed to determine the evolution of a material line during a deformation.[5] Recalling that the $i$th column of $\mathbf{F}(\boldsymbol{X}(t),t)$ is the basis vector $\boldsymbol{g}_i(t)$, we can rewrite (1.12) in the coordinate system defined by these vectors:

$$\Delta\boldsymbol{X}(t) = \Delta X_1(0)\boldsymbol{g}_1(t) + \Delta X_2(0)\boldsymbol{g}_2(t) + \Delta X_3(0)\boldsymbol{g}_3(t). \tag{1.13}$$

For $t > 0$ the basis vectors $\boldsymbol{g}_i(t)$ are not generally orthogonal. However, it is always possible to find a *reciprocal basis* comprised of vectors $\boldsymbol{h}_j(t)$ that satisfy the so-called *biorthogonality* condition $\boldsymbol{g}_i(t)\cdot\boldsymbol{h}_j(t) = \delta_{ij}$. For the parallelepiped whose edges are defined by the $\boldsymbol{g}_i$, the normal vectors to the faces of the parallelepiped are defined by the $\boldsymbol{h}_i$, as shown in Figure 1.5. Now thinking of coordinate planes that are embedded in the material and evolve with it, the $\boldsymbol{h}_i$ are basis vectors that are orthogonal to these planes and are said to be *covariant*. Since the basis vectors $\boldsymbol{g}_i(t)$ are contained in the columns of the deformation gradient tensor $\mathbf{F}(t)$, the columns of $(\mathbf{F}(t)^{\mathrm{T}})^{-1}$ will contain the reciprocal basis vectors $\boldsymbol{h}_j(t)$. These vectors satisfy

$$\frac{d\boldsymbol{h}_j}{dt} = -\mathbf{L}^{\mathrm{T}}\cdot\boldsymbol{h}_j. \tag{1.14}$$

Problem 1.4 illustrates one context in which this reciprocal basis arises.

Just as the vectors $\boldsymbol{g}_i$ comprise a codeforming basis set for representing vectors associated with a deforming material, the dyads $\boldsymbol{g}_i\boldsymbol{g}_j$ form a basis for representing second-order tensors. In index notation, where $g_{kj} = \boldsymbol{e}_k\cdot\boldsymbol{g}_j$, these dyads evolve as follows:

$$\begin{aligned}\left(\frac{d\boldsymbol{g}_i\boldsymbol{g}_j}{dt}\right)_{lk} = \frac{dg_{li}g_{kj}}{dt} &= \frac{dg_{li}}{dt}g_{kj} + g_{li}\frac{dg_{kj}}{dt}\\ &= L_{lm}g_{mi}g_{kj} + g_{li}L_{km}g_{mj}\\ &= L_{lm}g_{mi}g_{kj} + g_{li}g_{mj}L^{\mathrm{T}}_{mk},\end{aligned}$$

which we can rewrite

$$\frac{d\boldsymbol{g}_i\boldsymbol{g}_j}{dt} = \boldsymbol{g}_i\boldsymbol{g}_j\cdot\nabla\boldsymbol{v} + \nabla\boldsymbol{v}^{\mathrm{T}}\cdot\boldsymbol{g}_i\boldsymbol{g}_j. \tag{1.15}$$

[5] In mathematical terms, $\mathbf{F}$ is the *fundamental solution matrix* for (1.2).

More generally, we can define the *Green tensor*[6]

$$\mathbf{B}(\boldsymbol{X}(t), t) = \sum_{i=1}^{3}\sum_{i=1}^{3} \boldsymbol{g}_i \boldsymbol{g}_j. \tag{1.16}$$

By construction, at $t = 0$, $\mathbf{B} = \boldsymbol{\delta}$, where $\boldsymbol{\delta}$ is the identity tensor. Using (1.15), we can write that

$$\frac{d\mathbf{B}(\boldsymbol{X}(t), t)}{dt} - \left(\mathbf{B}(\boldsymbol{X}(t), t) \cdot \nabla \boldsymbol{v}(\boldsymbol{X}(t), t) + \nabla \boldsymbol{v}^{\mathrm{T}}(\boldsymbol{X}(t), t) \cdot \mathbf{B}(\boldsymbol{X}(t), t)\right) = \mathbf{0}. \tag{1.17}$$

If we think about $\mathbf{B}$ as a tensor field, i.e., as a function of position $\boldsymbol{x}$ in the flow field rather than as a tensor attached to a particular material point $\boldsymbol{X}(t)$, then the time derivative is replaced by a substantial derivative[7]

$$\frac{D}{Dt} = \frac{\partial}{\partial t} + \boldsymbol{v} \cdot \nabla,$$

in which case

$$\frac{D\mathbf{B}(\boldsymbol{x}, t)}{Dt} - \left(\mathbf{B}(\boldsymbol{x}, t) \cdot \nabla \boldsymbol{v}(\boldsymbol{x}, t) + \nabla \boldsymbol{v}^{\mathrm{T}}(\boldsymbol{x}, t) \cdot \mathbf{B}(\boldsymbol{x}, t)\right) = \mathbf{0}. \tag{1.18}$$

The quantity on the left-hand side of this expression is called the *contravariant convected derivative* or *upper convected derivative* of $\mathbf{B}$ and is denoted $\mathbf{B}_{(1)}$ or $\overset{\nabla}{\mathbf{B}}$. This derivative is the rate of change of a tensor relative to a coordinate system that is deforming with the material. The Green tensor is fundamentally important to the theory of elasticity and viscoelasticity, in part because the stress tensor for a simple model of a material called the neo-Hookean solid is proportional to it. We will see this object again in Section 8.6, where it arises naturally in a model of the dynamics of dilute polymer solutions. The Green tensor is related to $\mathbf{F}$ by

$$\mathbf{B}(\boldsymbol{X}(t), t) = \mathbf{F}(\boldsymbol{X}(t), t) \cdot \mathbf{F}^{\mathrm{T}}(\boldsymbol{X}(t), t). \tag{1.19}$$

(See Problem 1.2.)

[6] We use the nomenclature recommended by the International Union of Pure and Applied Chemistry (IUPAC) (Kaye et al. 1998). This tensor is also called the *left Cauchy–Green tensor* or sometimes the *Finger tensor*, although according to the IUPAC standard the latter term refers to the quantity $\left(\mathbf{F}^{\mathrm{T}} \cdot \mathbf{F}\right)^{-1}$. The *right Cauchy–Green tensor* $\mathbf{C}$, also called the *Cauchy tensor*, is given by $\mathbf{C} = \mathbf{F}^{\mathrm{T}} \cdot \mathbf{F}$.

[7] Consider a scalar field $f(\boldsymbol{x}(t), t)$. The rate of change of this field as measured by an observer moving with velocity $d\boldsymbol{x}/dt = \boldsymbol{v}$ (i.e., moving as a material point $\boldsymbol{X}(t)$) is given by the *substantial derivative* $Df/Dt$. In Cartesian coordinates, we can use the chain rule to write this as

$$\begin{aligned}\frac{Df(\boldsymbol{x}, t)}{Dt} &= \frac{\partial f}{\partial t} + \sum_{i=1}^{3} \frac{dx_i}{dt}\frac{\partial f}{\partial x_i} \\ &= \frac{\partial f}{\partial t} + \sum_{i=1}^{3} v_i \frac{\partial f}{\partial x_i} \\ &= \frac{\partial f}{\partial t} + \boldsymbol{v} \cdot \nabla f.\end{aligned}$$

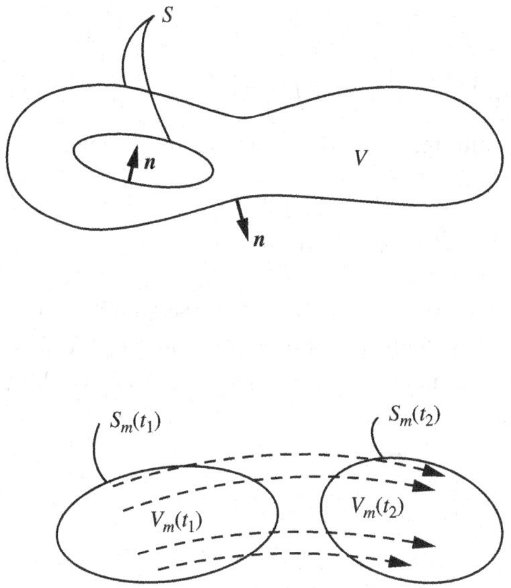


**Figure 1.6** Arbitrary stationary (top) and material (bottom) volumes within a material for derivation of conservation laws.

A number of other deformation tensors and convected derivatives also arise in studies of complex fluids and elastic solids. Section 9.4 illustrates the importance of these derivatives in the general context of constitutive models for stress in a material.

## 1.2 Conservation Equations

### 1.2.1 Conservation of Mass

Consider an arbitrarily chosen stationary volume $V$ with boundary $S$ within a material of mass density $\rho$ as shown in Figure 1.6. The outward unit normal vector for the volume is denoted $\boldsymbol{n}$. Given that the mass flux at any position $\boldsymbol{x}$ in the volume is $\rho \boldsymbol{v}$, the mass balance for this domain can be written as

$$\frac{d}{dt}\int_V \rho \, dV = -\int_S \boldsymbol{n} \cdot (\rho \boldsymbol{v}) \, dS. \tag{1.20}$$

This equation simply states that the rate of accumulation of mass in the domain is equal to the integral over the boundary of the mass flux into the volume $-\boldsymbol{n} \cdot (\rho \boldsymbol{v})$ – mass is neither created nor destroyed within the volume $V$. Applying the divergence theorem (Equation (A.10) in Appendix A.2) to the right-hand side and rearranging yields that

$$\frac{d}{dt}\int_V \rho \, dV + \int_V \boldsymbol{\nabla} \cdot (\rho \boldsymbol{v}) \, dV = 0. \tag{1.21}$$

Since the volume $V$ is stationary, this can be rewritten

$$\int_V \frac{\partial \rho}{\partial t} \, dV + \int_V \boldsymbol{\nabla} \cdot (\rho \boldsymbol{v}) \, dV = 0$$

or

$$\int_V \left( \frac{\partial \rho}{\partial t} + \nabla \cdot (\rho \boldsymbol{v}) \right) dV = 0.$$

The domain of integration is arbitrary, so the only way the integral can vanish is for the integrand to vanish pointwise:

$$\frac{\partial \rho}{\partial t} + \nabla \cdot (\rho \boldsymbol{v}) = 0 \tag{1.22}$$

at every point in the domain. This is the statement of conservation of mass at a point in a continuous medium and is known as the *continuity equation*. We will exclusively consider *incompressible flows* during which density changes are negligible and the continuity equation reduces to

$$\nabla \cdot \boldsymbol{v} = 0. \tag{1.23}$$

## 1.2.2 Conservation of Momentum

To address conservation of momentum, we take a slightly different approach than we did for mass conservation. Consider an arbitrary *material volume* $V_{\mathrm{M}}(t)$, by which we mean a volume whose elements move with the material velocity $\boldsymbol{v}(\boldsymbol{x},t)$ (Figure 1.6). Therefore points on the boundary of this volume $S_{\mathrm{M}}(t)$ are also moving with this velocity. The total momentum of this volume is

$$\int_{V_{\mathrm{M}}(t)} \rho \boldsymbol{v}\, dV$$

and Newton's second law applied to it becomes

$$\frac{d}{dt} \int_{V_{\mathrm{M}}(t)} \rho \boldsymbol{v}\, dV = \boldsymbol{F}_V + \boldsymbol{F}_S. \tag{1.24}$$

Here the left-hand side is the rate of change of momentum of the volume, $\boldsymbol{F}_V$ incorporates forces exerted on the material in the volume by external fields that act at each point in the material and $\boldsymbol{F}_S$ incorporates forces exerted across the boundary of the volume by the neighboring material – stresses.

Many body forces – gravitational, electrical, magnetic – can act at each point in a material in an external field. For the moment, we do not specify the nature of these forces but simply write that

$$\boldsymbol{F}_V = \int_{V_{\mathrm{M}}(t)} \boldsymbol{f}(\boldsymbol{x})\, dV, \tag{1.25}$$

where $\boldsymbol{f}(\boldsymbol{x})$ is an arbitrary position-dependent force density.

The net surface force is the integral of the stresses exerted across the boundary of $V$. Denoting the stress vector (sometimes called the traction vector) exerted at a point on the boundary by the neighboring material outside the boundary as $\boldsymbol{t}(\boldsymbol{x})$, this force can be written

$$\boldsymbol{F}_S = \int_{S_{\mathrm{M}}(t)} \boldsymbol{t}(\boldsymbol{x})\, dS.$$

Under very general circumstances, the traction vector has the form

$$\boldsymbol{t} = \boldsymbol{n} \cdot \boldsymbol{\sigma},$$

where $\boldsymbol{\sigma}$ is a second-order tensor called the *stress tensor* (Batchelor 1967, Aris 1989, Pozrikidis 1997). In Cartesian coordinates, the quantity $\sigma_{xy}$, for example, is the force per unit area in the $y$ direction exerted on a surface of constant $x$ by the material at greater $x$. Conservation of angular momentum yields, again under very general circumstances, that the stress tensor is symmetric: $\sigma_{ij} = \sigma_{ji}$ (see Problem 1.6). The stress tensor will be the subject of further discussion later in this chapter and elsewhere in the book.

Using the expressions for $\boldsymbol{F}_V$ and $\boldsymbol{F}_S$, we can rewrite (1.24) as

$$\frac{d}{dt}\int_{V_M(t)} \rho\boldsymbol{v}\, dV = \int_{V_M(t)} \boldsymbol{f}\, dV + \int_{S_M(t)} \boldsymbol{n}\cdot\boldsymbol{\sigma}\, dS.$$

Now Leibniz's rule, (A.15), can be applied to the left-hand side:

$$\int_{V_M(t)} \frac{\partial \rho\boldsymbol{v}}{\partial t}\, dV + \int_{S_M(t)} \rho\boldsymbol{v}(\boldsymbol{n}\cdot\boldsymbol{v})\, dS = \int_{V_M(t)} \boldsymbol{f}\, dV + \int_{S_M(t)} \boldsymbol{n}\cdot\boldsymbol{\sigma}\, dS.$$

Combining the surface integrals yields that

$$\int_{V_M(t)} \frac{\partial \rho\boldsymbol{v}}{\partial t}\, dV = \int_{S_M(t)} \boldsymbol{n}\cdot(-\rho\boldsymbol{v}\boldsymbol{v} + \boldsymbol{\sigma})\, dS + \int_{V_M(t)} \boldsymbol{f}\, dV,$$

where we have used the definition of a dyad: $(\boldsymbol{n}\cdot\boldsymbol{v})\boldsymbol{v} = \boldsymbol{n}\cdot(\boldsymbol{v}\boldsymbol{v})$. From this form of the equation, we can recognize that the quantity $\rho\boldsymbol{v}\boldsymbol{v} - \boldsymbol{\sigma}$ is the total momentum flux at a point in the material. The first term represents momentum carried by the bulk motion of the material and is called the *convective* or *advective momentum flux*, while the second represents momentum transported via microscopic mechanisms.

Applying the divergence theorem to the surface integral on the right-hand side converts it to a volume integral. Again, since the volume under consideration is arbitrary, we conclude that at every point in the domain the following equation holds:

$$\frac{\partial \rho\boldsymbol{v}}{\partial t} = -\nabla\cdot(\rho\boldsymbol{v}\boldsymbol{v}) + \nabla\cdot\boldsymbol{\sigma} + \boldsymbol{f}.$$

With some rearrangement and the use of the continuity equation, this can be written in more conventional form:

$$\rho\left(\frac{\partial \boldsymbol{v}}{\partial t} + \boldsymbol{v}\cdot\nabla\boldsymbol{v}\right) = \nabla\cdot\boldsymbol{\sigma} + \boldsymbol{f}, \tag{1.26}$$

which is called the *Cauchy momentum balance equation*. This equation is very broadly valid, independent of the material. Now we split the stress into an isotropic component that at equilibrium is just the thermodynamic pressure, and an additional component $\boldsymbol{\tau}$, sometimes called the *extra stress tensor*, that vanishes at equilibrium:

$$\boldsymbol{\sigma} = -p\boldsymbol{\delta} + \boldsymbol{\tau}.$$

With this definition, we have another common form of the Cauchy equation

$$\rho\left(\frac{\partial \boldsymbol{v}}{\partial t} + \boldsymbol{v}\cdot\nabla\boldsymbol{v}\right) = -\nabla p + \nabla\cdot\boldsymbol{\tau} + \boldsymbol{f}. \tag{1.27}$$

For an incompressible fluid, the pressure cannot be viewed as a thermodynamic quantity. It is best viewed as a Lagrange multiplier that enables the pointwise incompressibility constraint $\nabla \cdot \boldsymbol{v} = 0$ to be satisfied. Section 1.3.4 illustrates this point explicitly for a specific case.

Microscopically, the stress tensor $\boldsymbol{\sigma}$ can have many different origins. For example, it may be due to the fluctuating momentum fluxes associated with thermal motion of the molecules forming the material, or to forces exerted across the boundary due to chemical bonds that span it (Bird et al. 1987, Chaikin & Lubensky 1995, McQuarrie 2000, Bird, Stewart & Lightfoot 2002). Later sections of this book will shed light on the microscopic origins of the stresses associated with particles or macromolecules in solution, and we will eventually see that a complex fluid can be *viscoelastic*, having both a viscous contribution to the stress that is proportional to the rate of deformation and an elastic contribution that is proportional to the amplitude of deformation.

For this and the next several chapters, however, our focus will be on a specific class of materials, incompressible *Newtonian* fluids, in which density is constant and the contribution to the stress at a point in the fluid due to its deformation is a linear and isotropic function of the strain rate $\mathbf{E}$ at that point. The stress cannot depend on $\mathbf{W}$, because there is no material deformation associated with rotation (see Section 9.4 for further discussion). For such a fluid, the quantity $\boldsymbol{\tau}$ is called the *viscous stress* and has the form

$$\boldsymbol{\tau} = 2\eta\mathbf{E} = \eta\left(\nabla\boldsymbol{v} + \nabla\boldsymbol{v}^{\mathrm{T}}\right), \tag{1.28}$$

where $\eta$ is the dynamic viscosity (usually just called the viscosity) of the fluid – it is the proportionality constant between stress and strain rate. The related quantity $\nu = \eta/\rho$, known as the *kinematic viscosity*, plays the role of a diffusivity for momentum transport. Now the Cauchy momentum equation reduces to the *incompressible Navier–Stokes equation*

$$\rho\left(\frac{\partial \boldsymbol{v}}{\partial t} + \boldsymbol{v}\cdot\nabla\boldsymbol{v}\right) = -\nabla p + \eta\nabla^2\boldsymbol{v} + \boldsymbol{f}. \tag{1.29}$$

Supplemented by (1.23) for mass conservation and appropriate boundary conditions, this equation provides a complete description of flows of incompressible Newtonian fluids. Because we only consider incompressible flow, when we write "Navier–Stokes equation" (or "Stokes equation," discussed later), we will generally mean "Navier–Stokes and continuity equations for incompressible flow."

In many situations (e.g., single-phase flow with no free interfaces, multiphase flow of density-matched fluids), the only body force is gravity and its only effect is to induce a hydrostatic pressure gradient in the fluid that has no influence on the fluid motion. Therefore, it is often useful to write the body force as the gradient of the gravitational potential per unit volume:

$$\boldsymbol{f} = \rho\boldsymbol{g} = -\nabla\Phi_{\mathrm{g}}, \qquad \Phi_{\mathrm{g}} = \rho g(h - h_0), \tag{1.30}$$

where $\boldsymbol{g}$ is the gravity vector, $h$ is vertical height, and $h_0$ an arbitrary reference height. Now the pressure and gravitational terms can be combined by defining the *dynamic pressure* $\mathbb{P} = p + \Phi_g$. For a fluid at rest, $\mathbb{P}$ is constant. Now the momentum equation can be written

$$\rho\left(\frac{\partial \boldsymbol{v}}{\partial t} + \boldsymbol{v}\cdot\boldsymbol{\nabla}\boldsymbol{v}\right) = -\boldsymbol{\nabla}\mathbb{P} + \eta\nabla^2\boldsymbol{v}. \quad (1.31)$$

Where there is no danger of confusion, $\mathbb{P}$ is often replaced by $p$.

Now consider a situation in which the length scale over which we expect velocities to vary is $L$ and the characteristic variation in velocities is $U$. These quantities will vary from system to system; for a particle of size $a$ in a shear flow with shear rate $\dot{\gamma}$, we could take $L = a$ and $U = a\dot{\gamma}$. Additionally, there may be a time scale $T$ for imposed variations in velocity. Using these scales, along with the viscous stress scale $\eta U/L$ as a characteristic pressure scale, we can rewrite the momentum equation in nondimensional form:

$$\mathrm{Re}\left(\mathrm{Sr}^{-1}\frac{\partial \boldsymbol{v}}{\partial t} + \boldsymbol{v}\cdot\boldsymbol{\nabla}\boldsymbol{v}\right) = -\boldsymbol{\nabla}p + \nabla^2\boldsymbol{v} + \boldsymbol{f}, \quad (1.32)$$

where all variables are now nondimensionalized with their corresponding scales, and two dimensionless groups appear:

$$\mathrm{Re} = \frac{UL}{\nu} \quad (1.33)$$

and

$$\mathrm{Sr} = \frac{TU}{L}. \quad (1.34)$$

The Reynolds number Re estimates the ratio between the time scale for diffusion of momentum $(L^2/\nu)$ over the length scale $L$ and the time scale for flow $(L/U)$ over that scale. The Strouhal number Sr estimates the ratio between the imposed time scale $T$ and the flow time scale $L/U$. If there is no imposed time scale (e.g., if the flow is generated by a steady boundary motion or pressure gradient), then $T = L/U$ and $\mathrm{Sr} = 1$.

For water at room temperature, $\nu \approx 10^{-6}\mathrm{m}^2/\mathrm{s}$. If $L \approx 1\mu\mathrm{m}$, then $\mathrm{Re} \ll 1$ as long as $U \ll 1\mathrm{m/s}$, a condition that is satisfied for a wide variety of microscale processes. Thus, for these processes and under the condition $\mathrm{Sr} = O(1)$, the acceleration and convective terms in the Navier–Stokes equation can be neglected and the momentum balance reduces to the *Stokes equation*, which in dimensional form is given by

$$-\boldsymbol{\nabla}p + \eta\nabla^2\boldsymbol{v} + \boldsymbol{f} = \boldsymbol{0}. \quad (1.35)$$

Flow governed by the Stokes equation, or *Stokes flow*, is *inertialess* – the mass of the fluid plays no role in its dynamics. Taking the divergence of (1.35) and applying incompressibility shows that the pressure field in Stokes flow satisfies a Poisson equation:

$$\nabla^2 p = \boldsymbol{\nabla}\cdot\boldsymbol{f}. \quad (1.36)$$

Similarly, taking the Laplacian of the Stokes equation and applying (1.36) yields that

$$\eta\nabla^2\nabla^2\boldsymbol{v} = \boldsymbol{\nabla}\boldsymbol{\nabla}\cdot\boldsymbol{f} - \nabla^2\boldsymbol{f}. \quad (1.37)$$

If $\boldsymbol{f} = \boldsymbol{0}$, this reduces to a biharmonic equation for $\boldsymbol{v}$.

Another important case arises when $\mathrm{Re} \ll 1$, but when forcing or motion at very short time scales is present so that $\mathrm{ReSr}^{-1} = O(1)$. In this case, the convective term in the Navier–Stokes equation is negligible but the acceleration term is not, and we arrive at the *transient Stokes equation*:

$$\rho \frac{\partial \boldsymbol{v}}{\partial t} = -\boldsymbol{\nabla} p + \eta \nabla^2 \boldsymbol{v} + \boldsymbol{f}. \tag{1.38}$$

This equation is also sometimes called the *linearized Navier–Stokes equation*. The Stokes and transient Stokes equations are the central equations of microscale fluid dynamics. The remainder of this chapter, as well as Chapters 2 through 4, elaborate on properties and solutions of the Stokes equations, while Chapter 5 addresses some key effects of inertia.

### 1.2.3 Boundary Conditions

Because interfacial flows are not a central topic of this book, we provide only a cursory treatment of boundary conditions at interfaces. Consider an interfacial surface $S_I$ that forms the boundary between two materials or phases; the unit normal vector pointing from material 1 to material 2 will be denoted $\boldsymbol{n}_I$, and $\boldsymbol{v}|_1$ and $\boldsymbol{v}|_2$ will be the velocities of the two phases at the interface. For simplicity, we will only consider interfaces at which no phase change is occurring and take the interface to be a material surface. Mass conservation requires that

$$\boldsymbol{n}_I \cdot \boldsymbol{v}|_1 = \boldsymbol{n}_I \cdot \boldsymbol{v}|_2 \tag{1.39}$$

at each point on the surface. This is called the *no-penetration condition*. This condition provides no information about the tangential velocities at the interface, but experimental observations indicate that for gases and small-molecule liquids under a wide range of conditions, the tangential velocities of the two phases are the same. This is called the *no-slip boundary condition* and can be written

$$(\boldsymbol{\delta} - \boldsymbol{n}_I \boldsymbol{n}_I) \cdot \boldsymbol{v}|_1 = (\boldsymbol{\delta} - \boldsymbol{n}_I \boldsymbol{n}_I) \cdot \boldsymbol{v}|_2, \tag{1.40}$$

where $(\boldsymbol{\delta} - \boldsymbol{n}_I \boldsymbol{n}_I)$ is the orthogonal projection operator onto the plane locally parallel to the interface. The combination of (1.39) and (1.40) implies that

$$\boldsymbol{v}|_1 = \boldsymbol{v}|_2. \tag{1.41}$$

For a flat interface, or if interfacial tension effects are negligible, continuity of stress across the interface is required by the momentum balance:

$$\boldsymbol{n}_I \cdot \boldsymbol{\sigma}|_1 = \boldsymbol{n}_I \cdot \boldsymbol{\sigma}|_2. \tag{1.42}$$

If, for example, material 2 is a gas with very small viscosity compared to that of material 1, the viscous stress in the gas phase is often taken to be negligible, in which case the normal and tangential components of (1.42) simplify to

$$p|_1 = p|_2, \tag{1.43}$$

$$(\boldsymbol{\delta} - \boldsymbol{n}_I \boldsymbol{n}_I) \cdot (\boldsymbol{n}_I \cdot \boldsymbol{\tau}) = \boldsymbol{0}. \tag{1.44}$$

The latter equation expresses that there is no shear stress across the interface. For example, in Cartesian coordinates, if the interface is the surface $y = 0$, then this condition, after application of the no-penetration condition $v_y = 0$, becomes

$$\frac{\partial v_x}{\partial y} = \frac{\partial v_z}{\partial y} = 0.$$

For gas–liquid problems, this condition, rather than no-slip, is often applied on the liquid phase, with the motion in the gas phase simply neglected.

Finally, for an interface between two fluids with interfacial tension $\gamma$, the interfacial stress balance becomes

$$\boldsymbol{n}_I \cdot (\boldsymbol{\sigma}|_2 - \boldsymbol{\sigma}|_1) + \boldsymbol{\nabla}_S \gamma + 2\mathcal{H}\gamma \boldsymbol{n}_I = \boldsymbol{0}. \tag{1.45}$$

Here $\boldsymbol{\nabla}_S = (\boldsymbol{\delta} - \boldsymbol{n}_I \boldsymbol{n}_I) \cdot \boldsymbol{\nabla}$ is the surface gradient operator and

$$\mathcal{H} = -\frac{1}{2} \boldsymbol{\nabla}_S \cdot \boldsymbol{n}_I \tag{1.46}$$

is the *mean curvature* of the interface. If fluid 1 is a bubble or drop of radius $R$ at equilibrium in a second fluid, (1.45) reduces at equilibrium to the elementary result

$$p|_2 - p|_1 = -\frac{2\gamma}{R}. \tag{1.47}$$

The pressure in the bubble or drop is higher than in the surrounding fluid. Leal (2007) provides a detailed discussion of transport at interfaces.

## 1.3 General Properties of Stokes Flow

The Stokes equation (1.35) displays a number of important general properties that follow from the absence of inertial effects in the low Reynolds number or Stokes flow regime. These properties arise repeatly as we analyze and predict the dynamics of small-scale flow so we introduce them here.

### 1.3.1 Linearity and Reversibility

Consider flow in some domain $V_D$ with boundary conditions applied on $S_D$. Leaving aside for the moment the issue of boundary conditions, we observe that this system of equations is linear, so it is straightforward to verify by substitution into the equations the following properties:

1. Let $\boldsymbol{f} = 0$. If $(\boldsymbol{v}_1, p_1)$ and $(\boldsymbol{v}_2, p_2)$ are both solutions, then so is $(\alpha \boldsymbol{v}_1 + \beta \boldsymbol{v}_2, \alpha p_1 + \beta p_2)$, where $\alpha$ and $\beta$ are arbitrary constant scalars.
2. Now consider the case $\boldsymbol{f} \neq 0$. If $(\boldsymbol{v}_1, p_1)$ is a solution for $\boldsymbol{f} = \boldsymbol{f}_1$ and $(\boldsymbol{v}_2, p_2)$ is a solution for $\boldsymbol{f} = \boldsymbol{f}_2$, then $(\alpha \boldsymbol{v}_1 + \beta \boldsymbol{v}_2, \alpha p_1 + \beta p_2)$ is a solution for $\boldsymbol{f} = \alpha \boldsymbol{f}_1 + \beta \boldsymbol{f}_2$, again where $\alpha$ and $\beta$ are arbitrary constant scalars.

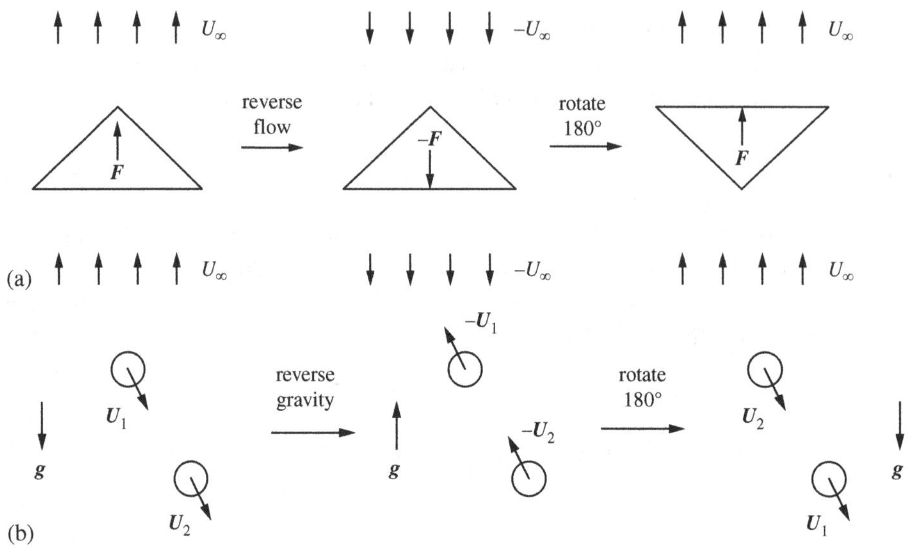


**Figure 1.7** (a) Drag on an object with triangular cross-section. (b) Sedimenting pair of identical rigid spheres.

Now turning to the issue of boundary conditions, for definiteness let us consider a no-slip condition

$$\boldsymbol{v} = \boldsymbol{v}_{\mathrm{D}} \quad \text{for } \boldsymbol{x} \in S_{\mathrm{D}}$$

where $\boldsymbol{v}_{\mathrm{D}}$ is the specified velocity of points on the boundary – e.g., $\boldsymbol{v}_{\mathrm{D}} = 0$ on a stationary wall. Again, linearity allows us to state the following result:

3. Let $\boldsymbol{f} = 0$. If $(\boldsymbol{v}_1, p_1)$ is a solution for $\boldsymbol{v}_{\mathrm{D}} = \boldsymbol{v}_{\mathrm{D}1}$ and $(\boldsymbol{v}_2, p_2)$ is a solution for $\boldsymbol{v}_{\mathrm{D}} = \boldsymbol{v}_{\mathrm{D}2}$, then $(\alpha\boldsymbol{v}_1 + \beta\boldsymbol{v}_2, \alpha p_1 + \beta p_2)$ is a solution for $\boldsymbol{v}_{\mathrm{D}} = \alpha\boldsymbol{v}_{\mathrm{D}1} + \beta\boldsymbol{v}_{\mathrm{D}2}$.

These properties are all specific cases of the superposition principle for linear equations. They will be very important as we construct solutions to the Stokes equation, as well as in deriving the reversibility property that we describe now.

From linearity it is simple to see the following:

1. In a given domain, if $\boldsymbol{v}_1$ is a flow driven by forcing $\boldsymbol{f}_1$, then the forcing $-\boldsymbol{f}_1$ drives the velocity $-\boldsymbol{v}_1$. I.e., the flow *reverses*.
2. Similarly, for $\boldsymbol{f} = 0$, if $\boldsymbol{v}_1$ is a flow driven by boundary motion $\boldsymbol{v}_{\mathrm{D}1}$, then the boundary motion $-\boldsymbol{v}_{\mathrm{D}1}$ drives the velocity $-\boldsymbol{v}_1$.

These simple statements summarize the *reversibility* principle for Stokes flow. Especially in conjunction with symmetries of the flow and geometry, this principle can be used to

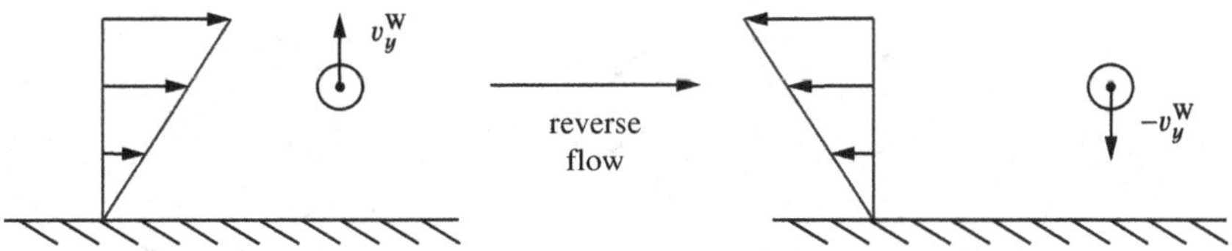


**Figure 1.8** Rigid sphere in shear flow near a plane rigid surface. By reversibility, the particle velocity in the right figure must be the negative of that in the left.

draw important qualitative conclusions about forces and motions involving Stokes flow. Here are some examples.

*Drag on an Object.*
Consider flow around a rigid object whose cross-section is a left-right symmetric isosceles triangle as shown in Figure 1.7(a). How does the drag change if we turn the object upside down? Consider first the leftmost drawing. The object is held stationary against a drag force $\boldsymbol{F}_{\text{drag}}$ in a flow that has the uniform velocity $\boldsymbol{v}_\infty = U_\infty \boldsymbol{e}_z$ far from the object. By reversibility, if we change the sign of $\boldsymbol{v}_\infty$, as shown in the center figure, the entire velocity field will change sign, as will the pressure field, and accordingly the drag force $\boldsymbol{F}_{\text{drag}}$ will change sign. Now note that the right figure is equivalent to the center figure viewed upside down and thus equivalent to the left figure except with the object turned upside down. Therefore, we can can conclude that in Stokes flow, the drag force on this object is the same when the object is upside down as when it is upright.

*Sedimentation of a Pair of Rigid Spheres.*
The left panel of Figure 1.7(b) shows a pair of identical rigid spheres sedimenting due to gravity with velocities $\boldsymbol{U}_1$ and $\boldsymbol{U}_2$ in an unbounded flow. Will these particles move relative to one another as they fall? If the sign of gravity is reversed ($\boldsymbol{g} \to -\boldsymbol{g}$), then Stokes flow reversibility implies that $\boldsymbol{U}_1 \to -\boldsymbol{U}_1$ and $\boldsymbol{U}_2 \to -\boldsymbol{U}_2$ as shown in the center panel. Now, rotating the entire system around the midpoint between the two particles yields the right panel. This configuration is identical to the left panel but with $\boldsymbol{U}_1$ replaced by $\boldsymbol{U}_2$ and vice versa. Therefore, $\boldsymbol{U}_1$ and $\boldsymbol{U}_2$ must be identical: the particles exhibit no relative motion as they sediment. However, if an initially collinear arrangement of three or more particles is considered, there is insufficient symmetry in the problem to conclude that the particles all fall at the same velocity and in fact they will not (Problem 2.5).

*Motion of a Rigid Sphere near a Wall.*
We commonly encounter situations in which a particle is in a flow near a rigid wall, as shown in Figure 1.8. Will a rigid sphere migrate in Stokes flow? Assume that in the flow shown on the left, the particle migrates at wall-normal speed $v_y^{\text{W}}$. In Stokes flow, when we reverse the imposed velocity field as shown on the right, the wall-normal velocity must become $v_y^{\text{W}\prime} = -v_y^{\text{W}}$ (so if the particle in the left figure is moving away from the wall, the particle in the right figure is moving toward it). But by symmetry, the left and right figures are equivalent (they are two views of the same process), so $v_y^{\text{W}\prime} = v_y^{\text{W}}$. The

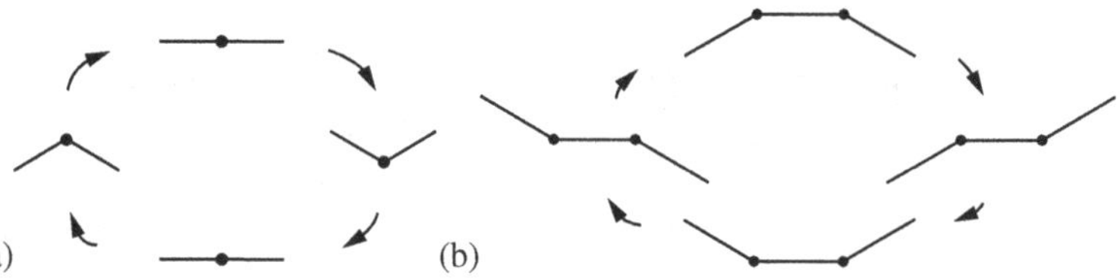


**Figure 1.9** (a) Cyclic motion of two rods connected by a motorized hinge. (b) Cyclic motion of three rods connected at two hinges.

only way for $v_y^W$ to satisfy both reversibility and symmetry is for it to be zero. By a very similar argument, a sphere sedimenting near to a vertical rigid wall in Stokes flow moves parallel to the wall. This argument relies both on reversibility and symmetry, the latter of which is lost if the particle is not spherical. Furthermore, due to deformability or inertial effects, particles that are spherical at rest *can* migrate toward or away from walls during flow, as we will describe in Sections 4.4 and 5.4.

*Locomotion of Linked Rigid Bodies.*
Consider the two rods connected by a hinge shown. The hinge is "motorized" so that it moves the rods relative to one another in a cyclic manner as shown in Figure 1.9(a). Can this motion generate locomotion in Stokes flow? Consider the cycle starting from the top figure in which the rods are aligned. Moving clockwise, the rods first fold upward toward each other, then back down, passing through the initial conformation as they fold downward. Then they fold back up to the straight initial conformation, completing one cycle of motion. In Stokes flow, any net motion of the object as the rods move upward away from the initial straight conformation will be exactly undone when the rods move back down into the initial configuration and likewise when the rods move down. Since in Stokes flow the fluid velocity is linearly proportional to the boundary velocity, this result holds even if the rods are moved quickly upward and slowly downward as long as both motions are slow enough that the inertia of the fluid remains negligible. More generally, *reciprocal motions* of boundaries, that is, cyclic motions that appear the same running forward or backward in time, lead to no net motion in Stokes flow (Purcell 1977, Lauga & Powers 2009). On the other hand, for the three-rod object undergoing the sequence of motions shown in Figure 1.9(b), no part of this cycle reverses the motion of the object induced by an earlier part. With this nonreciprocal motion, Stokes flow reversibility does not preclude net motion after one cycle. Indeed, this object will display a net translation through the fluid, a crude form of locomotion.

### 1.3.2 Stress Equilibrium

We noted earlier that at low Reynolds number, the transport of momentum by viscous diffusion is very rapid compared to transport by convection. An important consequence of this fact is known as *stress equilibrium.* To elucidate this mathematically, we observe that in the absence of body forces, the Stokes equation (1.35) is equivalent to the statement

$$\nabla \cdot \boldsymbol{\sigma} = \mathbf{0}. \tag{1.48}$$

(The Stokes equation in its usual form is recovered by inserting the Newtonian constitutive equation $\boldsymbol{\sigma} = -p\boldsymbol{\delta} + 2\eta\mathbf{E}$.) Integrating this equation over an arbitrary volume $V$ in the fluid and then applying the divergence theorem yields that

$$\int_S \boldsymbol{n} \cdot \boldsymbol{\sigma}\, dS = \int_S \boldsymbol{t}\, dS = \boldsymbol{F}_S = 0. \tag{1.49}$$

That is, the sum of forces on the boundary of any domain in the fluid is always zero – the stresses are in equilibrium. If body forces are present, then

$$\nabla \cdot \boldsymbol{\sigma} + \boldsymbol{f} = 0, \tag{1.50}$$

and repeating the preceding steps yields that

$$\int_S \boldsymbol{t}\, dS + \int_V \boldsymbol{f}\, dV = 0. \tag{1.51}$$

The sum of the stresses and the body forces is zero: any body forces exerted within the fluid are instantaneously transmitted to the boundary or, if the domain is unbounded, to infinity. This point will be revisited in Section 5.1.

### 1.3.3 Lorentz Reciprocal Relations

The divergence theorem provides an important link between behavior in the interior of a domain and at its boundaries. Generalizations of this theorem provide such relationships for multiple functions. For example, given two scalar fields $\phi'$ and $\phi''$, Green's second identity, (A.17), states that

$$\int_V \left(\phi'\nabla^2\phi'' - \phi''\nabla^2\phi'\right) dV = \int_S \left(\phi'\boldsymbol{n} \cdot \nabla\phi'' - \phi''\boldsymbol{n} \cdot \nabla\phi'\right) dS.$$

Letting $\boldsymbol{u} = -\nabla\phi$, this becomes

$$\int_V \left(\phi'\nabla \cdot \boldsymbol{u}'' - \phi''\nabla \cdot \boldsymbol{u}'\right) dV = \int_S \left(\phi'\boldsymbol{n} \cdot \boldsymbol{u}'' - \phi''\boldsymbol{n} \cdot \boldsymbol{u}'\right) dS. \tag{1.52}$$

If both $\phi'$ and $\phi''$ are solutions to Laplace's equation, $\nabla^2\phi = 0$, then this identity gives us a relationship between $\phi'$ and $\phi''$:

$$\int_S \left(\phi'\boldsymbol{n} \cdot \boldsymbol{u}'' - \phi''\boldsymbol{n} \cdot \boldsymbol{u}'\right) dS = 0. \tag{1.53}$$

An analogous set of results applies to the Stokes equation. Consider velocity fields $\boldsymbol{v}'$ and $\boldsymbol{v}''$ and their corresponding stress tensors $\boldsymbol{\sigma}'$ and $\boldsymbol{\sigma}''$. Beginning with an expression analogous to the left-hand side of (1.52), we can use the product rule and divergence theorem to show that[8]

$$\int_V (\boldsymbol{v}' \cdot (\nabla \cdot \boldsymbol{\sigma}'') - \boldsymbol{v}'' \cdot (\nabla \cdot \boldsymbol{\sigma}'))\ dV = \int_S (\boldsymbol{v}' \cdot (\boldsymbol{n} \cdot \boldsymbol{\sigma}'') - \boldsymbol{v}'' \cdot (\boldsymbol{n} \cdot \boldsymbol{\sigma}'))\ dS - \int_V (\nabla \boldsymbol{v}' : \boldsymbol{\sigma}'' - \nabla \boldsymbol{v}'' : \boldsymbol{\sigma}')\, dS. \tag{1.54}$$

Inserting the Newtonian form of the stress tensor, (1.28), the last term on the right-hand side can be written

$$\int_V (\nabla \boldsymbol{v}' : \boldsymbol{\sigma}'' - \nabla \boldsymbol{v}'' : \boldsymbol{\sigma}')\, dV = \int_V \left( \nabla \boldsymbol{v}' : (-p''\boldsymbol{\delta} + \eta(\nabla \boldsymbol{v}'' + \nabla \boldsymbol{v}''^{\mathrm{T}})) - \nabla \boldsymbol{v}'' : (-p'\boldsymbol{\delta} + \eta(\nabla \boldsymbol{v}' + \nabla \boldsymbol{v}'^{\mathrm{T}})\right) dV.$$

The terms containing pressure vanish, because $\nabla \boldsymbol{v} : \boldsymbol{\delta} = \nabla \cdot \boldsymbol{v} = 0$, and the other terms cancel because $\nabla \boldsymbol{v}' : \nabla \boldsymbol{v}''^{\mathrm{T}} = \nabla \boldsymbol{v}'^{\mathrm{T}} : \nabla \boldsymbol{v}''$. Therefore, this entire integral vanishes, reducing (1.54) to

$$\int_V (\boldsymbol{v}' \cdot (\nabla \cdot \boldsymbol{\sigma}'') - \boldsymbol{v}'' \cdot (\nabla \cdot \boldsymbol{\sigma}'))\ dV = \int_S (\boldsymbol{v}' \cdot (\boldsymbol{n} \cdot \boldsymbol{\sigma}'') - \boldsymbol{v}'' \cdot (\boldsymbol{n} \cdot \boldsymbol{\sigma}'))\ dS. \tag{1.55}$$

This is the most fundamental and general of the *Lorentz reciprocal relations*; it is valid for *any incompressible Newtonian flow*. The other Lorentz reciprocal relations, which we present now, are special cases of (1.55) for solutions of the Stokes equation with no imposed body force: $\nabla \cdot \boldsymbol{\sigma}' = \nabla \cdot \boldsymbol{\sigma}'' = 0$. Now the left-hand side of (1.55) vanishes, so

$$\int_S (\boldsymbol{v}' \cdot (\boldsymbol{n} \cdot \boldsymbol{\sigma}'') - \boldsymbol{v}'' \cdot (\boldsymbol{n} \cdot \boldsymbol{\sigma}'))\ dS = 0. \tag{1.56}$$

The quantity $\boldsymbol{n} \cdot \boldsymbol{\sigma}$ is the traction vector $\boldsymbol{t}$, so (1.56) can be rewritten

$$\int_S (\boldsymbol{v}' \cdot \boldsymbol{t}'' - \boldsymbol{v}'' \cdot \boldsymbol{t}')\ dS = 0. \tag{1.57}$$

Applying the divergence theorem to (1.56) leads to the volumetric form

$$\int_S (\nabla \cdot (\boldsymbol{\sigma}'' \cdot \boldsymbol{v}') - \nabla \cdot (\boldsymbol{\sigma}' \cdot \boldsymbol{v}''))\ dV = 0. \tag{1.58}$$

Equations (1.55)–(1.58) are widely used in situations where the primed problem has a known solution and we seek information about a related problem, which is taken to be the double-primed problem. Examples of such applications are found in Sections 3.5, 4.1, 4.6, and 6.4.

### 1.3.4 Mechanical Energy Balance and the Minimum Dissipation Principle

Many examples can be found in which physical laws can be stated as variational principles. For example, thermodynamic equilibrium is the state that maximizes entropy subject to various constraints (Robertson 1993), and Newton's equations of motion can be derived from Hamilton's principle of least action (Goldstein 1980). In the context of

[8] Equation (A.4) defines the double dot product as used in this book.

fluid dynamics, a natural question to ask is, "what is the flow that minimizes the energy dissipation rate subject to the constraint of incompressibility?"

To address this question, we first consider the mechanical energy balance for an incompressible continuum, obtained by taking the dot product of the Cauchy momentum equation with the velocity $\boldsymbol{v}$ and rearranging:

$$\frac{D}{Dt}\left(\frac{1}{2}\rho v^2\right) = \boldsymbol{v} \cdot \boldsymbol{\nabla} \cdot \boldsymbol{\sigma} + \boldsymbol{v} \cdot \boldsymbol{f}. \tag{1.59}$$

Here $v^2 = \boldsymbol{v} \cdot \boldsymbol{v}$. The left-hand side of this expression is simply the rate of change of kinetic energy of a material element. Integrating over the entire domain $V_{\mathrm{D}}$ of flow and applying the divergence theorem, we find that

$$\int_{V_{\mathrm{D}}} \frac{D}{Dt}\left(\frac{1}{2}\rho v^2\right) dV = \int_{S_{\mathrm{D}}} \boldsymbol{n} \cdot \boldsymbol{\sigma} \cdot \boldsymbol{v}\, dS + \int_{V_{\mathrm{D}}} \boldsymbol{v} \cdot \boldsymbol{f}\, dV - \Phi_v, \tag{1.60}$$

where

$$\Phi_v = \int_{V_{\mathrm{D}}} \boldsymbol{\sigma} : \boldsymbol{\nabla}\boldsymbol{v}\, dV. \tag{1.61}$$

The left-hand side of (1.60) is the total rate of change of kinetic energy of the material. The first two terms on the right-hand side are the rates of work on the material exerted across the domain boundary and via the body forces within the material. The last term is the one of primary interest here: it is the rate of interconversion between kinetic and internal energy (Bird et al. 2002). For an incompressible Newtonian fluid, the expression for $\Phi_v$ reduces to

$$\Phi_v = \int_V \boldsymbol{\tau} : \boldsymbol{\nabla}\boldsymbol{v}\, dV = \int_V 2\eta\mathbf{E} : \boldsymbol{\nabla}\boldsymbol{v}\, dV = \int_V \eta\left(\boldsymbol{\nabla}\boldsymbol{v} + \boldsymbol{\nabla}\boldsymbol{v}^{\mathrm{T}}\right) : \boldsymbol{\nabla}\boldsymbol{v}\, dV, \tag{1.62}$$

and it is straightforward to show that this is always nonnegative. (For a material with some elasticity, $\Phi_v$ can be negative, in which case stored elastic energy is being converted to mechanical energy.)

To address the previously posed question, we wish to minimize $\Phi_v$ subject to the incompressibility constraint, and we will consider the case of a given velocity on the boundary. Since incompressibility must apply at every point in the fluid, the Lagrange multiplier associated with that constraint must be a function of position. Denoting this as $\lambda(\boldsymbol{x})$, we write the objective functional to be extremized as

$$I = \int_V \left(\eta\left(\boldsymbol{\nabla}\boldsymbol{v} + \boldsymbol{\nabla}\boldsymbol{v}^{\mathrm{T}}\right) : \boldsymbol{\nabla}\boldsymbol{v} - \lambda(\boldsymbol{x})\boldsymbol{\nabla} \cdot \boldsymbol{v}\right) dV, \tag{1.63}$$

subject to incompressibility and the boundary condition

$$\boldsymbol{v} = \boldsymbol{v}_{\mathrm{D}} \tag{1.64}$$

on the boundary $S_{\mathrm{D}}$. This is a standard application of the calculus of variations (Greenberg 1978).

It will be convenient to work in index notation. We define a comparison function

$$V_i = v_i + \epsilon w_i,$$

where $w_i = 0$ on the boundary but is otherwise arbitrary, and $v_i$ is the velocity field that yields the extremum. Since $w_i = 0$ on the boundary, $V_i = v_i = v_{Di}$ there. We seek the equation that must be satisfied by $v_i$ such that

$$I(\epsilon) = \int_V \left( \eta \left( \frac{\partial V_i}{\partial x_j} + \frac{\partial V_j}{\partial x_i} \right) \frac{\partial V_i}{\partial x_j} - \lambda(\boldsymbol{x}) \frac{\partial V_i}{\partial x_i} \right) dV$$

is an extremum. This occurs when

$$\left. \frac{dI}{d\epsilon} \right|_{\epsilon=0} = 0.$$

Substituting in the definition of $V_i$, this equation becomes

$$\int_V \left[ 2\eta \left( \frac{\partial v_i}{\partial x_j} + \frac{\partial v_j}{\partial x_i} \right) \frac{\partial w_i}{\partial x_j} - \lambda \frac{\partial w_i}{\partial x_i} \right] dV = 0.$$

The first term can be written using the product rule:

$$\int_V \left[ \frac{\partial}{\partial x_j} \left( \left( \frac{\partial v_j}{\partial x_i} + \frac{\partial v_i}{\partial x_j} \right) w_i \right) - w_i \left( \frac{\partial}{\partial x_j} \left( \frac{\partial v_j}{\partial x_i} + \frac{\partial v_i}{\partial x_j} \right) \right) \right] dV,$$

and applying the divergence theorem yields:

$$\int_S n_j \left( \frac{\partial v_j}{\partial x_i} + \frac{\partial v_i}{\partial x_j} \right) w_i \, dS - \int_V w_j \left( \frac{\partial}{\partial x_j} \left( \frac{\partial v_j}{\partial x_i} + \frac{\partial v_i}{\partial x_j} \right) \right) dV.$$

Similarly, the second term becomes

$$-\int_S n_j w_j \lambda \, dS + \int_V w_j \frac{\partial \lambda}{\partial x_j} dV.$$

The boundary terms vanish because $w_i = 0$ there, so substituting these expressions into the extremality condition and applying incompressibility yields that

$$\int_V \left[ -2\eta \frac{\partial^2 v_i}{\partial x_j^2} + \frac{\partial \lambda}{\partial x_i} \right] w_i \, dV = 0.$$

Since $w_i$ is arbitrary, the term in the square brackets must be zero everywhere in the domain. Letting $\lambda = 2p$, we therefore have that

$$= \eta \frac{\partial^2 v_i}{\partial x_j^2} - \frac{\partial p}{\partial x_i} = 0.$$

This is the Stokes equation, with pressure appearing as the Lagrange multiplier needed to satisfy incompressibility. Finally, to show that the solution to the Stokes equation yields the *minimum* dissipation, it suffices to show that any other velocity field that satisfies the boundary conditions has greater dissipation (Problem 1.8). Thus the *minimum dissipation principle* states that the incompressible Newtonian flow satisfying the boundary condition (1.64) that has the least dissipation is the solution to the Stokes flow problem in that geometry.

It is important to note that this minimum dissipation result does not imply, for example, that a flowing particle suspension minimizes dissipation, because in that problem the boundaries are moving with time and the preceding statement applies only to an instant of time – the aforementioned minimization statement does not incorporate the time-dependent motion of the particles, so it says nothing about the suspension dynamics as a whole.

## Problems

**1.1** Find the eigenvalues and eigenvectors of (1.7) for arbitrary $\alpha \in [-1, 1]$ and use the results to show that Figure 1.4 is correct.

**1.2** Let $\boldsymbol{A}(\boldsymbol{X}(t), t) = \mathbf{F}(\boldsymbol{X}(t), t) \cdot \mathbf{F}(\boldsymbol{X}(t), t)^{\mathrm{T}}$. Show that

$$\frac{d\boldsymbol{A}(\boldsymbol{X}(t),t)}{dt} = \left(\boldsymbol{A}(\boldsymbol{X}(t),t) \cdot \nabla\boldsymbol{v}(\boldsymbol{X}(t),t) + \nabla\boldsymbol{v}^{\mathrm{T}}(\boldsymbol{X}(t),t) \cdot \boldsymbol{A}(\boldsymbol{X}(t),t)\right).$$

Since $\boldsymbol{A}(\boldsymbol{X}(0), 0) = \mathbf{F}(\boldsymbol{X}(0), 0) \cdot \mathbf{F}(\boldsymbol{X}(0), 0)^{\mathrm{T}} = \boldsymbol{\delta} = \mathbf{B}(\boldsymbol{X}(0), 0)$, this is equivalent to deriving (1.19).

**1.3** The inverse $\mathbf{B}^{-1}$ of the Green tensor is called the *Piola tensor*. Show that the evolution equation for $\mathbf{B}^{-1}$, which defines the *lower* or *covariant* convected derivative, is given by

$$\frac{D\mathbf{B}^{-1}(\boldsymbol{x},t)}{Dt} + \left(\mathbf{B}^{-1}(\boldsymbol{x},t) \cdot \nabla\boldsymbol{v}^{\mathrm{T}}(\boldsymbol{x},t) + \nabla\boldsymbol{v}(\boldsymbol{x},t) \cdot \mathbf{B}^{-1}(\boldsymbol{x},t)\right) = \left(\mathbf{B}^{-1}\right)^{(1)} = \overset{\Delta}{\mathbf{B}^{-1}} = 0. \tag{1.65}$$

Hint: $D(\mathbf{B} \cdot \mathbf{B}^{-1})/Dt = 0$.

**1.4** In the absence of molecular diffusion, the evolution of the concentration $c(\boldsymbol{x}, t)$ of a solute in an incompressible fluid is given by

$$\frac{\partial c}{\partial t} + \boldsymbol{v} \cdot \nabla c = 0.$$

Consider a linear flow field $\boldsymbol{v} = \mathbf{L} \cdot \boldsymbol{x}$ and an initial concentration field that varies sinusoidally with position, so it can be written as $c(\boldsymbol{x}, 0) = c_0 e^{i\boldsymbol{k}\cdot\boldsymbol{x}} + \text{c.c.}$ Seeking a solution $c(\boldsymbol{x}, t) = c_0 e^{i\boldsymbol{k}(t)\cdot\boldsymbol{x}} + \text{c.c.}$, find the evolution equation for the time-dependent wavevector $\boldsymbol{k}(t)$. How is this vector related to $\boldsymbol{g}_i(t)$ and/or $\boldsymbol{h}_j(t)$?

**1.5** (a) For the general shear flow case (no eigenvalues of $\mathbf{L}$ with positive real parts), we can write

$$\mathbf{L} = \begin{bmatrix} 0 & a & 0 \\ 0 & 0 & b \\ 0 & 0 & 0 \end{bmatrix}.$$

If $a$ and $b$ are constant, how fast can a material line stretch in this flow?

(b) Now, allowing a general time-dependent velocity gradient, use (1.11) to find an expression for the time averaged stretch rate

$$\sigma = \lim_{t\to\infty} \frac{1}{t} \ln\left(\frac{||\Delta\boldsymbol{X}(t)||}{||\Delta\boldsymbol{X}(0)||}\right)$$

in terms of $\mathbf{F}(t)$ and $\Delta\boldsymbol{X}(0)$. If $\sigma > 0$ for some initial material line orientation $\Delta\boldsymbol{X}(0)$, then the material line stretches (and nearby fluid elements diverge) exponentially fast, on average. For each trajectory in a flow, there are three independent values of $\sigma$; they are called *Lyapunov exponents*.

(c) Find the Lyapunov exponents for uniaxial extensional flow with extension rate $\dot{\epsilon}$, for which there is an orthogonal coordinate system in which

$$\mathbf{L} = \dot{\epsilon} \begin{bmatrix} 1 & 0 & 0 \\ 0 & -\frac{1}{2} & 0 \\ 0 & 0 & -\frac{1}{2} \end{bmatrix}.$$

(d) The end-to-end vector $\boldsymbol{Q}$ for a simple model of polymer molecule in flow (Section 8.6) satisfies

$$\frac{d\boldsymbol{Q}}{dt} = \mathbf{L} \cdot \boldsymbol{Q} - \frac{2}{\lambda}\boldsymbol{Q}(t) + \boldsymbol{\xi}(t),$$

where $\lambda$ is a relaxation time for the polymer and $\boldsymbol{\xi}(t)$ is a rapidly fluctuating term that comes from Brownian motion (Chapter 6). Neglecting this term (which does not affect the final result), find a criterion relating $\sigma$ and $\lambda$ that determines whether or not a polymer molecule will stretch indefinitely along a trajectory. Hint: Use an integrating factor.

**1.6** The angular momentum of a material volume $V_{\mathrm{M}}$ with respect to some fixed origin is

$$\int_{V_{\mathrm{M}}} \boldsymbol{x} \times (\rho \boldsymbol{v})\, dV.$$

If we allow external torques to be imposed at points within a material (this could occur, for example, by application of an external electric field to a fluid of polar molecules), then the conservation of angular momentum in this volume can be written in index notation as

$$\frac{d}{dt}\int_{V_{\mathrm{M}}} \epsilon_{ijk} x_j (\rho v_k)\, dV = \int_{V_{\mathrm{M}}} \epsilon_{ijk} x_j (\rho g_k)\, dV + \int_{S_{\mathrm{M}}} \epsilon_{ijk} x_j (n_l \sigma_{lk})\, dS + \int_{V_{\mathrm{M}}} c_i\, dV, \tag{1.66}$$

where $\boldsymbol{c}(\boldsymbol{x})$ is the torque density field (also called a couple density field or a body torque field).

(a) Use the divergence theorem to show that

$$\int_{S_{\mathrm{M}}} \epsilon_{ijk} x_j (n_l \sigma_{lk})\, dS = \int_{V_{\mathrm{M}}} \left( \epsilon_{ilk} \sigma_{lk} + \epsilon_{ijk} x_j \left( \frac{\partial \sigma_{lk}}{\partial x_l} \right) \right) dV.$$

(b) Use this result along with (1.66) to show that

$$\epsilon_{ilk} \sigma_{lk} + c_i = 0. \tag{1.67}$$

If there are no couples, $c_i = 0$ and the only general way to satisfy this equation is to take $\sigma_{lk} = \sigma_{kl}$. Therefore, in the absence of external torques, the stress tensor must be symmetric. Furthermore, multiplying by $\epsilon_{imn}$ and using (A.8) yields that

$$\sigma_{mn} - \sigma_{nm} + \epsilon_{imn} c_i = 0, \tag{1.68}$$

so even in the presence of external torques, the stress tensor can only be asymmetric if it depends on $\boldsymbol{c}$ in the specific manner shown here.

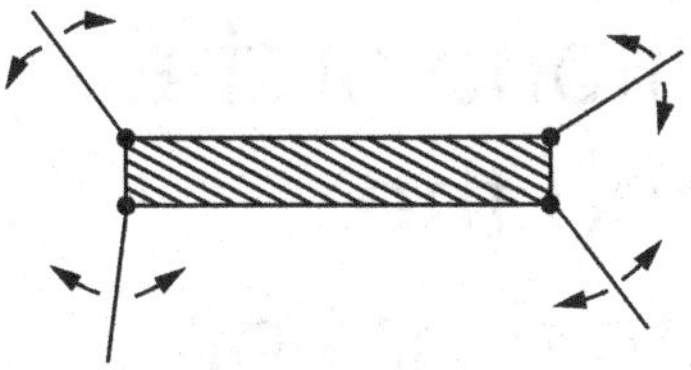

**Figure 1.10** A microswimmer. Each limb can pivot around its junction with the body in the plane of the page.

**1.7** A friend of yours who is a whiz at nanofabrication has constructed a small "swimming" device made of rods connected by hinges such that the "limbs" of the device can be moved by tiny motors – see Figure 1.10. Each limb can only move in the plane of the paper. But alas, your friend doesn't know any fluid mechanics – the protocol that was designed to sequentially move the limbs doesn't work, and the "swimmer" flops pathetically back and forth in the fluid. Design a limb motion protocol (i.e., a repeating sequence of limb motions) that will move the swimmer in one direction on average at zero Reynolds number. Also, give an example of a protocol that will *not* lead to any net motion.

**1.8** (a) Show that for viscous incompressible flow the rate of conversion of mechanical to internal energy

$$\Phi_v = \int_{V_D} \sigma : \nabla \boldsymbol{v} \, dV$$

is nonnegative.

(b) Show that for an incompressible flow

$$\int_{V_D} \mathbf{E} : \nabla \boldsymbol{v} \, dV = \int_{V_D} \mathbf{E} : \mathbf{E} \, dV,$$

where **E** is the deformation rate. It will be easier to work with the second expression in the next part of the problem.

(c) Now consider incompressible velocity fields $\boldsymbol{v}$ and $\boldsymbol{v}'$, where $\boldsymbol{v}$ solves the Stokes equation and $\boldsymbol{v}'$ does not (but both $\boldsymbol{v}$ and $\boldsymbol{v}'$ satisfy the same boundary conditions). Show that the dissipation rate for $\boldsymbol{v}'$ is always greater than or equal to that for $\boldsymbol{v}$. Start with the identity that

$$\int_{V_D} \mathbf{E}' : \mathbf{E}' \, dV - \int_{V_D} \mathbf{E} : \mathbf{E} \, dV = \int_{V_D} (\mathbf{E}' - \mathbf{E}) : (\mathbf{E}' - \mathbf{E}) \, dV + 2 \int_{V_D} (\mathbf{E}' - \mathbf{E}) : \mathbf{E} \, dV$$

and show that the right-hand side is nonnegative. This completes the proof that the Stokes equation yields the flow that minimizes dissipation.

Hint: Notice that

$$\frac{\partial (v_i - v_i')}{\partial x_j} E_{ij} = \frac{\partial}{\partial x_j} \left( (v_i - v_i') E_{ij} \right) - (v_i - v_i') \frac{\partial E_{ij}}{\partial x_j}.$$

The first of these terms is a divergence and the second contains $\frac{\partial E_{ij}}{\partial x_j} = \frac{1}{2} \nabla^2 v_i$.

# 2 Fundamental Solutions of the Stokes Equation and the Point-Particle Approximation

The linearity of the Stokes equation allows solutions to complex problems to be put together as sums of solutions to simpler problems. Perhaps the most important of these simpler problems is the flow generated by a point force (a delta function force density) exerted at some position in the flow domain.[1] Physically, one can imagine a small particle settling due to gravity in the fluid. That particle exerts a force on the fluid as it settles, and if we are far enough away from the particle, it appears as a point force. In general, the solution to a linear differential equation with a delta function forcing is called a *Green's function*. With this solution, a number of other *fundamental solutions* can be generated. We develop these here for unbounded flow domains and illustrate how they form a basis for computing and understanding the dynamics of particles in flow.

## 2.1 Free-Space Green's Function: The Stokeslet

Consider Stokes flow in an unbounded domain driven by a point force $\boldsymbol{F}$ exerted at the origin:

$$-\boldsymbol{\nabla} p + \eta\nabla^2\boldsymbol{v} + \boldsymbol{F}\delta(\boldsymbol{x}) = \boldsymbol{0}, \tag{2.1}$$

$$\boldsymbol{\nabla}\cdot\boldsymbol{v} = 0, \tag{2.2}$$

with $p \to p_\infty$ and $\boldsymbol{v} \to 0$ as $r = |\boldsymbol{x}| \to \infty$. For problems without a given reference pressure, we can take $p_\infty = 0$, since the velocity only depends on the pressure gradient. Taking the divergence of (2.1) and applying continuity yields that

$$-\nabla^2 p + \boldsymbol{F}\cdot\boldsymbol{\nabla}\delta(\boldsymbol{x}) = 0. \tag{2.3}$$

Introducing a function $\phi$ that satisfies

$$\nabla^2\phi = -\delta(\boldsymbol{x}) \tag{2.4}$$

(this function is in fact the free-space Green's function for the Laplace equation), we can rewrite (2.3) as

$$\begin{aligned} -\nabla^2 p &= \boldsymbol{F}\cdot\boldsymbol{\nabla}\nabla^2\phi \\ &= \nabla^2\boldsymbol{F}\cdot\boldsymbol{\nabla}\phi. \end{aligned}$$

[1] Properties of the delta function and its derivatives are described in Appendix A.4.

The solution to this equation is

$$p = p_\infty - \boldsymbol{F} \cdot \boldsymbol{\nabla}\phi. \tag{2.5}$$

Inserting this solution back into the Stokes equation and rearranging yields that

$$\eta\nabla^2\boldsymbol{v} = -\boldsymbol{F} \cdot (\boldsymbol{\nabla\nabla}\phi + \boldsymbol{\delta}\delta(\boldsymbol{x}))\,.$$

We can use (2.4) to rewrite this as

$$\eta\nabla^2\boldsymbol{v} = -\boldsymbol{F} \cdot \left(\boldsymbol{\nabla\nabla}\phi - \boldsymbol{\delta}\nabla^2\phi\right). \tag{2.6}$$

Now we introduce a new function $\zeta$ that satisfies

$$\nabla^2\zeta = \phi. \tag{2.7}$$

Substituting this into the first term on the right-hand side of (2.6) and rearranging yields that

$$\eta\nabla^2\boldsymbol{v} = -\nabla^2\left(\boldsymbol{F} \cdot (\boldsymbol{\nabla\nabla}\zeta - \boldsymbol{\delta}\phi)\right).$$

From this equation, it is clear that we can write an expression for the velocity field as

$$\boldsymbol{v} = -\frac{1}{\eta}\left(\boldsymbol{F} \cdot (\boldsymbol{\nabla\nabla}\zeta - \boldsymbol{\delta}\phi)\right) + \boldsymbol{a},$$

where $\boldsymbol{a}$ is any divergence-free function that satisfies $\nabla^2\boldsymbol{a} = \boldsymbol{0}$ everywhere. The condition that the velocity vanishes at infinity leads us to set $\boldsymbol{a} = \boldsymbol{0}$.

Now it simply remains to determine $\phi$ and $\zeta$. Seeking spherically symmetric solutions to (2.4) and (2.7), respectively, and discarding solutions that will lead to a velocity that does not vanish at infinity, we find that

$$\phi = \frac{1}{4\pi r}$$

and

$$\zeta = \frac{r}{8\pi}.$$

Using these results and the identity

$$\boldsymbol{\nabla} f(r) = \frac{df}{dr}\boldsymbol{\nabla} r = \frac{df}{dr}\frac{\boldsymbol{x}}{r}, \tag{2.8}$$

we find that

$$\boldsymbol{v} = \frac{1}{8\pi\eta r}\left(\boldsymbol{\delta} + \frac{\boldsymbol{xx}}{r^2}\right) \cdot \boldsymbol{F}. \tag{2.9}$$

Now letting the point force be exerted at an arbitrary position $\boldsymbol{x}_0$, we can write this expression as

$$\boldsymbol{v}(\boldsymbol{x}) = \mathbf{G}(\boldsymbol{x} - \boldsymbol{x}_0) \cdot \boldsymbol{F}, \tag{2.10}$$

where

$$\mathbf{G}(\boldsymbol{x} - \boldsymbol{x}_0) = \frac{1}{8\pi\eta|\boldsymbol{x} - \boldsymbol{x}_0|}\left(\boldsymbol{\delta} + \frac{(\boldsymbol{x} - \boldsymbol{x}_0)(\boldsymbol{x} - \boldsymbol{x}_0)}{|\boldsymbol{x} - \boldsymbol{x}_0|^2}\right) \tag{2.11}$$

is the free space Green's function for the Stokes equation. This tensor is also commonly

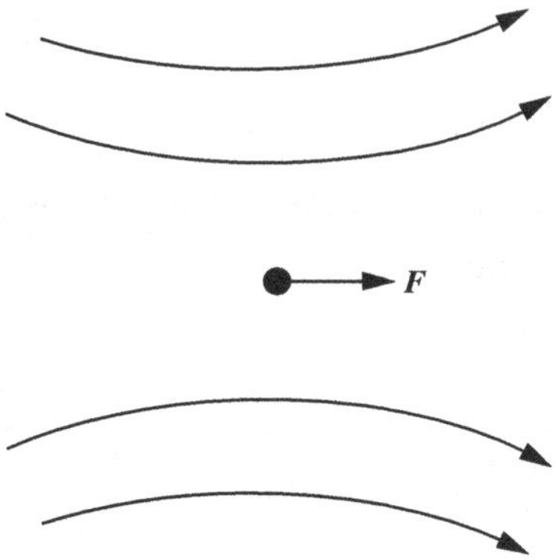


**Figure 2.1** Stokeslet velocity field.

called the *Stokeslet* or *Oseen–Burgers tensor*. Because the velocity field is divergence-free, the Green's function is as well:

$$\nabla \cdot \mathbf{G} = \mathbf{0}. \tag{2.12}$$

(Recall that the independent variable is $\boldsymbol{x}$, not $\boldsymbol{x}_0$: $\nabla \equiv \frac{\partial}{\partial \boldsymbol{x}}$.) Similarly, (1.37) implies that

$$\nabla^2 \nabla^2 \mathbf{G} = \mathbf{0}. \tag{2.13}$$

Given the Stokeslet velocity field (see Figure 2.1), the pressure and stress fields can be easily obtained:

$$p(\boldsymbol{x}) = p_\infty + \mathbf{\Pi}(\boldsymbol{x} - \boldsymbol{x}_0) \cdot \boldsymbol{F}, \quad \mathbf{\Pi}(\boldsymbol{x}) = \frac{1}{4\pi} \frac{\boldsymbol{x}}{r^3}, \tag{2.14}$$

$$\boldsymbol{\sigma}(\boldsymbol{x}) = \mathbf{T}(\boldsymbol{x} - \boldsymbol{x}_0) \cdot \boldsymbol{F}, \quad \mathbf{T}(\boldsymbol{x}) = -\frac{3}{4\pi} \frac{\boldsymbol{xxx}}{r^5}. \tag{2.15}$$

The quantities $\mathbf{G}$ and $\mathbf{\Pi}$ satisfy this form of the Stokes equation:

$$-\nabla \mathbf{\Pi} + \eta \nabla^2 \mathbf{G} + \boldsymbol{\delta} \delta(\boldsymbol{x} - \boldsymbol{x}_0) = \mathbf{0}. \tag{2.16}$$

The stress tensor satisfies

$$\nabla \cdot \boldsymbol{\sigma} + \boldsymbol{F} \delta(\boldsymbol{x} - \boldsymbol{x}_0) \tag{2.17}$$

so

$$\nabla \cdot \mathbf{T} + \boldsymbol{\delta} \delta(\boldsymbol{x} - \boldsymbol{x}_0) = \mathbf{0}. \tag{2.18}$$

The Stokeslet velocity field is axisymmetric with respect to an axis collinear with $\boldsymbol{F}$ and also satisfies a fore-aft symmetry with respect to the direction of $\boldsymbol{F}$, as required by reversibility: if the sign of $\boldsymbol{F}$ changes, so does the sign of $\boldsymbol{v}$. A very important aspect of the Stokeslet, especially with regard to particle–particle and particle–wall interactions, is that[2] $\mathbf{G}(\boldsymbol{x} - \boldsymbol{x}_0) \sim |\boldsymbol{x} - \boldsymbol{x}_0|^{-1}$; the velocity field generated by a point force decays

[2] When comparing mathematical expressions, we use the symbol "$\sim$" to mean "scales as," generally in some limiting case. That is, if $a \sim b$, then $a/b = O(1)$ in some limit. Later we will also use the symbol to

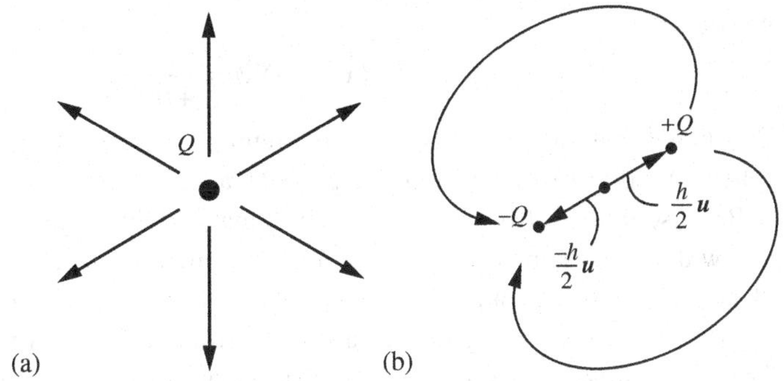


**Figure 2.2** Point source (a) and source dipole (b) flow fields.

very slowly with distance. The consequences of this fact arise throughout the field of microhydrodynamics.

## 2.2 Point Source and Point Source Dipole

In the construction of the Stokeslet, we implicitly required that the flow field be incompressible everywhere and that no sources or sinks of mass arose. Imagine for the moment the case where a point source of fluid mass with volumetric flow rate $Q$ exists at a position $\boldsymbol{x}_0$, which for the moment we take to be the origin. (This case can be approximately realized by injecting fluid into a large fluid-filled domain through a long needle.) In this case, the Stokes equation is unchanged, but the continuity equation becomes

$$-\nabla \cdot \boldsymbol{v} + s(\boldsymbol{x}) = 0 \tag{2.19}$$

with the volumetric source density

$$s(\boldsymbol{x}) = Q\delta(\boldsymbol{x}).$$

This has a simple solution of the form

$$\boldsymbol{v} = -Q\nabla\phi, \tag{2.20}$$

which leads to the equation

$$\nabla^2\phi = -\delta(\boldsymbol{x}). \tag{2.21}$$

(A velocity field that is proportional to the gradient of a function is called a potential flow.) We found the solution $\phi = \frac{1}{4\pi r}$ to (2.21) in the previous section; here it leads to the velocity field

$$\boldsymbol{v}(\boldsymbol{x}) = \boldsymbol{S}(\boldsymbol{x})Q, \tag{2.22}$$

denote that a random variable has a given probability distribution (see Appendix A.5). Usage will be clear from context.

where

$$\boldsymbol{S}(\boldsymbol{x}) = -\boldsymbol{\nabla}\phi = \frac{\boldsymbol{x}}{4\pi r^3}. \tag{2.23}$$

This flow is entirely radial, as shown in Figure 2.2(a), and the total flow rate across any sphere of radius $r$ centered at the origin is $Q$, as required.

Because it arises from a point source of mass, which cannot occur in the interior of a flow domain, the preceding solution is of limited direct relevance. The combination of an equal and opposite source and sink, however, leads to no net fluid entering or leaving the domain and generates a flow field that arises in a variety of applications. This *source dipole* has a source density $s(\boldsymbol{x})$ that consists of point sources of magnitude $\pm Q$ separated by a distance $h$ along a line in the $\boldsymbol{u}$ direction, where $\boldsymbol{u}$ is an arbitrary unit vector, as shown in Figure 2.2(b):

$$s(\boldsymbol{x}) = Q\left[\delta(\boldsymbol{x} - (\boldsymbol{x}_0 + \boldsymbol{u}\ h/2)) - \delta(\boldsymbol{x} - (\boldsymbol{x}_0 - \boldsymbol{u}\ h/2))\right]. \tag{2.24}$$

Because (2.19) is linear, we can superimpose point source solutions to find the velocity field for the dipole as

$$\boldsymbol{v}(x) = Q\left[\boldsymbol{S}(\boldsymbol{x} - (\boldsymbol{x}_0 + \boldsymbol{u}\ h/2)) - \boldsymbol{S}(\boldsymbol{x} - (\boldsymbol{x}_0 - \boldsymbol{u}\ h/2))\right]. \tag{2.25}$$

Multiplying and dividing by $h$, we can put (2.24) and (2.25) in a familiar form:

$$s(\boldsymbol{x}) = Qh\frac{\delta(\boldsymbol{x} - (\boldsymbol{x}_0 + \boldsymbol{u}\ h/2)) - \delta(\boldsymbol{x} - (\boldsymbol{x}_0 - \boldsymbol{u}\ h/2))}{h}$$

and

$$\boldsymbol{v}(x) = Qh\frac{\boldsymbol{S}(\boldsymbol{x} - (\boldsymbol{x}_0 + \boldsymbol{u}\ h/2)) - \boldsymbol{S}(\boldsymbol{x} - (\boldsymbol{x}_0 - \boldsymbol{u}\ h/2))}{h}.$$

Taking the limit $h \to 0$ while keeping the product $Qh \equiv Q_{\mathrm{d}}$ constant yields the point source dipole density

$$s_{\mathrm{d}}(\boldsymbol{x}) = Q_{\mathrm{d}}\boldsymbol{u} \cdot \boldsymbol{\nabla}_0\delta(\boldsymbol{x} - \boldsymbol{x}_0)$$

and corresponding velocity field

$$\boldsymbol{v}(x) = Q_{\mathrm{d}}\boldsymbol{u} \cdot \boldsymbol{\nabla}_0\boldsymbol{S}(\boldsymbol{x} - \boldsymbol{x}_0),$$

where $\boldsymbol{\nabla}_0 = \frac{\partial}{\partial \boldsymbol{x}_0}$. Now note that for any function $f(\boldsymbol{x} - \boldsymbol{x}_0)$,

$$\boldsymbol{\nabla}_0 f(\boldsymbol{x} - \boldsymbol{x}_0) = -\boldsymbol{\nabla} f(\boldsymbol{x} - \boldsymbol{x}_0).$$

Using this fact and (2.23) for $\boldsymbol{S}$, we have that

$$\boldsymbol{v}(x) = Q_{\mathrm{d}}\boldsymbol{u} \cdot \boldsymbol{\nabla}_0\boldsymbol{S} = -Q_{\mathrm{d}}\boldsymbol{u} \cdot \boldsymbol{\nabla}\boldsymbol{S}(\boldsymbol{x} - \boldsymbol{x}_0) = (\boldsymbol{\nabla\nabla}\frac{1}{4\pi r}) \cdot Q_{\mathrm{d}}\boldsymbol{u} \equiv \boldsymbol{S}_{\mathrm{d}} \cdot (Q_{\mathrm{d}}\boldsymbol{u}), \tag{2.26}$$

where

$$\boldsymbol{S}_{\mathrm{d}} = -\boldsymbol{\nabla}\boldsymbol{S} = (\boldsymbol{\nabla\nabla}\frac{1}{4\pi r}) = \frac{1}{4\pi r^3}\left(-\boldsymbol{\delta} + 3\frac{\boldsymbol{xx}}{r^2}\right). \tag{2.27}$$

Because this flow is just the derivative of the potential flow field found for a point source, it is often called a *potential dipole* flow. We will see in Section 3.3.1 that this flow arises when a particle moves in a fluid, displacing fluid from ahead of it to behind it.

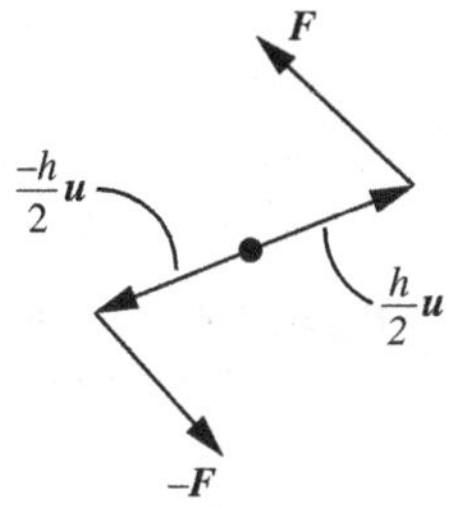


**Figure 2.3** Force dipole.

## 2.3 Force Dipole Solutions: Stresslet and Rotlet

By analogy with the simple potential dipole solution just presented, we can also construct *force dipole* solutions by superposition of Stokeslets. Consider the force density arising from equal and opposite forces applied at points $\boldsymbol{x}_0 \pm \boldsymbol{u}\ h/2$ as shown in Figure 2.3:

$$\boldsymbol{f}_{\mathrm{d}}(\boldsymbol{x}) = \boldsymbol{F}\left[\delta(\boldsymbol{x} - (\boldsymbol{x}_0 + \boldsymbol{u}\ h/2)) - \delta(\boldsymbol{x} - (\boldsymbol{x}_0 - \boldsymbol{u}\ h/2))\right].$$

This density yields a velocity field

$$\boldsymbol{v}(\boldsymbol{x}) = \left[\mathbf{G}(\boldsymbol{x} - (\boldsymbol{x}_0 + \boldsymbol{u}\ h/2)) - \mathbf{G}(\boldsymbol{x} - (\boldsymbol{x}_0 - \boldsymbol{u}\ h/2))\right] \cdot \boldsymbol{F}.$$

Following what we did previously with the source dipole, we recognize that in the limit $h \to 0$ with $\boldsymbol{F}_{\mathrm{d}} = \lim_{h\to 0} \boldsymbol{F} h$ held constant, this velocity field can be written

$$\boldsymbol{v}(\boldsymbol{x} - \boldsymbol{x}_0) = \boldsymbol{u} \cdot \boldsymbol{\nabla}_0 \mathbf{G}(\boldsymbol{x} - \boldsymbol{x}_0) \cdot \boldsymbol{F}_{\mathrm{d}}. \tag{2.28}$$

In index notation:

$$v_i = \left(\frac{\partial}{\partial x_{0j}} G_{ik}\right) F_{\mathrm{d},k} u_j. \tag{2.29}$$

This flow field is sometimes called the *Stokeslet doublet*. Based on the structure of (2.28), we can see that this flow results from a force density

$$\begin{aligned} \boldsymbol{f}_{\mathrm{d}}(\boldsymbol{x} - \boldsymbol{x}_0) &= \boldsymbol{u} \cdot \boldsymbol{\nabla}_0 \delta(\boldsymbol{x} - \boldsymbol{x}_0) \boldsymbol{F}_{\mathrm{d}} \\ &= \boldsymbol{F}_{\mathrm{d}} \boldsymbol{u} \cdot \boldsymbol{\nabla}_0 \delta(\boldsymbol{x} - \boldsymbol{x}_0) \\ &= -\boldsymbol{F}_{\mathrm{d}} \boldsymbol{u} \cdot \boldsymbol{\nabla} \delta(\boldsymbol{x} - \boldsymbol{x}_0) \\ &= \boldsymbol{\nabla} \cdot (-\boldsymbol{u} \boldsymbol{F}_{\mathrm{d}} \delta(\boldsymbol{x} - \boldsymbol{x}_0)). \end{aligned} \tag{2.30}$$

This force density has the structure of the divergence of a singular stress. This is an important observation, so we will elaborate on it here. Let the force dipole tensor $\mathbf{D}$ be defined as

$$\mathbf{D} = -\boldsymbol{F}_{\mathrm{d}} \boldsymbol{u}, \tag{2.31}$$

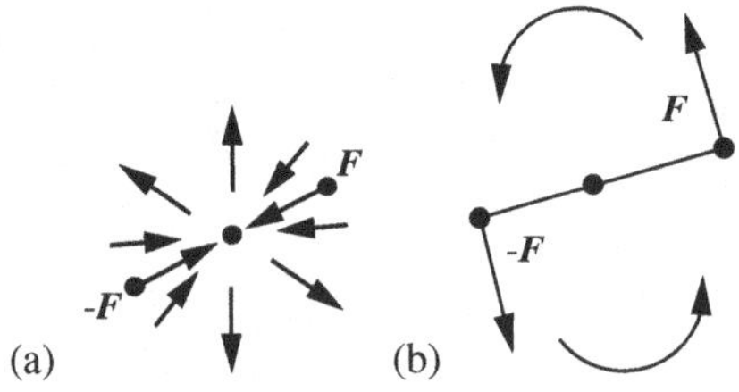


**Figure 2.4** (a) Stresslet and (b) rotlet force dipoles and their resulting velocity fields.

so we can rewrite (2.29) in index notation as

$$v_{\mathrm{d},i} = \left(\frac{\partial}{\partial x_j} G_{ik}\right) D_{kj}, \tag{2.32}$$

where

$$\frac{\partial}{\partial x_j} G_{ik} = -\frac{1}{8\pi\eta}\left(\frac{1}{r^3}\left(\delta_{ik}x_j - \delta_{ij}x_k - \delta_{kj}x_i\right) + 3\frac{x_i x_j x_k}{r^5}\right).$$

Note that for each value of $i$, (2.32) has the form $A_{jk}B_{kj}$. Now it is straightforward to show that $A_{jk}B_{kj} = A^{\mathrm{S}}_{jk}B^{\mathrm{S}}_{kj} + A^{\mathrm{A}}_{jk}B^{\mathrm{A}}_{kj}$, where superscripts $S$ and $A$ represent symmetric and antisymmetric parts respectively. (E.g., $A^{\mathrm{S}}_{jk} = \frac{1}{2}(A_{jk} + A_{kj})$, $A^{\mathrm{A}}_{jk} = \frac{1}{2}(A_{jk} - A_{kj})$.) Therefore, we can write $\boldsymbol{v} = \boldsymbol{v}^{\mathrm{S}} + \boldsymbol{v}^{\mathrm{A}}$, where

$$v^{\mathrm{S}}_i = \frac{1}{2}\left(\frac{\partial G_{ik}}{\partial x_j} + \frac{\partial G_{ij}}{\partial x_k}\right) D^{\mathrm{S}}_{kj} \tag{2.33}$$

and

$$v^{\mathrm{A}}_i = \frac{1}{2}\left(\frac{\partial G_{ik}}{\partial x_j} - \frac{\partial G_{ij}}{\partial x_k}\right) D^{\mathrm{A}}_{kj}. \tag{2.34}$$

The isotropic part of $\mathbf{D}$ drives no motion because $\boldsymbol{\nabla}\cdot\mathbf{G} = \mathbf{0}$ (see Problem 2.2), so it is appropriate to just consider the flow due to the traceless part of the force dipole tensor:

$$\mathbf{D} - \frac{1}{3}\,\mathrm{Tr}\,\mathbf{D}\boldsymbol{\delta}.$$

Furthermore, we define the symmetric tensor

$$\mathbf{S} = \mathbf{D}^{\mathrm{S}} - \frac{1}{3}\,\mathrm{Tr}\,\mathbf{D}^{\mathrm{S}}\boldsymbol{\delta} = \mathbf{D}^{\mathrm{S}} - \frac{1}{3}\,\mathrm{Tr}\,\mathbf{D}\boldsymbol{\delta} \tag{2.35}$$

and antisymmetric tensor

$$\mathbf{R} = \mathbf{D}^{\mathrm{A}} - \frac{1}{3}\,\mathrm{Tr}\,\mathbf{D}^{\mathrm{A}}\boldsymbol{\delta} = \mathbf{D}^{\mathrm{A}} \tag{2.36}$$

so that $\mathbf{D}$ can be written

$$\mathbf{D} = \mathbf{S} + \mathbf{R} + \frac{1}{3}\,\mathrm{Tr}\,\mathbf{D}\boldsymbol{\delta}. \tag{2.37}$$

Figure 2.4 shows the force dipoles and resulting flows associated with $\mathbf{S}$ and $\mathbf{R}$. From

Figure 2.4(b), it is apparent that $-\mathbf{R}$ corresponds to a point torque exerted on the fluid. Using these definitions, we can rewrite (2.33)–(2.34) as

$$v_i^{\mathrm{STR}} = T_{jik}^{\mathrm{STR}} S_{kj} \tag{2.38}$$

and

$$v_i^{\mathrm{ROT}} = T_{jik}^{\mathrm{ROT}} R_{kj}, \tag{2.39}$$

where STR stands for "stresslet" and ROT stands for "rotlet." The origins of these names will become clearer momentarily. The two third-order tensors are, respectively,

$$T_{jik}^{\mathrm{STR}} = -\frac{1}{8\pi\eta}\frac{3x_i x_j x_k}{r^5}, \tag{2.40}$$

and

$$T_{jik}^{\mathrm{ROT}} = -\frac{1}{8\pi\eta}\frac{\delta_{ik}x_j - \delta_{ij}x_k}{r^3}. \tag{2.41}$$

Note that $\mathbf{T}^{\mathrm{STR}}$ differs only by a scalar factor from the stress tensor $\mathbf{T}$ corresponding to the Stokelet $\mathbf{G}$.

Returning to the expression (2.30), we can rewrite this as

$$\begin{aligned} \boldsymbol{f}_{\mathrm{d}}(\boldsymbol{x}) &= \boldsymbol{\nabla}\cdot\left(\mathbf{D}^T\delta(\boldsymbol{x}-\boldsymbol{x}_0)\right) \\ &= \boldsymbol{\nabla}\cdot\left((\mathbf{S}+\mathbf{R}^T)\delta(\boldsymbol{x}-\boldsymbol{x}_0)\right) + \boldsymbol{\nabla}\left(\frac{1}{3}\operatorname{Tr}\mathbf{D}\delta(\boldsymbol{x}-\boldsymbol{x}_0)\right). \end{aligned}$$

The antisymmetric tensor $\mathbf{R}$ can be written as a pseudovector:

$$R_{ij} = -\frac{1}{2}\epsilon_{ijk}M_k, \tag{2.42}$$

where $\boldsymbol{M} = -\boldsymbol{T}$ is the torque exerted on the dipole (so $\boldsymbol{T}$ is the torque exerted by the dipole on the fluid). Now we can rewrite (2.39) as

$$\begin{aligned} v_i &= -T_{jik}^{\mathrm{ROT}}\frac{1}{2}\epsilon_{kjl}M_l \\ &= \frac{1}{16\pi\eta}\frac{\delta_{ik}x_j - \delta_{ij}x_k}{r^3}\epsilon_{kjl}M_l \\ &= \frac{1}{8\pi\eta r^3}\epsilon_{ijl}x_j M_l. \end{aligned} \tag{2.43}$$

With these results, we can rewrite (2.30) as

$$\boldsymbol{f}_{\mathrm{d}}(\boldsymbol{x}) = \boldsymbol{\nabla}\cdot(\mathbf{S}\delta(\boldsymbol{x}-\boldsymbol{x}_0)) - \frac{1}{2}\boldsymbol{\nabla}\times(\boldsymbol{M}\delta(\boldsymbol{x}-\boldsymbol{x}_0)) + \boldsymbol{\nabla}\left(\frac{1}{3}\operatorname{Tr}\mathbf{D}\delta(\boldsymbol{x}-\boldsymbol{x}_0)\right).$$

Interpretation of these terms is as follows: $\mathbf{S}$ is the symmetric part of the force dipole exerted *by* the fluid *on* the dipole – it arises in the equation as a point (symmetric) stress and is often called the *stresslet*; $\boldsymbol{M}$ is the torque exerted *by* the fluid *on* the dipole, and it or its equivalent tensor $\mathbf{R}$ is often called the *rotlet*; $\operatorname{Tr}\mathbf{D}$ is a uniform radially inward stress exerted *by* the fluid *on* the dipole – to enforce incompressibility, this will be counterbalanced by a contribution to the pressure field. By Newton's third law, taking

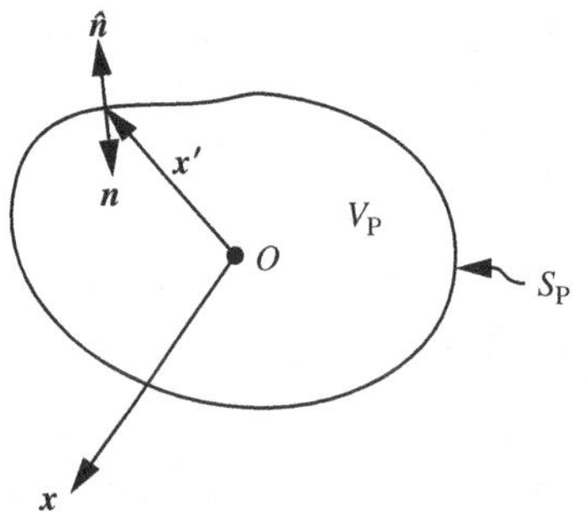


**Figure 2.5** Particle in an unbounded fluid.

the negatives of these yields the symmetric force dipole and the torque exerted *by* the dipole *on* the fluid. The following section elaborates on how the singularity solutions we have just derived are relevant to the motion of a particle in a fluid.

## 2.4 Multipole Expansion and Average Stress in a Suspension

The point force solution described in the preceding section was introduced by imagining a small particle sedimenting in a fluid, exerting a force on it. We return now to this idea to consider in general a particle with characteristic dimension $a$ in an unbounded fluid in the Stokes flow regime. The flow induced by the presence of the particle is generated by the forces that the particle exerts on the fluid: it is an integral over the Stokeslets generated by the force that each infinitesimal element of the particle surface exerts on the fluid. If the particle is centered at the origin, occupies the domain $V_P$, and has surface $S_P$ as shown in Figure 2.5, then this statement can be written as follows:

$$v_i(\boldsymbol{x}) - v_{\infty,i}(\boldsymbol{x}) = -\int_{S_P} G_{ij}(\boldsymbol{x}-\boldsymbol{x}')\left(\hat{n}_k\sigma_{kj}(\boldsymbol{x}')\right) dS(\boldsymbol{x}'). \tag{2.44}$$

Here $\boldsymbol{v}_\infty$ is the velocity the fluid would have in the absence of the particle, $G_{ij}$ is the free-space Green's function derived previously, $\hat{\boldsymbol{n}}$ is the unit normal pointing from the particle surface into the fluid, and $\left(\hat{n}_k\sigma_{kj}(\boldsymbol{x}')\right) dS(\boldsymbol{x}')$ is the force by the fluid exerted on the area element of particle surface located at position $\boldsymbol{x}'$. We use the hat on the normal vector to distinguish it from the usual normal, e.g., as appears in the divergence theorem, which is *outward* with respect to the fluid domain. Thus $\hat{\boldsymbol{n}} = -\boldsymbol{n}$.

Now consider a point $\boldsymbol{x}$ that is very far from the particle, so that $|\boldsymbol{x}| \gg |\boldsymbol{x}'|$ for all points $\boldsymbol{x}'$ on the particle surface ($|\boldsymbol{x}'| \lesssim a$). In this case, we can Taylor expand the Green's function:

$$G_{ij}(\boldsymbol{x}-\boldsymbol{x}') = G_{ij}|_{\boldsymbol{x}'=0} - x'_l \left.\frac{\partial G_{ij}}{\partial x_l}\right|_{\boldsymbol{x}'=0} + \frac{1}{2}x'_l x'_m \left.\frac{\partial}{\partial x_m}\frac{\partial G_{ij}}{\partial x_l}\right|_{\boldsymbol{x}'=\boldsymbol{0}} + O\left(\left(\frac{a}{|\boldsymbol{x}|}\right)^4\right),$$

and substitute this into (2.44) to yield:

$$
\begin{aligned}
v_i(\boldsymbol{x}) - v_{\infty,i}(\boldsymbol{x}) = &- G_{ij}|_{\boldsymbol{x}'=0} \int_{S_P} \left(\hat{n}_k \sigma_{kj}(\boldsymbol{x}')\right) dS(\boldsymbol{x}') \\
&+ \left.\frac{\partial G_{ij}}{\partial x_l}\right|_{\boldsymbol{x}'=0} \int_{S_P} x'_l \left(\hat{n}_k \sigma_{kj}(\boldsymbol{x}')\right) dS(\boldsymbol{x}') \\
&- \frac{1}{2} \frac{\partial}{\partial x_m} \left.\frac{\partial G_{ij}}{\partial x_l}\right|_{\boldsymbol{x}'=0} \int_{S_P} x'_l x'_m \left(\hat{n}_k \sigma_{kj}(\boldsymbol{x}')\right) dS(\boldsymbol{x}') \\
&+ O\left(\left(\frac{a}{|\boldsymbol{x}|}\right)^4\right).
\end{aligned} \tag{2.45}
$$

This expression is called the *multipole expansion* of the velocity field. We now examine its structure.

Consider the integral in the first term on the right-hand side of this expression, which we will denote $\boldsymbol{F}_{\text{drag}}$:

$$
F_{\text{drag},j} = \int_{S_P} \left(\hat{n}_k \sigma_{kj}(\boldsymbol{x}')\right) dS(\boldsymbol{x}'). \tag{2.46}
$$

This is simply the stress exerted *on* the particle *by* the fluid integrated over the surface of the particle – the total drag force exerted on the particle by the fluid. By Newton's third law, this is simply the negative of the force $\boldsymbol{F}$ that that the particle exerts on the fluid. It multiplies the free-space Green's function, giving a Stokeslet contribution to the velocity field that just arises from the total force exerted by the particle on the fluid.

The next term contains the gradient of the Green's function, which as we have seen is associated with point stresses and torques exerted on the fluid – it is the Stokeslet doublet. We denote the integral in this term as $\mathbf{D}$:

$$
D_{jl} = \int_{S_P} x'_l \left(\hat{n}_k \sigma_{kj}(\boldsymbol{x}')\right) dS(\boldsymbol{x}'). \tag{2.47}
$$

Note that this contains in its integrand the same structure as in (2.31), a force multiplied by a position vector. It is indeed the dipole tensor associated with the particle. Now we can write the multipole expansion in terms of the Stokeslet, stresslet, and rotlet solutions previously introduced:

$$
v_i(\boldsymbol{x}) - v_{\infty,i} = -G_{ij}(\boldsymbol{x}) F_{\text{drag},j} + \frac{\partial G_{ij}}{\partial x_l}(\boldsymbol{x}) D_{jl} + O\left(\left(\frac{a}{|\boldsymbol{x}|}\right)^3\right) \tag{2.48}
$$

$$
= -G_{ij}(\boldsymbol{x}) F_{\text{drag},j} + T^{\text{STR}}_{lij} S_{jl} + T^{\text{ROT}}_{lij} R_{jl} + O\left(\left(\frac{a}{|\boldsymbol{x}|}\right)^3\right). \tag{2.49}
$$

Now consider a particle that is sufficiently small that its inertia is negligible, so any external forces or torques exerted on it are balanced by forces and torques the particle exerts in turn on the fluid.[3] For a sedimenting particle, the first term dominates at long

[3] We see in Section 6.1 that suspended particles have a characteristic inertial-viscous relaxation time – on time scales larger than this, particle inertia is negligible. For very small particles, this time scale is very short compared to most time scales of interest.

distances because of the slow $1/r$ decay of the Stokeslet velocity field. For a neutrally buoyant particle, or more generally a particle that is not subjected to an external force, the first term vanishes and the second, which decays as $1/r^2$, dominates. Furthermore, if there are no external torques exerted on the particle, then the particle will exert no net torque on the fluid. Therefore, **D** will be symmetric and only a stresslet flow field will be generated. Thus a neutrally buoyant particle subjected to no external torques generates a velocity field:

$$v_i(\boldsymbol{x}) - v_{\infty,i} = T^{\text{STR}}_{lij} S_{jl} + \left( \left( \frac{a}{|\boldsymbol{x}|} \right)^3 \right).$$

Because it is symmetric, the particle stresslet **S** can be written in terms of its eigenvalues $\lambda_i$ and orthonormal eigenvectors $\boldsymbol{w}_i$ as a sum of dyads:

$$\mathbf{S} = \lambda_1 \boldsymbol{w}_1 \boldsymbol{w}_1 + \lambda_2 \boldsymbol{w}_2 \boldsymbol{w}_2 + \lambda_3 \boldsymbol{w}_3 \boldsymbol{w}_3. \tag{2.50}$$

Thus the stresslet for an arbitrary particle is the sum of three orthogonal stresslet dyads: i.e., three flows of the form shown in Figure 2.4(a), with the directions of the arrows determined by the value of $\lambda_i$. (The figure shows the case $\lambda_i < 0$.)

Note that in the absence of particle inertia, the no-net-force, no-net-torque conditions hold even if the particle is self-propelled, like a microorganism that swims using a flagellum. In this case, the backward thrust exerted by the flagellum on the fluid is balanced by the forward force exerted by the cell body on the fluid as it translates, generating a stresslet. If the flagella rotate, exerting a torque on the fluid, the body of a cell propelled by a rotating flagellum will counterrotate to exert an equal and opposite torque. This can be modeled as a torque dipole, which can be constructed as were the source and force dipoles previously described – it is a dipole of force dipoles. This will involve the second derivative of **G** and is thus part of the third term in (2.45), which is called the *quadrupole*. We will not treat this term in detail other than to note that in addition to the torque dipole it includes a potential dipole associated with the fluid that a particle displaces from in front of it to behind it as it moves through fluid. This term arises when considering the flow driven by a translating sphere, which we do in Sections 3.3.1 and 3.4.

Now consider a suspension of many neutrally buoyant force- and torque-free particles in a Newtonian fluid. If the externally imposed deformations have a scale much larger than the distance between particles, then we can define effective properties of the suspension by considering quantities that are averaged over a volume $V$ that is large compared to the distance between particles but small compared to the overall size of the flow system (Batchelor 1970*b*, Landau & Lifschitz 1984). Let there be $N$ particles in the volume so the number density $n = N/V$. Each particle is given a label $\alpha = 1, 2, 3, \ldots, N$, and particle $\alpha$ occupies volume $V_\alpha$ with surface $S_\alpha$. Once again, $\hat{\boldsymbol{n}}$ denotes the unit normal pointing from a particle into the fluid. We assume that the boundary of the volume does not cut through any particles. The volume-averaged velocity gradient is

$$\Gamma_{ji} = \frac{1}{V} \int_V \frac{\partial v_i}{\partial x_j} \, dV. \tag{2.51}$$

The integral can be split into a contribution from the suspending fluid and one from the particles:

$$\Gamma_{ji} = \frac{1}{V}\int_{V-\sum_{\alpha=1}^{N} V_\alpha} \frac{\partial v_i}{\partial x_j}\, dV + \frac{1}{V}\int_{\sum_{\alpha=1}^{N} V_\alpha} \frac{\partial v_i}{\partial x_j}\, dV$$
$$= \frac{1}{V}\int_{V-\sum_{\alpha=1}^{N} V_\alpha} \frac{\partial v_i}{\partial x_j}\, dV + \frac{1}{V}\sum_{\alpha=1}^{N}\int_{S_\alpha} v_i \hat{n}_j\, dS. \tag{2.52}$$

Similarly, the volume-averaged stress tensor is

$$\Sigma_{ij} = \frac{1}{V}\int_V \sigma_{ij}\, dV \tag{2.53}$$
$$= \frac{1}{V}\int_{V-\sum_{\alpha=1}^{N} V_\alpha} \left[-p\delta_{ij} + \eta\left(\frac{\partial v_i}{\partial x_j} + \frac{\partial v_j}{\partial x_i}\right)\right] dV + \frac{1}{V}\sum_{\alpha=1}^{N}\int_{V_\alpha} \sigma_{ij}\, dV. \tag{2.54}$$

The second integral in this expression contains the stress inside the particles. While we cannot determine this without further information about the particles, we do know that if inertia is negligible within the particles, then the stress there satisfies $\nabla\cdot\boldsymbol{\sigma} = \mathbf{0}$. Furthermore, based on the results of Sections 2.3 and 2.4, we can anticipate the importance of the first moment of the stress, $\boldsymbol{\sigma x}$. Observe that

$$\frac{\partial}{\partial x_i}\sigma_{ij}x_k = \frac{\partial \sigma_{ij}}{\partial x_i}x_k + \sigma_{ij}\delta_{ki} = \frac{\partial \sigma_{ij}}{\partial x_i}x_k + \sigma_{kj}.$$

Invoking $\nabla\cdot\boldsymbol{\sigma} = \mathbf{0}$ reduces this equation to

$$\frac{\partial}{\partial x_i}\sigma_{ij}x_k = \sigma_{kj}. \tag{2.55}$$

Therefore,

$$\int_{V_\alpha} \sigma_{ij}\, dV = \int_{V_\alpha} \frac{\partial}{\partial x_k}\sigma_{kj}x_i\, dV$$
$$= \int_{S_\alpha} \hat{n}_k \sigma_{kj} x_i\, dS$$
$$= D_{ij}^{\alpha} \tag{2.56}$$

where $D_{ij}^{\alpha}$ is precisely the dipole tensor for particle $\alpha$, which we have already encountered in the multipole expansion; see (2.47). Incorporating (2.52) and (2.56) in (2.54) yields that

$$\Sigma_{ij} = -p_{\text{eff}}\delta_{ij} + \eta\left(\Gamma_{ij} + \Gamma_{ji}\right) + \Sigma_{\text{P},ij}. \tag{2.57}$$

Here

$$p_{\text{eff}} = \frac{1}{V}\int_{V-\sum_{\alpha=1}^{N} V_\alpha} p\, dV \tag{2.58}$$

is an effective pressure for the suspension, so the first two terms of (2.57) simply comprise a Newtonian stress based on the average velocity gradient. The last term, given by

$$\Sigma_{P,ij} = \frac{1}{V}\sum_{\alpha} D_{ij}^{\alpha} - \frac{1}{V}\eta\sum_{\alpha}\int_{S_\alpha}\left(u_i\hat{n}_j + u_j\hat{n}_i\right)\,dS$$
$$= n\left[\frac{1}{N}\sum_{\alpha} D_{ij}^{\alpha}\right] - n\eta\left[\frac{1}{N}\sum_{\alpha}\int_{S_\alpha}\left(u_i\hat{n}_j + u_j\hat{n}_i\right)\,dS\right], \qquad (2.59)$$

is the particle-phase contribution to the average stress. The second term here cancels out the "extra" stress that comes from integrating the fluid stress over the particle volume. It can be rewritten:

$$-n\eta\left[\frac{1}{N}\sum_{\alpha}\int_{S_\alpha}\left(u_i\hat{n}_j + u_j\hat{n}_i\right)\,dS\right] = -n\eta\left[\frac{1}{N}\sum_{\alpha}\int_{V_\alpha}\left(\frac{\partial v_i}{\partial x_j} + \frac{\partial v_j}{\partial x_i}\right)\,dV\right]$$
$$= -n\eta\left[\frac{1}{N}\sum_{\alpha}\int_{V_\alpha} 2E_{ij}\,dV\right].$$

If the particles are rigid, then within them the rate of strain tensor $E_{ij} = 0$, in which case this term vanishes and the particle stress reduces to

$$\Sigma_{P,ij} = n\left[\frac{1}{N}\sum_{\alpha} D_{ij}^{\alpha}\right]. \qquad (2.60)$$

This is simply the number density of particles times the average force dipole per particle. Thus we can understand the stress in a suspension of rigid particles as if the particles were simply symmetric force dipoles – stresslets. Equations (2.59) and (2.60) can be rewritten with $\mathbf{D}^{\alpha}$ replaced with $\mathbf{S}^{\alpha}$, with the understanding that the isotropic part of $\mathbf{D}$ will make a contribution to the pressure.

The preceding results are valid for any concentration of particles as long as inertia is negligible in the particles. If the suspension is not dilute, then each of the particles generates a fluid motion that affects the velocity field experienced by the other particles, resulting in a complex mathematical problem that does not generally have a simple solution. If the suspension is dilute, however, then we can evaluate the expressions as if the particles were in isolation, which we will do for suspension of rigid spheres in Section 3.3.3 and a polymer molecule in dilute solution in Section 8.5.

## 2.5 Stokes's Law, Hydrodynamic Interactions, and the Mobility of a System of Point Particles

By the linearity of the Stokes equation, an isolated particle moving with velocity $\boldsymbol{U}$ exerts a force $\boldsymbol{F}$ on the fluid that is linearly proportional to the relative velocity of the particle and the surrounding fluid. For an isotropic object such as a rigid sphere or spherical drop, we can write

$$\boldsymbol{F} = \zeta\left(\boldsymbol{U} - \boldsymbol{v}(\boldsymbol{R})\right). \qquad (2.61)$$

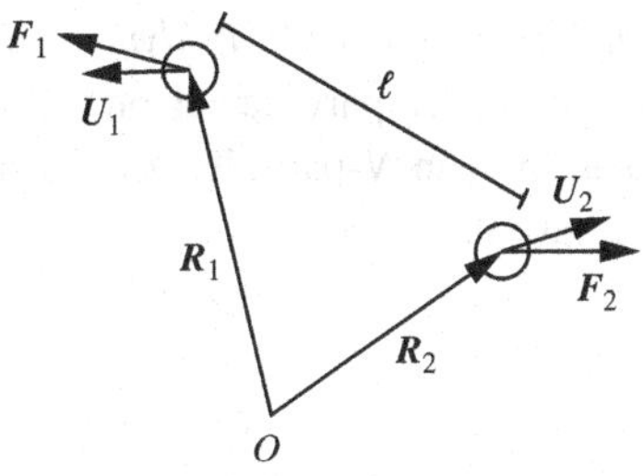


**Figure 2.6** Two hydrodynamically interacting spheres in a fluid.

Here $\boldsymbol{R}$ is the particle position, $\boldsymbol{v}(\boldsymbol{R})$ is the fluid velocity at that position in the absence of the particle and $\zeta$ is the *friction coefficient* for the particle. The result is called *Stokes's law*; formulas for $\zeta$ for spheres and for flexible polymer molecules in their equilibrium conformations will be derived in Sections 3.3.1 and 8.3, respectively. For a sphere of radius $a$, $\zeta = 6\pi\eta a$. For a nonspherical particle such as a fiber or for an initially isotropic object that has been deformed by flow, the friction coefficient is a tensor, as described in Section 3.7. The *mobility* of a particle is the inverse of its friction coefficient – for a nonspherical particle, it is also a tensor.

A question of broad interest is how the motion of an object in fluid (cell, macromolecule, particle segment) is affected by the presence and motion of other objects in the fluid. Imagine a pair of isotropic particles, each with characteristic dimension $a$, that are at positions $\boldsymbol{R}_1$ and $\boldsymbol{R}_2$ a distance $\ell$ apart in an unbounded fluid, as illustrated in Figure 2.6. Each of these spheres obeys Stokes's law and additionally, each generates a flow as it moves. If $a \ll \ell$, then we can approximate this flow using just the first term of the multipole expansion: i.e., as a Stokeslet. We will call this the point-particle approximation. For example, particle 1 exerts a force $\boldsymbol{F}_1$ on the fluid given based on the difference between its velocity and the fluid velocity in its absence:

$$\boldsymbol{F}_1 = \zeta(\boldsymbol{U}_1 - \boldsymbol{v}_1),$$

where

$$\boldsymbol{v}_1 = \boldsymbol{v}_\infty(\boldsymbol{R}_1) + \mathbf{G}(\boldsymbol{R}_1 - \boldsymbol{R}_2) \cdot \boldsymbol{F}_2.$$

Here $\boldsymbol{v}_\infty$ is the velocity field in the absence of any particles and $\mathbf{G}(\boldsymbol{R}_1 - \boldsymbol{R}_2) \cdot \boldsymbol{F}_2$ is the velocity induced at position $\boldsymbol{R}_1$ by the force exerted by the particle at $\boldsymbol{R}_2$. Corresponding results hold for $\boldsymbol{v}_2$ and $\boldsymbol{F}_2$. Combining these results yields that

$$\begin{bmatrix} \boldsymbol{U}_1 \\ \boldsymbol{U}_2 \end{bmatrix} = \begin{bmatrix} \boldsymbol{v}_\infty(\boldsymbol{R}_1) \\ \boldsymbol{v}_\infty(\boldsymbol{R}_2) \end{bmatrix} + \begin{bmatrix} \frac{1}{\zeta}\boldsymbol{\delta} & \mathbf{G}(\boldsymbol{R}_1 - \boldsymbol{R}_2) \\ \mathbf{G}(\boldsymbol{R}_2 - \boldsymbol{R}_1) & \frac{1}{\zeta}\boldsymbol{\delta} \end{bmatrix} \cdot \begin{bmatrix} \boldsymbol{F}_1 \\ \boldsymbol{F}_2 \end{bmatrix}.$$

Letting $\mathbf{U} = \begin{bmatrix} \boldsymbol{U}_1 \\ \boldsymbol{U}_2 \end{bmatrix}$, etc., this can be succinctly rewritten:

$$\mathbf{U} = \mathbf{V}_\infty + \mathbf{M} \cdot \mathbf{F}. \tag{2.62}$$

The off-diagonal blocks of the *mobility tensor* **M** represent the *hydrodynamic interactions* between the particles, the effect that the motion of each particle has on the other.

We can generalize this discussion and notation to an $N$-particle system, as long as all the particles are far apart. For particle $\alpha$, we have that

$$\boldsymbol{F}_\alpha = \zeta(\boldsymbol{U}_\alpha - \boldsymbol{v}(\boldsymbol{R}_\alpha)), \tag{2.63}$$

where

$$\boldsymbol{v}(\boldsymbol{R}_\alpha) = \boldsymbol{v}_\infty(\boldsymbol{R}_\alpha) + \sum_{\beta \neq \alpha} \mathbf{G}(\boldsymbol{R}_\alpha - \boldsymbol{R}_\beta) \cdot \boldsymbol{F}_\beta.$$

Now (2.62) holds, with

$$\mathsf{M}_{\alpha\beta} = \frac{1}{\zeta}\delta\delta_{\alpha\beta} + (1 - \delta_{\alpha\beta})\mathbf{G}(\boldsymbol{R}_{\alpha\beta}), \tag{2.64}$$

where $\mathsf{M}_{\alpha\beta}$ is the block of **M** corresponding to the velocity induced at particle $\alpha$ by the force that particle $\beta$ exerts on the fluid and $\boldsymbol{R}_{\alpha\beta} = \boldsymbol{R}_\alpha - \boldsymbol{R}_\beta$. Equation (2.62) is the generalization of Stokes's law for a system of $N$ point particles. This point is perhaps more clear if we define the *resistance tensor* $\mathsf{M}^{-1}$ and rewrite (2.62) as

$$\mathsf{F} = \mathsf{M}^{-1} \cdot (\mathsf{U} - \mathsf{V}_\infty). \tag{2.65}$$

Finally, letting $\mathsf{R} = \begin{bmatrix} \boldsymbol{R}_1^{\mathrm{T}} & \boldsymbol{R}_2^{\mathrm{T}} & \dots & \boldsymbol{R}_N^{\mathrm{T}} \end{bmatrix}^{\mathrm{T}}$, (2.62) can be rewritten

$$\frac{d\mathsf{R}}{dt} = \mathsf{V}_\infty + \mathsf{M} \cdot \mathsf{F}. \tag{2.66}$$

This is an evolution equation for the particle positions. The quantities $\mathsf{V}_\infty$, **M**, and **F** are often nonlinear functions of the particle positions, rendering (2.66) a nonlinear equation with potentially very complex solutions.

The specific expressions here are valid only if the particles are far apart. For rigid spheres, we will develop a more accurate approximation in Section 3.6. Nevertheless, the general form given by (2.62) remains the same. The mobility **M** is always symmetric, a result that follows from (4.3). Another important result comes from energetic considerations. The rate of work $\dot{W}$ that the particles perform as they move through an otherwise stationary fluid is

$$\dot{W} = \sum_{\alpha=1}^{N} \boldsymbol{F}_\alpha \cdot \boldsymbol{U}_\alpha = \mathsf{F} \cdot \mathsf{U} = \mathsf{F} \cdot \mathsf{M} \cdot \mathsf{F}.$$

This must be positive for any nonzero **F** (the particles cannot extract energy from an otherwise stationary fluid), so

$$\mathsf{F} \cdot \mathsf{M} \cdot \mathsf{F} > 0. \tag{2.67}$$

Thus **M** is positive definite.

A brief example further illuminates the properties of **M**. Consider a pair of spheres that at some instant are at positions $\boldsymbol{R}_1 = \ell/2\, \boldsymbol{e}_1 = [\ell/2 \quad 0 \quad 0]^{\mathrm{T}}$ and $\boldsymbol{R}_2 = -\ell/2\, \boldsymbol{e}_1 = [-\ell/2 \quad 0 \quad 0]^{\mathrm{T}}$. Letting $\mathsf{V}_\infty = \mathbf{0}$, the instantaneous pair velocity **U** is given by

$$\mathsf{U} = \mathsf{M} \cdot \mathsf{F}. \tag{2.68}$$

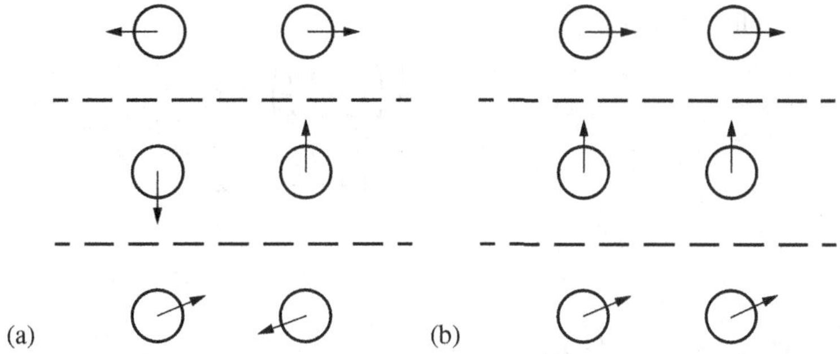

**Figure 2.7** Competitive (a) and cooperative (b) modes of relative motion for a pair of particles in Stokes flow.

At this instant, **M** is given from (2.64) as

$$\mathbf{M} = \frac{1}{\zeta}\begin{bmatrix} 1 & 0 & 0 & \frac{3}{2}\frac{a}{\ell} & 0 & 0 \\ 0 & 1 & 0 & 0 & \frac{3}{4}\frac{a}{\ell} & 0 \\ 0 & 0 & 1 & 0 & 0 & \frac{3}{4}\frac{a}{\ell} \\ \frac{3}{2}\frac{a}{\ell} & 0 & 0 & 1 & 0 & 0 \\ 0 & \frac{3}{4}\frac{a}{\ell} & 0 & 0 & 1 & 0 \\ 0 & 0 & \frac{3}{4}\frac{a}{\ell} & 0 & 0 & 1 \end{bmatrix}. \tag{2.69}$$

Substantial insight into the relation between **U**, **M**, and **F** can be gained by expressing **M** in terms of its eigenvalues and eigenvectors; i.e., solutions $\lambda_i$ and $\mathbf{Z}_i$ to the problem $\mathbf{M}\cdot\mathbf{Z} = \lambda\mathbf{Z}$. Because **M** is symmetric, the eigenvectors $\mathbf{Z}_i$ are orthogonal modes of forcing that lead to independent modes of collective motion of the particles: i.e., if the force vector $\mathbf{F} = \mathbf{Z}_i$, then the corresponding pair velocity is $\mathbf{U}_i = \lambda_i\mathbf{Z}_i$.

If the particles are infinitely far apart, the eigenvalues (mobilities) are all $\frac{1}{\zeta}$ and any six orthogonal vectors in $\mathbb{R}^6$ form a set of eigenvectors. If $a/\ell$ is finite, there are two families of eigenvalues and eigenvectors. The first family is

$$\lambda_1 = \frac{1}{\zeta}\left(1 - \frac{3}{2}\frac{a}{\ell}\right), \quad \mathbf{Z}_1 = \begin{bmatrix} -\boldsymbol{e}_1 \\ \boldsymbol{e}_1 \end{bmatrix},$$

$$\lambda_2 = \frac{1}{\zeta}\left(1 - \frac{3}{4}\frac{a}{\ell}\right), \quad \mathbf{Z}_2 = \begin{bmatrix} -\boldsymbol{e}_2 \\ \boldsymbol{e}_2 \end{bmatrix},$$

$$\lambda_3 = \lambda_2, \quad \mathbf{Z}_3 = \begin{bmatrix} -\boldsymbol{e}_3 \\ \boldsymbol{e}_3 \end{bmatrix},$$

and the second

$$\lambda_4 = \frac{1}{\zeta}\left(1 + \frac{3}{2}\frac{a}{\ell}\right), \quad \mathbf{Z}_4 = \begin{bmatrix} \boldsymbol{e}_1 \\ \boldsymbol{e}_1 \end{bmatrix},$$

$$\lambda_5 = \frac{1}{\zeta}\left(1 + \frac{3}{4}\frac{a}{\ell}\right), \quad \mathbf{Z}_5 = \begin{bmatrix} \boldsymbol{e}_2 \\ \boldsymbol{e}_2 \end{bmatrix},$$

$$\lambda_6 = \lambda_5, \quad \mathbf{Z}_6 = \begin{bmatrix} \boldsymbol{e}_3 \\ \boldsymbol{e}_3 \end{bmatrix}.$$

Thus, for example, $\mathbf{U}_1 = \lambda_1 \mathbf{Z}_1$, or

$$\begin{bmatrix} \boldsymbol{U}_1 \\ \boldsymbol{U}_2 \end{bmatrix}_1 = \frac{1}{\zeta}\left(1 - \frac{3}{2}\frac{a}{\ell}\right)\begin{bmatrix} \boldsymbol{F}_1 \\ \boldsymbol{F}_2 \end{bmatrix},$$

where

$$\begin{bmatrix} \boldsymbol{F}_1 \\ \boldsymbol{F}_2 \end{bmatrix} = \mathbf{Z}_1 = \begin{bmatrix} -\boldsymbol{e}_1 \\ \boldsymbol{e}_1 \end{bmatrix}.$$

Eigenvector $\mathbf{Z}_1$ corresponds to forces $\boldsymbol{F}_1$ and $\boldsymbol{F}_2$ that drive the two particles together or apart along the $x$-direction – imagine that a repulsive force is pushing the particles away from one another. The corresponding particle velocities are proportional to $\lambda_1$, which is smaller than $1/\zeta$, indicating that the motion of each particle occurs more slowly in the presence of the other particle than in its absence. This occurs because each particle drives a velocity field that is opposite to the direction the other particle is moving. Similar arguments hold for modes 2 and 3, so all three are "competitive modes" of collective motion. These are shown in Figure 2.7(a).

Turning to the second family of modes, $\mathbf{Z}_4$ corresponds to a force such as gravity that drives both particles in the same direction along the $x$-axis. The corresponding eigenvalue $\lambda_4$ is greater than $1/\zeta$ so the motions of particles 1 and 2 are "cooperative." The force particle 1 exerts on the fluid pushes particle 2 in the same direction. Modes 5 and 6 are also cooperative. Modes 4, 5, and 6 all correspond to the sedimentation of the pair of particles along one of the coordinate directions. Since $\lambda > 1/\zeta$ for all three of these modes, we see that two particles sediment faster than one. These motions are shown in Figure 2.7(b). An important example of hydrodynamic cooperativity is presented in Section 8.3 in the context of diffusion of long flexible polymer molecules in solution. Sections 3.6, 3.10, and 3.11 discuss aspects of the general $N$-particle mobility problem that move beyond the point-particle approximation.

## 2.6 Regularized Stokeslets

In the preceding example, observe that if the particles become too close together, then $\lambda_1$, $\lambda_2$, and $\lambda_3$ become negative (with $\lambda_1$ doing so first, at $\ell = 3a/2$), violating the positive definiteness condition on $\mathbf{M}$. This occurs because the point-particle approximation becomes invalid when the particles are not far apart. In fact, if the particles are too close together, this approximation allows particle 2 to be moved through the fluid by the flow

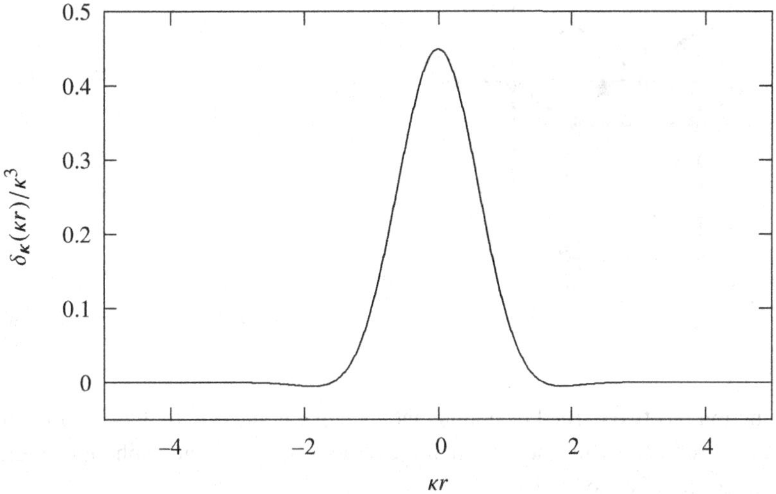


**Figure 2.8** Regularized delta function (2.70) plotted as $\delta_\kappa(\kappa r)/\kappa^3$ *vs.* $\kappa r$.

generated by particle 1 at a speed higher than particle 1 is itself moving. This is clearly unphysical and arises because of the $1/r$ singularity in the Stokeslet velocity field. A simple "fix" to this problem is to *regularize* the Stokeslet, modifying it to eliminate the singularity. One way to do this is to repeat the solution procedure described in Section 2.1, with the delta function $\delta(\boldsymbol{x})$ replaced by a regularized (three-dimensional) delta function such as this one (Figure 2.8):

$$\delta_\kappa(\boldsymbol{x}) = \frac{\kappa^3}{\sqrt{\pi}^3} \exp\left(-\kappa^2 r^2\right)\left[\frac{5}{2} - \kappa^2 r^2\right]. \tag{2.70}$$

This becomes a delta function in the limit $\kappa \to \infty$, and we can view $\kappa^{-1}$ as the length scale over which the delta function has been smeared out. The resulting velocity field (Problem 2.1) is

$$\boldsymbol{v}(\boldsymbol{x}) = \mathbf{G}_\kappa(\boldsymbol{x}) \cdot \boldsymbol{F}, \tag{2.71}$$

where $\mathbf{G}_\kappa(\boldsymbol{x})$ is a regularized Green's function:

$$\mathbf{G}_\kappa(\boldsymbol{x}) = \frac{\operatorname{erf}(\kappa r)}{8\pi\eta r}\left(\boldsymbol{\delta} + \frac{\boldsymbol{x}\boldsymbol{x}}{r^2}\right) + \frac{\kappa e^{-\kappa^2 r^2}}{4\pi^{3/2}\eta}\left(\boldsymbol{\delta} - \frac{\boldsymbol{x}\boldsymbol{x}}{r^2}\right). \tag{2.72}$$

As $\kappa r \to \infty$, $\operatorname{erf}(\kappa r) \to 1$ and $e^{-\kappa^2 r^2} \to 0$, so $\mathbf{G}_\kappa(\boldsymbol{x}) \to \mathbf{G}(\boldsymbol{x})$. Observe that $\mathbf{G}_\kappa$ does not diverge at the origin. Replacing $\mathbf{G}(\boldsymbol{R}_{\alpha\beta})$ with $\mathbf{G}_\kappa(\boldsymbol{R}_{\alpha\beta})$ in (2.64) yields a positive definite pair mobility as long as $\kappa^{-1} > 3a/\sqrt{\pi}$ (Hernández-Ortiz, de Pablo, & Graham 2007) (see Problem 2.4). Similar results can also be derived using other regularized delta functions (Cortez, Fauci, & Medovikov 2005, Nguyen & Cortez 2014).

Another approach is to directly replace the Stokeslet with an approximate version. The *Rotne–Prager–Yamakawa* (RPY) tensor $\mathbf{G}^{\mathrm{RPY}}(\boldsymbol{x})$ (Rotne & Prager 1969, Yamakawa 1970) is one such function that is widely used in simulation of polymer dynamics. This

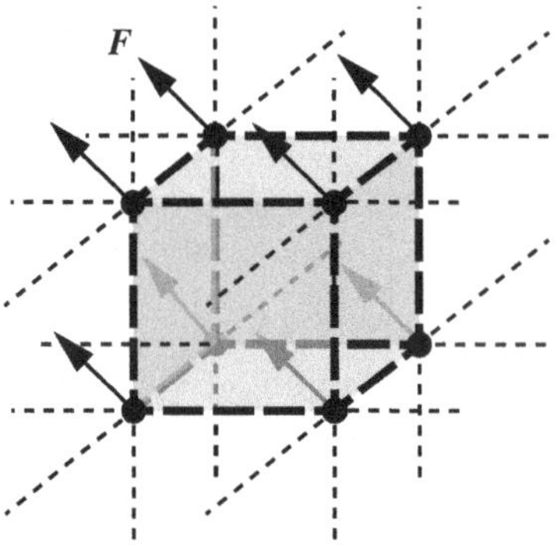


**Figure 2.9** A periodic array of point forces.

function has two parts. For points that are separated by more than $2a$, the Stokeslet flow field is replaced with the flow field generated by an isolated sphere of radius $a$ (which we derive in Section 3.3):

$$\mathbf{G}^{\mathrm{RPY}}(\boldsymbol{x}) = \left(1 + \frac{a^2}{6}\nabla^2\right)\mathbf{G}(\boldsymbol{x}), \quad \text{if } |\boldsymbol{x}| \geq 2a. \tag{2.73}$$

More explicitly,

$$\mathbf{G}^{\mathrm{RPY}}(\boldsymbol{x}) = \frac{1}{8\pi\eta|\boldsymbol{x}|}\left(C_1\boldsymbol{\delta} + C_2\frac{\boldsymbol{x}\boldsymbol{x}}{|\boldsymbol{x}|^2}\right). \tag{2.74}$$

When $|\boldsymbol{x}| < 2a$, we have instead

$$\mathbf{G}^{\mathrm{RPY}}(\boldsymbol{x}) = \frac{1}{6\pi\eta a}\left(C_1'\boldsymbol{\delta} + C_2'\frac{\boldsymbol{x}\boldsymbol{x}}{|\boldsymbol{x}|^2}\right), \quad \text{if } |\boldsymbol{x}| < 2a. \tag{2.75}$$

The functions $C_1$, etc. are given by

$$C_1 = 1 + \frac{2a^2}{3|\boldsymbol{x}|^2}, \quad C_2 = 1 - \frac{2a^2}{|\boldsymbol{x}|^2}, \tag{2.76}$$

$$C_1' = 1 - \frac{9|\boldsymbol{x}|}{32a}, \quad C_2' = \frac{3|\boldsymbol{x}|}{32a}. \tag{2.77}$$

Replacing $\mathbf{G}(\boldsymbol{R}_{\alpha\beta})$ with $\mathbf{G}^{\mathrm{RPY}}(\boldsymbol{R}_{\alpha\beta})$ in (2.64) yields a positive definite mobility for all particle separations (Rotne & Prager 1969).

## 2.7 Periodic Array of Point Forces

In all of the preceding cases, we have considered a single point forcing or a single particle in an unbounded domain. We now take our first steps away from this idealization to consider the flow generated by a point force in a periodic domain, or equivalently an infinite set of point forces exerted at points of a lattice in an unbounded domain (Hasimoto 1959) as shown in Figure 2.9. In developing this solution, some of the subtleties and technicalities associated with multiparticle systems will be seen for the first time.

Consider a set of point forces of strength $\boldsymbol{F}$ located at positions on a periodic lattice

$$\boldsymbol{x}_{\boldsymbol{m}} = m_1 \boldsymbol{a}^{(1)} + m_2 \boldsymbol{a}^{(2)} + m_3 \boldsymbol{a}^{(3)}, \quad (m_1, m_2, m_3 = 0, \pm 1, \pm 2, \ldots),$$

where $\boldsymbol{m} = (m_1, m_2, m_3)$ and the vectors $\boldsymbol{a}^{(i)}$ are the basis vectors defining the unit cell of the lattice. The volume of the unit cell is $V = \boldsymbol{a}^{(1)} \cdot (\boldsymbol{a}^{(2)} \times \boldsymbol{a}^{(3)})$. The position $\boldsymbol{x}_{(0,0,0)}$ is the origin and for a cubic lattice with side $L$, the basis vectors are simply $\boldsymbol{a}^{(i)} = L\boldsymbol{e}_i$. The force density is

$$\boldsymbol{f}(\boldsymbol{x}) = \boldsymbol{F} \sum_{\boldsymbol{m}} \delta(\boldsymbol{x} - \boldsymbol{x}_{\boldsymbol{m}}) = \boldsymbol{F} \sum_{m_1=-\infty}^{\infty} \sum_{m_2=-\infty}^{\infty} \sum_{m_3=-\infty}^{\infty} \delta(\boldsymbol{x} - \boldsymbol{x}_{(m_1,m_2,m_3)}). \tag{2.78}$$

Since this is a superposition of point forces, we begin by naively writing the resulting velocity field as a superposition of Stokeslets:

$$\boldsymbol{v}(\boldsymbol{x}) = \left( \sum_{\boldsymbol{m}} \mathbf{G}(\boldsymbol{x} - \boldsymbol{x}_{\boldsymbol{m}}) \right) \cdot \boldsymbol{F}. \tag{2.79}$$

A simple estimate of the magnitude of $\boldsymbol{v}$ begins to indicate why our naive approach is inadequate. Let $\boldsymbol{x}_0$ be any point that is not on the lattice, $r$ be the distance from that point to an arbitrary point in the domain, and $r_0$ be the distance from $\boldsymbol{x}$ to the nearest lattice point. A crude estimate of the magnitude of the velocity at $\boldsymbol{x}$ can be obtained by noting that $\mathbf{G} \sim 1/8\pi\eta r$ and that far enough away from $\boldsymbol{x}$, we can approximate the force distribution as a uniform density $\boldsymbol{f}$. We will approach our estimate by considering the contributions from point forces within a radius $L$ of $\boldsymbol{x}_0$ and then taking $L$ to $\infty$. Approximating the sum in (2.79) as an integral and using the uniform force density approximation, we conclude that

$$|\boldsymbol{v}(\boldsymbol{x}_0)| \sim \lim_{L\to\infty} \int_{r_0}^{L} \frac{1}{8\pi\eta r} |\boldsymbol{f}| \, r^2 \, dr. \tag{2.80}$$

As we take the limit of infinite domain size $L$, this integral diverges as $L^2$, suggesting that the velocity should diverge everywhere, because $\boldsymbol{x}_0$ is arbitrary. (One can make a more careful estimate and the result remains the same.) Although the contribution of distant point forces decays as $1/r$, the number of point forces within the volume grows as $r^2$. A similar divergence arises in considering the Poisson equation (e.g., for the electrostatic energy in a system of point charges or the gravitational potential energy in a system of point masses), for the same reason. See also Problem 2.6.

A resolution to this divergence is to counterbalance the momentum introduced by the point forces with a uniform pressure gradient that makes the net rate of introduction of momentum zero. One way to "implement" this pressure gradient would be to "put a no-flux boundary at infinity" so that the average velocity through the system is zero.

To mathematically formulate this approach, it is best to consider the problem in Fourier space, in which case the average velocity is proportional to the zero-wavenumber component of the Fourier-transformed velocity. To begin, we provide some definitions. A function $f(\boldsymbol{x})$ that is periodic over the domain $V$ can be written

$$f(\boldsymbol{x}) = \sum_{\boldsymbol{k}} \check{f}(\boldsymbol{k}) e^{i\boldsymbol{k}\cdot\boldsymbol{x}}, \tag{2.81}$$

where $\check{f}(\boldsymbol{k})$ is the finite Fourier transform of the function, defined as

$$\check{f}(\boldsymbol{k}) = \frac{1}{V}\int_V f(\boldsymbol{x})e^{-i\boldsymbol{k}\cdot\boldsymbol{x}}\, dV. \tag{2.82}$$

Here the permissible wavevectors $\boldsymbol{k}$ are given by

$$\boldsymbol{k} = 2\pi\left(m_1\boldsymbol{b}^{(1)} + m_2\boldsymbol{b}^{(2)} + m_3\boldsymbol{b}^{(3)}\right),$$

where the $\boldsymbol{b}^{(i)}$ are the basis vectors of the reciprocal lattice:

$$\boldsymbol{b}^{(1)} = \frac{\boldsymbol{a}^{(2)}\times\boldsymbol{a}^{(3)}}{V},\ \boldsymbol{b}^{(2)} = \frac{\boldsymbol{a}^{(3)}\times\boldsymbol{a}^{(1)}}{V},\ \boldsymbol{b}^{(3)} = \frac{\boldsymbol{a}^{(1)}\times\boldsymbol{a}^{(2)}}{V}.$$

For a cubic lattice, $\boldsymbol{b}^{(i)} = (1/L)\boldsymbol{e}_i$. The sum in (2.81) is over the $\boldsymbol{k}$ values corresponding to all values of $\boldsymbol{m}$. These wavevectors satisfy

$$\boldsymbol{k}\cdot\boldsymbol{a}^{(j)} = 2\pi m_j \quad (j = 1, 2, 3).$$

Finally, for a function $g(\boldsymbol{x})$ of the form

$$g(\boldsymbol{x}) = \sum_{\boldsymbol{m}} g_I(\boldsymbol{x} - \boldsymbol{x}_{\boldsymbol{m}}), \tag{2.83}$$

where $g_{\mathrm{I}}(\boldsymbol{x})$ is a *nonperiodic* function, the following identity can be established:

$$\check{g}(\boldsymbol{k}) = \frac{1}{V}\hat{g}_{\mathrm{I}}(\boldsymbol{k}), \tag{2.84}$$

where

$$\hat{g}_{\mathrm{I}}(\boldsymbol{k}) = \int g_{\mathrm{I}}(\boldsymbol{x})e^{-i\boldsymbol{k}\cdot\boldsymbol{x}}\, dV \tag{2.85}$$

is the three-dimensional Fourier transform of $g_{\mathrm{I}}(\boldsymbol{x})$ (i.e., the integral is over all space; information on Fourier transforms is given in Appendix A.3).

Now the velocity field is given by

$$\boldsymbol{v}(\boldsymbol{x}) = \sum_{\boldsymbol{k}} \check{\boldsymbol{v}}(\boldsymbol{k})e^{i\boldsymbol{k}\cdot\boldsymbol{x}}. \tag{2.86}$$

The pressure *gradient* will be taken to be spatially periodic with a nonzero (but as yet undetermined) mean value $\boldsymbol{g}$ that will be chosen to counteract the introduction of momentum by the point forces. Thus the pressure itself can be written

$$p(\boldsymbol{x}) = \boldsymbol{g}\cdot\boldsymbol{x} + \sum_{\boldsymbol{k}} \check{p}(\boldsymbol{k})e^{i\boldsymbol{k}\cdot\boldsymbol{x}}. \tag{2.87}$$

Applying the finite Fourier transform to the momentum and continuity equations leads to

$$-\boldsymbol{g}\delta_{\boldsymbol{k}\boldsymbol{0}} - i\boldsymbol{k}\check{p}(\boldsymbol{k}) - \eta k^2\check{\boldsymbol{v}}(\boldsymbol{k}) + \frac{1}{V}\boldsymbol{F} = \boldsymbol{0}, \tag{2.88}$$

$$i\boldsymbol{k}\cdot\check{\boldsymbol{v}}(\boldsymbol{k}) = 0. \tag{2.89}$$

Here and elsewhere, $k^2 = \boldsymbol{k}\cdot\boldsymbol{k}$ and we have applied (2.84) to (2.78). The quantity $\delta_{\boldsymbol{k}\boldsymbol{0}}$ is unity if $\boldsymbol{k} = \boldsymbol{0}$ and zero otherwise.

To solve (2.88)–(2.89), we consider first the zero-wavenumber case ($\boldsymbol{k} = \boldsymbol{0}$). Note that $\boldsymbol{g}$ is the mean pressure gradient; if this were not present, there would be no bounded solutions for $\check{\boldsymbol{v}}(\boldsymbol{k} = \boldsymbol{0})$ and $\check{p}(\boldsymbol{k} = \boldsymbol{0})$ – this is the divergence that was revealed by our previous naive analysis. However, the pressure gradient *is* present, and we can set it to balance the imposed force:

$$\boldsymbol{g} = \frac{1}{V}\boldsymbol{F}.$$

Now the velocity at $\boldsymbol{k} = \boldsymbol{0}$ is arbitrary, a consequence of the *Galilean invariance* of the Stokes (and Navier–Stokes) equations: for every solution $\boldsymbol{v}$ to the equations, $\boldsymbol{v} + \boldsymbol{U}$, where $\boldsymbol{U}$ is a constant is also a solution. This is a general result of classical mechanics – the equations of motion are unchanged if the reference frame moves at a constant velocity. In any case, if we impose the condition of no net flow through the domain, then $\check{\boldsymbol{v}}(\boldsymbol{k} = \boldsymbol{0}) = \boldsymbol{0}$. Similarly, we can take $\check{p}(\boldsymbol{k} = \boldsymbol{0}) = 0$.

Turning to the $\boldsymbol{k} \neq 0$ terms, we note that taking the divergence of the Stokes equation is equivalent in Fourier space to taking the dot product with $i\boldsymbol{k}$, which yields

$$\boldsymbol{0} = k^2 \check{p}(\boldsymbol{k}) + i\frac{1}{V}\boldsymbol{k} \cdot \boldsymbol{F},$$

where we have used the continuity equation $i\boldsymbol{k} \cdot \check{\boldsymbol{v}}(\boldsymbol{k}) = 0$. Thus

$$\check{p}(\boldsymbol{k}) = -\frac{1}{V}\frac{i\boldsymbol{k} \cdot \boldsymbol{F}}{k^2},$$

and we can use this result to solve for the velocity field (for $\boldsymbol{k} \neq 0$):

$$\check{\boldsymbol{v}}(\boldsymbol{k}) = \frac{1}{\eta V k^2}\left(\boldsymbol{\delta} - \frac{1}{k^2}\boldsymbol{k}\boldsymbol{k}\right) \cdot \boldsymbol{F}. \tag{2.90}$$

Note that $\left(\boldsymbol{\delta} - \frac{1}{k^2}\boldsymbol{k}\boldsymbol{k}\right)$ is the projection operator onto the plane perpendicular to $\boldsymbol{k}$: in Fourier space, the continuity equation requires that the velocity field be orthogonal to $\boldsymbol{k}$, and this operator enforces that orthogonality. Inserting this expression into (2.86) yields an expression for the velocity field.

The introduction of the pressure gradient $\boldsymbol{g}$ (or equivalently the no-flux condition at infinity) resolves the divergence problem. Unfortunately, because of the $1/k^2$ dependence in (2.90), the Fourier series (2.86), converges too slowly to be directly useful in general. This issue can be remedied, however, by recasting the solution in a form comprised of two rapidly converging sums. This *Ewald sum* approach originated in the context of the Poisson equation for electrostatics (Allen & Tildesley 1987) and was developed by Hasimoto for the Stokes equation (Hasimoto 1959). There are two ways to derive the desired result: one is to work directly with the solution in Fourier space (2.90) (Hasimoto 1959, Sierou & Brady 2001); the other is to begin with the Stokes equation and recast the form of the delta-function forcing (Hernández-Ortiz et al. 2007, Kumar & Graham 2012, Zhang, de Pablo & Graham 2012).

We choose the latter path. The slow decay of the velocity field with $k$ originates in the fact that the velocity is driven by a delta-function forcing, whose Fourier transform does not decay *at all* as $k$ increases. On the other hand, for any smooth function $f(x)$ that

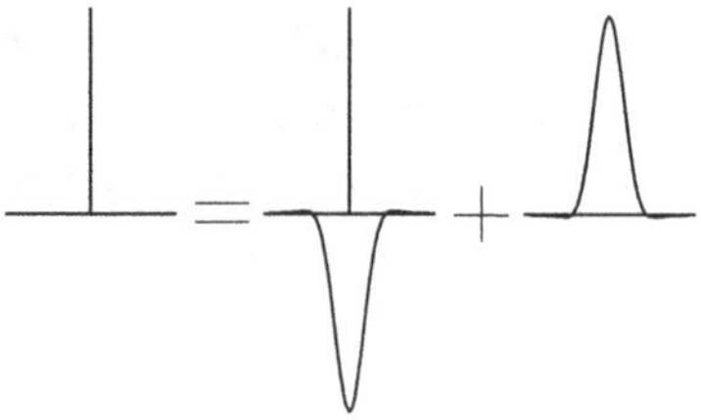

**Figure 2.10** Splitting of a delta function $\delta(\boldsymbol{x})$ (left) into the sum of $\delta(\boldsymbol{x}) - \delta_\kappa(\boldsymbol{x})$ (center) and $\delta_\kappa(\boldsymbol{x})$ (right).

decays rapidly as $|x| \to \infty$, its Fourier transform decays rapidly as $|k| \to \infty$ (Gasquet & Witomski 1998). Therefore, if we replace the delta function $\delta(\boldsymbol{x})$ with an appropriate regularization $\delta_\kappa(\boldsymbol{x})$ where $\kappa^{-1}$ measures the width of the peak, we would recover a rapidly converging Fourier series for the velocity. Of course, that would only be an approximate solution to the problem with true delta-function forcing. The resolution to this issue is to split the delta-function forcing into two pieces by addition and subtraction, as shown in Figure 2.10. That is, we write the force density (2.78) as the sum of a "local" part $\boldsymbol{f}_{\rm l}(\boldsymbol{x})$ and a "global" part $\boldsymbol{f}_{\rm g}(\boldsymbol{x})$: i.e.,

$$\boldsymbol{f}(\boldsymbol{x}) = \boldsymbol{f}_{\rm l}(\boldsymbol{x}) + \boldsymbol{f}_{\rm g}(\boldsymbol{x}), \tag{2.91}$$

where

$$\boldsymbol{f}_{\rm l}(\boldsymbol{x}) = \boldsymbol{F} \sum_{\boldsymbol{m}} (\delta(\boldsymbol{x} - \boldsymbol{x_m}) - \delta_\kappa(\boldsymbol{x} - \boldsymbol{x_m})), \tag{2.92}$$

$$\boldsymbol{f}_{\rm g}(\boldsymbol{x}) = \boldsymbol{F} \sum_{\boldsymbol{m}} \delta_\kappa(\boldsymbol{x} - \boldsymbol{x_m}). \tag{2.93}$$

The velocity field will now also be the sum of a local and a global part:

$$\boldsymbol{v}(\boldsymbol{x}) = \boldsymbol{v}_{\rm l}(\boldsymbol{x}) + \boldsymbol{v}_{\rm g}(\boldsymbol{x}), \tag{2.94}$$

where $\boldsymbol{v}_{\rm g}$ is driven by $\boldsymbol{f}_{\rm g}$ and $\boldsymbol{v}_{\rm l}$ by $\boldsymbol{f}_{\rm l}$. The quantity $\kappa$ is called the *splitting parameter.*

To proceed, we choose $\delta_\kappa(\boldsymbol{x})$ to be the quasi-Gaussian function (2.70), which leads to a solution equivalent to that of Hasimoto (1959). In an unbounded domain, the Stokes flow velocity field driven by $\boldsymbol{F}\delta_\kappa(\boldsymbol{x})$ is

$$\boldsymbol{v}_\kappa(\boldsymbol{x}) = \mathbf{G}_\kappa(\boldsymbol{x}) \cdot \boldsymbol{F}, \tag{2.95}$$

where $\mathbf{G}_\kappa(\boldsymbol{x})$ is the regularized Green's function given by (2.72).

The local velocity field results from the local force density and is given by

$$\boldsymbol{v}_{\rm l}(\boldsymbol{x}) = \sum_{\boldsymbol{m}} (\mathbf{G}(\boldsymbol{x} - \boldsymbol{x_m}) - \mathbf{G}_\kappa(\boldsymbol{x} - \boldsymbol{x_m})) \cdot \boldsymbol{F}. \tag{2.96}$$

This sum converges very rapidly, because when $\kappa r \gg 1$, $\mathbf{G}(\boldsymbol{x}) - \mathbf{G}_\kappa(\boldsymbol{x})$ decays faster than $e^{-\kappa^2 r^2}$. That is, the local velocity field at a given position $\boldsymbol{x}$ is affected only by the point forces within a distance of several times $\kappa^{-1}$. Thus the local part is also called the "short-ranged" part and $\delta_\kappa(\boldsymbol{x})$ as it is used in (2.92) is called a *screening function,* because its effect is to counteract or screen out the delta-function forcing to yield a solution that is virtually zero for $\kappa r \gg 1$.

The global or "long-ranged" velocity field $\boldsymbol{v}_g(\boldsymbol{x})$ is driven by a periodic array of regularized delta functions. In Fourier space, the governing equations as given in (2.88)–(2.89), but with $\hat{\delta}(\boldsymbol{k})(=1)$ replaced by $\hat{\delta}_\kappa(\boldsymbol{k})$, are as follows:

$$-\boldsymbol{g}\delta_{\boldsymbol{k}0} - i\boldsymbol{k}\check{p}_g(\boldsymbol{k}) - \eta k^2 \check{\boldsymbol{v}}_g(\boldsymbol{k}) + \frac{1}{V}\boldsymbol{F}\hat{\delta}_\kappa(\boldsymbol{k}) = \boldsymbol{0}, \tag{2.97}$$

$$\boldsymbol{k}\cdot\check{\boldsymbol{v}}_g(\boldsymbol{k}) = 0. \tag{2.98}$$

Because $\hat{\delta}_\kappa(\boldsymbol{0}) = 1$ (as must be the case for any regularized delta function), the $\boldsymbol{k} = \boldsymbol{0}$ solution is as in the original solution approach: $\check{\boldsymbol{v}}_g(\boldsymbol{0}) = \boldsymbol{0}$, $\boldsymbol{g} = \frac{1}{V}\boldsymbol{F}$. For $\boldsymbol{k} \neq 0$, the solution is similar to (2.90):

$$\check{\boldsymbol{v}}_g(\boldsymbol{k}) = \frac{\hat{\delta}_\kappa(\boldsymbol{k})}{\eta V k^2}\left(\boldsymbol{\delta} - \frac{1}{k^2}\boldsymbol{k}\boldsymbol{k}\right)\cdot\boldsymbol{F}. \tag{2.99}$$

To determine $\hat{\delta}_\kappa(\boldsymbol{k})$, the following observation is useful:[4]

$$r^2 e^{-\kappa^2 r^2} = -\frac{1}{2\kappa}\frac{\partial}{\partial\kappa}e^{-\kappa^2 r^2}.$$

Therefore, we can write

$$\delta_\kappa(\boldsymbol{x}) = \frac{\kappa^3}{2\sqrt{\pi^3}}\left(5 + \kappa\frac{\partial}{\partial\kappa}\right)e^{-\kappa^2 r^2},$$

the Fourier transform of which is

$$\begin{aligned}\hat{\delta}_\kappa(\boldsymbol{k}) &= \frac{\kappa^3}{2\sqrt{\pi^3}}\left(5 + \kappa\frac{\partial}{\partial\kappa}\right)\left(\frac{\pi}{\kappa^2}\right)^{3/2} e^{-k^2/4\kappa^2},\\ &= \left(1 + \frac{k^2}{4\kappa^2}\right)e^{-k^2/4\kappa^2}.\end{aligned}$$

Thus

$$\check{\boldsymbol{v}}_g(\boldsymbol{k}) = \frac{1}{\eta V k^2}\left(1 + \frac{k^2}{4\kappa^2}\right)e^{-k^2/4\kappa^2}\left(\boldsymbol{\delta} - \frac{1}{k^2}\boldsymbol{k}\boldsymbol{k}\right)\cdot\boldsymbol{F}. \tag{2.100}$$

This decays extremely rapidly as $k \to \infty$, resulting in a rapidly converging Fourier series. Finally, putting the local and global pieces together, we find that for an array of point forces, the velocity field is given by

$$\begin{aligned}\boldsymbol{v}(\boldsymbol{x}) = \Bigg[&\sum_{\boldsymbol{m}}\left(\mathbf{G}(\boldsymbol{x} - \boldsymbol{x}_{\boldsymbol{m}}) - \mathbf{G}_\kappa(\boldsymbol{x} - \boldsymbol{x}_{\boldsymbol{m}})\right)\\ &+ \sum_{\boldsymbol{k}\neq 0}\frac{1}{\eta V k^2}\left(1 + \frac{k^2}{4\kappa^2}\right)e^{-k^2/4\kappa^2}e^{i\boldsymbol{k}\cdot\boldsymbol{x}}\left(\boldsymbol{\delta} - \frac{1}{k^2}\boldsymbol{k}\boldsymbol{k}\right)\Bigg]\cdot\boldsymbol{F}.\end{aligned} \tag{2.101}$$

The quantity in brackets is the *Ewald sum* representation of the Green's function for a periodic array of point forces, in a form slightly different from that originally given

[4] Identities like this are immensely valuable in evaluating Fourier transforms involving exponential and Gaussian functions.

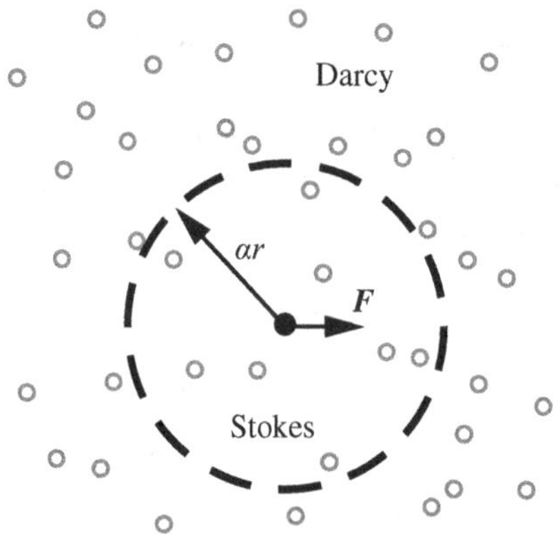


**Figure 2.11** Schematic of a fluid-saturated porous medium, showing flow regimes relative to the screening length $\alpha^{-1}$.

by Hasimoto (1959). Especially in combination with fast Fourier transform algorithms, it is widely used in simulations of hydrodynamically interacting particles in a periodic domain (see, e.g., Metsi 2000, Sierou & Brady 2001, Banchio & Brady 2003, Saintillan, Darve, & Shaqfeh 2005, Stoltz, de Pablo, & Graham 2006, Zhang et al. 2012). The same idea of local and global solutions can be applied to nonperiodic domains as well (Hernández-Ortiz et al. 2007, Hernández-Ortiz et al. 2008, Kumar & Graham 2012, Zhu et al. 2014, Beams, Olson, & Freund 2016). The local solution remains the same (aside from any changes due to a different choice of $\delta_\kappa(\boldsymbol{x})$), while the global solution is determined numerically, with boundary conditions on $\boldsymbol{v}_g(\boldsymbol{x})$ chosen so that $\boldsymbol{v}_l(\boldsymbol{x})+\boldsymbol{v}_g(\boldsymbol{x})$ satisfies the correct boundary conditions for the overall problem of interest.

## 2.8 Flow in a Porous Medium and Hydrodynamic Screening

In addition to the case of a suspension of free particles that we addressed in Section 2.4, we are also often interested in flow through a set of stationary obstacles, a fluid-saturated porous medium. Consider the following idealized situation: a nominally uniform flow with velocity $\boldsymbol{v}_\infty$ through a set of randomly and isotropically located fixed spherical obstacles of radius $a$ and number density $n$, as shown in Figure 2.11. The volume fraction $\phi$ of spheres is the number density times the volume of each sphere:

$$\phi = \frac{4}{3}\pi a^3 n.$$

If $\phi \ll 1$, then the distance between the obstacles (which scales as $n^{-1/3}$) is much larger than their size, so to a first approximation we neglect hydrodynamic interactions between them. Many gels, for example, satisfy the diluteness condition (Larson 1999, Rubinstein & Colby 2003). Each obstacle then exerts a force $-\zeta\boldsymbol{v}_\infty$ on the fluid, giving a volume-averaged force density

$$\boldsymbol{f} = -\zeta n\boldsymbol{v}_\infty = -\frac{9}{2}\frac{\eta}{a^2}\phi\boldsymbol{v}_\infty. \quad (2.102)$$

Here we have used the Stoke's law result $\zeta = 6\pi\eta a$. The *Brinkman model* of a porous medium imagines that this force density is spread uniformly through the fluid so the forced Stokes equation

$$-\nabla p + \eta \nabla^2 \boldsymbol{v} + \boldsymbol{f} = \boldsymbol{0}$$

becomes the *Brinkman equation*

$$-\nabla p + \eta \nabla^2 \boldsymbol{v} - \frac{\eta}{k}\boldsymbol{v} = \boldsymbol{0}, \tag{2.103}$$

where

$$k = \frac{2}{9}\frac{a^2}{\phi} \tag{2.104}$$

is the *permeability* of the porous medium. The continuity equation is unchanged. Observe that, because $k$ has units of length$^2$, a new length scale $k^{1/2}$ has entered the flow problem. In a uniform flow, the Laplacian term vanishes, yielding *Darcy's law*, a widely used model for bulk flow through porous media:

$$\boldsymbol{v} = -\frac{k}{\eta}\nabla p. \tag{2.105}$$

Even if the flow is not uniform, the Brinkman equation reduces to Darcy's law in the limit $k/L^2 \to 0$, where $L$ is a characteristic length scale for velocity variations. Taking the divergence of (2.105) and applying continuity gives that

$$\nabla^2 p = 0.$$

Recalling the discussion of Section 2.2, we see that flow governed by Darcy's law is a form of potential flow, with the velocity potential being the pressure.

The permeability is a function of the structure of the medium, not the fluid flowing through it. By the linearity of Stokes flow, even if the medium is not dilute and hydrodynamic interactions cannot be neglected, there will still be a linear relationship between the averaged body force and the mean velocity, so while the specific expression (2.104) is limited to the dilute limit, the Brinkman and Darcy models are not. Corrections to $k$ due to hydrodynamic interactions at finite concentration are addressed in a number of studies (Hinch 1977, Kim & Russel 1985, Durlofsky & Brady 1987) and many empirical studies report the dependence of $k$ on the structure and volume fraction of the porous medium (Bird et al. 2002).

Now that we have a model for flow in a porous medium, we can address the question of how flow is affected by the presence of the obstacles. A simple way to approach this question is to return to the situation that opened this chapter, that of the flow driven by a point force. Now Brinkman's equation becomes

$$-\nabla p + \eta \nabla^2 \boldsymbol{v} - \frac{\eta}{k}\boldsymbol{v} + \boldsymbol{F}\delta(\boldsymbol{x}) = \boldsymbol{0}. \tag{2.106}$$

The solution to this equation, sometimes called the *Brinkmanlet*, can be found by the same approach used to find the Stokeslet, and the solution (Exercise 2.7) is

$$\boldsymbol{v}(\boldsymbol{x}) = \mathbf{B}(\boldsymbol{x}) \cdot \boldsymbol{F}, \tag{2.107}$$

where

$$\mathbf{B}(\boldsymbol{x}) = \frac{1}{4\pi\eta\alpha^2 r^3}\left[2\left(1-(1+\alpha r)\,e^{-\alpha r}\right)\frac{\boldsymbol{x}\boldsymbol{x}}{r^2} + \left(\left(1+\alpha r+\alpha^2 r^2\right)e^{-\alpha r}-1\right)\left(\boldsymbol{\delta}-\frac{\boldsymbol{x}\boldsymbol{x}}{r^2}\right)\right], \tag{2.108}$$

and $\alpha = k^{-1/2}$ (Howells 2006). This quantity is the inverse of the new length scale that we noted previously. For $\alpha r \ll 1$, $\mathbf{B}(x)$ reduces to $\mathbf{G}(\boldsymbol{x})$, so on length scales smaller than $k^{1/2}$, the flow looks like Stokes flow. For $\alpha r \gg 1$, the terms containing the exponentials vanish and

$$\mathbf{B} \approx \frac{1}{4\pi\eta\alpha^2 r^3}\left(3\frac{\boldsymbol{x}\boldsymbol{x}}{r^2} - \boldsymbol{\delta}\right). \tag{2.109}$$

This is in fact the Green's function for the Darcy equation (2.105) with point forcing:

$$\boldsymbol{v} = -\frac{k}{\eta}\nabla p + \frac{k}{\eta}\boldsymbol{F}\delta(\boldsymbol{x}). \tag{2.110}$$

Referring back to (2.27), we recognize that (2.109) leads to a potential dipole flow, consistent with the general observation that Darcy flow is a form of potential flow.

Observe that when $\alpha r \gg 1$, the velocity field decays as $1/r^3$ in contrast to the $1/r$ of the Stokeslet; consequently, the hydrodynamic interaction between distant positions in a porous medium is much weaker than in a pure fluid. This phenomenon is called *Brinkman screening*; it is a specific case of the more general phenomenon of *hydrodynamic screening*. This effect occurs because the stationary obstacles in the porous medium remove momentum from the fluid that in their absence would have to diffuse out to infinity, and thus lead to faster decay of the velocity. The length scale $\alpha^{-1} = k^{1/2}$, which determines the distance at which this momentum absorption becomes appreciable, is called the *Brinkman screening length*. In the dilute limit, $\alpha^{-1} \sim a\phi^{-1/2}$, which is much larger than the obstacle spacing $n^{-1/3}$.

Having these results in hand, we can ask whether the hydrodynamic interaction between distant particles exerting forces on the fluid in a porous medium is long-ranged, in the sense that the sum of the interactions diverges as the volume of the domain becomes infinite. Revisiting our naive estimate (2.80) for forces exerted on a pure fluid and replacing $1/8\pi\eta r$ by $1/4\pi\eta\alpha^2 r^3$ yields

$$|\boldsymbol{v}| \sim \lim_{L\to\infty}\int_{r_0}^{L}\frac{1}{4\pi\eta\alpha^2 r^3}\,|\boldsymbol{f}|\,r^2\,dr \sim \ln L,$$

which diverges weakly as $L \to \infty$. However, in contrast to a Stokeslet flow, which is always primarily in the direction of the imposed force, the Darcy flow is a potential dipole flow, which in some parts of the domain is in the opposite direction to the imposed force (see Figure 2.2). Therefore, let us make a more careful estimate considering the velocity in the $z$-direction at some point in the domain due to a force density exerted in that direction. Recall that the idea is to add up the contributions to the velocity at a given point due to the forces exerted at distant points. Now our estimate becomes

$$v_z \sim \lim_{L\to\infty} 4\pi\int_{r_0}^{L}\int_0^{\pi}\frac{1}{4\pi\eta\alpha^2 r^3}(3\cos^2\theta - 1)r^2\sin\theta\,d\theta\,dr = 0.$$

The integral vanishes upon integration over the polar angle $\theta$ – the contributions for $\theta < \pi/2$ and $\theta > \pi/2$ cancel out by symmetry. Here the naive prediction that the integral weakly diverges is incorrect because of the symmetry of the flow field. (The other velocity components vanish by axisymmetry.) This result has been called "screening by symmetry" (Tlusty 2006). Thus we do not expect long-range interactions between particles exerting forces on the fluid in a porous medium. No such cancellation occurs for the free-space Stokeslet, which always drives flow in the direction of the imposed force.

## Problems

**2.1** The Stokeslet blows up at the origin. This singularity is a consequence of using the delta-function idealization and can be problematic so it is of interest to derive regularized Stokeslets that arise from a force density that is slightly smeared out in space. Repeat the derivation of the Stokeslet given in Section 2.1, replacing the delta function with the nonsingular function given by (2.70). Hint: For problems involving $\nabla^2 u$ in spherical coordinates, the substitution $u = w/r$ is often convenient.

**2.2** The flow driven by the symmetric part of the force dipole **D** is given by

$$v_i = \frac{1}{2}\left(\frac{\partial G_{ik}}{\partial x_j} + \frac{\partial G_{ij}}{\partial x_k}\right) D^{\mathrm{S}}_{kj}.$$

By simplifying the term in parentheses, show that this velocity field can be written as the superposition of a point *source* flow and a stresslet flow. Confirm that this quantity is consistent with the result $\boldsymbol{\nabla} \cdot \mathbf{G} = \mathbf{0}$. By writing $D^{\mathrm{S}}_{kj} = S_{kj} + \frac{1}{3} D^{\mathrm{S}}_{ll} \delta_{kj}$ and recalling that the stresslet tensor **S** is traceless, show that the point source flow cancels out (as it must), yielding (2.38).

**2.3** Consider the torque associated with a force density:

(a) The total torque (with respect to the origin) exerted by a force density on a fluid is given by

$$\boldsymbol{T} = \int_V \boldsymbol{x} \times \boldsymbol{f}(\boldsymbol{x})\, dV.$$

For the antisymmetric force density $\boldsymbol{f}(\boldsymbol{x}) = -\frac{1}{2}\nabla \times (\boldsymbol{M}\delta(\boldsymbol{x}))$, show that the total torque exerted on the fluid by this force density is $-\boldsymbol{M}$.

(b) Verify this result by going back to the picture of the force dipole itself. Use the fact that the total torque exerted by the forces on the fluid is

$$\sum_i \boldsymbol{x}_i \times \boldsymbol{F}_i;$$

let $\boldsymbol{x}_1 = -\boldsymbol{u}\frac{h}{2}$ and $\boldsymbol{F}_1 = -\boldsymbol{F}$ and $\boldsymbol{x}_2 = +\boldsymbol{u}\frac{h}{2}$ and $\boldsymbol{F}_2 = +\boldsymbol{F}$. I.e., show that this expression reduces to $-\boldsymbol{M}$.

**2.4** Use the regularized Green's function from Problem 2.1 (which is given by (2.72)) in the mobility equation (2.64) to find a modified form of (2.69) for a pair of particles aligned on the $x$-axis. Find the eigenvalue corresponding to the competitive mode $\mathbf{Z} = [-1\ 0\ 0\ 1\ 0\ 0]^{\mathrm{T}}$ and derive the condition given in the text that $\kappa$ must satisfy for this eigenvalue to be positive.

**2.5** Consider a system of three regularized point particles that are each subjected to an external force $\boldsymbol{F}$ in otherwise stationary fluid ($\boldsymbol{v}_\infty = \mathbf{0}$). A force balance on each particle yields that $\boldsymbol{F}_\alpha = \boldsymbol{F}$. Noting that

$$\boldsymbol{U}_\alpha = \frac{d\boldsymbol{R}_\alpha}{dt},$$

where $\boldsymbol{R}_\alpha$ is the position of particle $\alpha$, write a code to time-integrate the dynamics of this system of particles. Let the particles each have a friction coefficient $\zeta = 6\pi\eta a$ and set $\kappa^{-1} = 3a/\sqrt{\pi}$. Natural length and time scales with which to nondimensionalize the problem are $a$ and $a\zeta/|\boldsymbol{F}|$, in which case the dimensionless magnitude of the force is unity. Show the evolution of a number of initial configurations, including a line of particles oriented parallel and perpendicular to $\boldsymbol{F}$.

**2.6** Show that the universe is finite in space and/or time. More specifically, show that a universe of point sources of photons (stars) infinite in space and time would result in a bright night sky, not a dark one. The photon flux from a star decays as $1/r^2$ (by conservation of photons). Construct the integral that is analogous to the expression for the Stokes flow velocity field due to an infinite lattice of point forces and show that it diverges for an infinite universe.

**2.7** Derive (2.108) from (2.106) and (2.109) from (2.110).

# 3 Beyond Point Particles

The singularity solutions presented in the preceding chapter provide important idealizations of real objects in a flow, a point made explicit through the multipole expansion. Now we build on these results to develop a more detailed description and understanding of spherical and nonspherical particles in unbounded flow as well as how hydrodynamic interactions are altered as we move beyond the point particle approximation.

## 3.1 General Solution to the Stokes Equation

In this section, we present a general solution to the Stokes equation and show its relationship first with singularity solutions and the multipole expansion and then to the important particular case of a spherical particle. We begin by recalling from Section 1.2.2 that the pressure field for the Stokes equation satisfies a Poisson equation (1.36), which in the absence of a force density $\boldsymbol{f}$ reduces to the Laplace equation

$$\nabla^2 p = 0. \tag{3.1}$$

Solutions of this equation are said to be *harmonic*. Observing that the Stokes equation itself also contains the Laplacian operator $\nabla^2$ motivates us to seek a solution to the Stokes equation as the sum $\boldsymbol{v} = \boldsymbol{v}^{\mathrm{H}} + \boldsymbol{v}^{\mathrm{P}}$ of a homogeneous solution $\boldsymbol{v}^{\mathrm{H}}$ that satisfies the vector Laplace equation

$$\nabla^2 \boldsymbol{v}^{\mathrm{H}} = \boldsymbol{0} \tag{3.2}$$

and a particular solution $\boldsymbol{v}^{\mathrm{P}}$ that satisfies

$$\nabla^2 \boldsymbol{v}^{\mathrm{P}} = \frac{1}{\eta} \boldsymbol{\nabla} p. \tag{3.3}$$

Taking the pressure at infinity to approach a constant, $p_\infty$, the latter solution can be taken to be

$$\boldsymbol{v}^{\mathrm{P}} = \frac{1}{2\eta} \boldsymbol{x} (p - p_\infty), \tag{3.4}$$

and incompressibility requires that

$$\boldsymbol{\nabla} \cdot \boldsymbol{v}^{\mathrm{H}} = -\boldsymbol{\nabla} \cdot \boldsymbol{v}^{\mathrm{P}} = -\frac{1}{2\eta} \left( 3(p - p_\infty) + \boldsymbol{x} \cdot \boldsymbol{\nabla} p \right). \tag{3.5}$$

Since $\boldsymbol{v}^{\mathrm{H}}$ and $p$ both obey the Laplace equation, it is clear that we can use solutions to this equation to construct solutions to the Stokes equation. We outline this approach here; more details of this solution can be found in Lamb (1932), Batchelor (1967), Kim & Karrila (1991), and Leal (2007).

Decaying solutions to the Laplace equation

$$\nabla^2 f = 0$$

can be found easily. A spherically symmetric one, valid for any $\boldsymbol{x} \neq \mathbf{0}$, is simply

$$f = \frac{1}{r}. \tag{3.6}$$

This is actually the solution to

$$\nabla^2 f = -4\pi\delta(\boldsymbol{x}), \tag{3.7}$$

i.e., it is a multiple of the Green's function for the Laplacian operator. Then we observe that if $f$ is a solution to (3.1), then so is $\boldsymbol{\nabla} f$, since

$$\boldsymbol{\nabla}\nabla^2 f = \nabla^2\boldsymbol{\nabla} f = -4\pi\boldsymbol{\nabla}\delta(\boldsymbol{x}). \tag{3.8}$$

This observation can be applied ad infinitum yielding the *decaying vector spherical harmonics*

$$\boldsymbol{\nabla}^n\frac{1}{r}, \qquad n = 0, 1, 2, \ldots, \infty. \tag{3.9}$$

In index notation, the Cartesian components of the first several of these are

$$\boldsymbol{\nabla}^0\frac{1}{r} = \frac{1}{r} \tag{3.10}$$

$$\left(\boldsymbol{\nabla}\frac{1}{r}\right)_i = \frac{\partial}{\partial x_i}\frac{1}{r} = -\frac{x_i}{r^3}, \tag{3.11}$$

$$\left(\boldsymbol{\nabla\nabla}\frac{1}{r}\right)_{ij} = \frac{\partial}{\partial x_i}\frac{\partial}{\partial x_j}\frac{1}{r} = 3\left(\frac{x_i x_j}{r^5} - \frac{\delta_{ij}}{3r^3}\right), \tag{3.12}$$

$$\left(\boldsymbol{\nabla\nabla\nabla}\frac{1}{r}\right)_{ijk} = \frac{\partial}{\partial x_i}\frac{\partial}{\partial x_j}\frac{\partial}{\partial x_k}\frac{1}{r} = -15\left(\frac{x_i x_j x_k}{r^7} - \frac{x_i\delta_{jk} + x_j\delta_{ki} + x_k\delta_{ij}}{5r^5}\right). \tag{3.13}$$

These functions may look vaguely familiar from the preceding sections, and are in fact the first terms of the multipole expansion of the Laplace equation.

Observe that because these functions are formed from repeated gradient operations, which commute, they are invariant under exchange of any two indices. For example,

$$\left((\boldsymbol{\nabla\nabla\nabla})\frac{1}{r}\right)_{ijk} = \left((\boldsymbol{\nabla\nabla\nabla})\frac{1}{r}\right)_{ikj} = \left((\boldsymbol{\nabla\nabla\nabla})\frac{1}{r}\right)_{jik}.$$

Furthermore, since they are all solutions to the Laplace equation, they are traceless with respect to any pair of indices. For example,

$$\left((\boldsymbol{\nabla\nabla\nabla})\frac{1}{r}\right)_{iik} = \left((\boldsymbol{\nabla}\cdot\boldsymbol{\nabla\nabla})\frac{1}{r}\right)_k = \left(\nabla^2\boldsymbol{\nabla}\frac{1}{r}\right)_k = \left(\boldsymbol{\nabla}\nabla^2\frac{1}{r}\right)_k = 0.$$

In short, the second- and higher-order vector spherical harmonics are symmetric and traceless with respect to any pair of indices.

All of the preceding solutions decay as $r \to \infty$. There are also *growing* vector spherical harmonics, which have the form

$$r^{2n+1}\nabla^n \frac{1}{r}, \qquad n = 0, 1, 2, \ldots, \infty. \tag{3.14}$$

Note that the solution for $n = 0$ is simply a constant. The general solution to the Laplace equation is the sum of the decaying and growing harmonics each multiplied by an arbitrary constant $c^{(n)}$ or $d^{(n)}$, respectively, of the appropriate tensorial order:

$$\begin{aligned} f(\boldsymbol{x}) = {} & \left(c^{(0)} + rd^{(0)}\right)\frac{1}{r} + \left(c_i^{(1)} + r^3 d_i^{(1)}\right)\frac{\partial}{\partial x_i}\frac{1}{r} \\ & + \left(c_{ij}^{(2)} + r^5 d_{ij}^{(2)}\right)\frac{\partial}{\partial x_i}\frac{\partial}{\partial x_j}\frac{1}{r} + \left(c_{ijk}^{(3)} + r^7 d_{ijk}^{(3)}\right)\frac{\partial}{\partial x_i}\frac{\partial}{\partial x_j}\frac{\partial}{\partial x_k}\frac{1}{r} + \ldots. \end{aligned} \tag{3.15}$$

This representation is closely related to the perhaps more familiar general solution in spherical coordinates:

$$f(r, \theta, \phi) = \sum_{n=0}^{\infty}\sum_{m=0}^{l}\left(a_{nm}r^n + b_{nm}r^{-(n+1)}\right)P_{nm}(\cos\theta)e^{im\phi}, \tag{3.16}$$

where $\theta$ and $\phi$ are here the polar and azimuthal angles, respectively, and $P_{nm}(\cos\theta)$ are the *associated Legendre polynomials*. Noting that

$$\nabla^n \frac{1}{r} \propto r^{-(n+1)},$$

we recognize that the angular part of this function corresponds to the terms $P_{nm}(\cos\theta)e^{im\phi}$ in (3.16).

The general solution to the Stokes equation can be presented in a form known as *Lamb's general solution* that is analogous to (3.16) (Lamb 1932, Kim & Karrila 1991). This will not be required for our purposes here. Rather, knowing that $p$ and the components of $\boldsymbol{v}^{\mathrm{H}}$ have the form given by (3.15), we will construct solutions for a number of important specific cases. For flows that vanish as $r \to \infty$, we need only keep the decaying terms.

## 3.2 The Stokeslet Revisited

As an initial example of applying the preceding results, consider again the flow driven by a point force that we first addressed in Section 2.1. Except at the origin where the point force is exerted, the flow satisfies the unforced Stokes equation. The pressure field is a scalar that must decay to a constant value $p_\infty$ as $r \to \infty$ and, since the Stokes equation is linear, the field must depend linearly on the force $\boldsymbol{F}$ exerted on the fluid. There are no other input parameters on which the pressure can depend. The only way that such a function can be constructed from the decaying vector harmonics is as the combination

$$p - p_\infty = \alpha \boldsymbol{F} \cdot \nabla \frac{1}{r} = -\alpha F_i \frac{x_i}{r^3}, \tag{3.17}$$

where $\alpha$ is an as-yet arbitrary constant.[1]

Similarly, the homogeneous part of the velocity field must be a vector that is a linear function of $\boldsymbol{F}$. Any linear relationship between $\boldsymbol{v}^{\mathrm{H}}$ and $\boldsymbol{F}$ can be expressed by multiplication by a scalar or by a second-order tensor, leading us to the form

$$v_i^{\mathrm{H}} = \beta \frac{1}{r} F_i + \gamma \left( \frac{\partial}{\partial x_i} \frac{\partial}{\partial x_j} \frac{1}{r} \right) F_j, \tag{3.18}$$

where $\beta$ and $\gamma$ are undetermined. The term multiplying $\gamma$ behaves as $r^{-3}$. This is overly singular at the origin, corresponding to a forcing $\boldsymbol{\nabla\nabla}\delta(\boldsymbol{x})$ rather than $\delta(\boldsymbol{x})$, so we must take $\gamma = 0$. Thus the solution has the form

$$v_i = \frac{1}{2\eta} x_i (p - p_\infty) + v_i^{\mathrm{H}} = -\frac{1}{2\eta} \alpha \frac{x_i x_j}{r^3} F_j + \frac{\beta}{r} F_i. \tag{3.19}$$

The incompressibility condition (3.5) requires that $\beta = -\alpha/2\eta$. Furthermore, the solution we seek must satisfy

$$\boldsymbol{\nabla} \cdot \boldsymbol{\sigma} + \boldsymbol{F}\delta(\boldsymbol{x}) = \mathbf{0}.$$

Integrating this equation over any volume containing the origin and applying the divergence theorem yields that

$$\int_S \boldsymbol{n} \cdot \boldsymbol{\sigma} \, dS + \boldsymbol{F} = \mathbf{0}.$$

Choosing for convenience a spherical volume centered on the origin, one finds that $\alpha = -1/4\pi$. Therefore,

$$v_i = \frac{1}{8\pi\eta} \frac{x_i x_j}{r^3} F_j + \frac{1}{8\pi\eta r} F_i = \frac{1}{8\pi\eta r} \left( \delta_{ij} + \frac{x_i x_j}{r^2} \right) F_j. \tag{3.20}$$

As required, we have recovered the Stokeslet result of Section 2.1. We now see that the free-space Green's function $\mathbf{G}(\boldsymbol{x})$ can be written in terms of vector spherical harmonics as

$$G_{ij}(\boldsymbol{x}) = \frac{1}{8\pi\eta} \left( \delta_{ij} - x_i \frac{\partial}{\partial x_j} \right) \frac{1}{r}. \tag{3.21}$$

Since the terms in the multipole expansion of Section 2.4 are derivatives of $\mathbf{G}(\boldsymbol{x})$, this expression provides the connection between those derivatives and the harmonics.

[1] One might ask why a term of the form

$$F_i \delta_{jk} \frac{\partial}{\partial x_i} \frac{\partial}{\partial x_j} \frac{\partial}{\partial x_k} \frac{1}{r}$$

cannot appear – this is a scalar function that is linear in $\boldsymbol{F}$ and depends on the isotropic tensor $\boldsymbol{\delta}$. Indeed, a term like this might appear, only to vanish as a consequence of the symmetric traceless property of the harmonics; i.e.,

$$\delta_{jk} \frac{\partial}{\partial x_i} \frac{\partial}{\partial x_j} \frac{\partial}{\partial x_k} \frac{1}{r} = \frac{\partial}{\partial x_i} \frac{\partial}{\partial x_j} \frac{\partial}{\partial x_j} \frac{1}{r} = \frac{\partial}{\partial x_i} \nabla^2 \frac{1}{r} = 0.$$

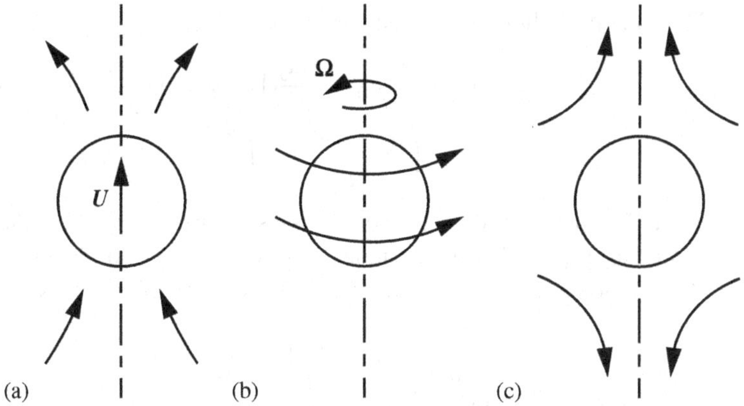


**Figure 3.1** (a) Sphere translating in stationary fluid, (b) sphere rotating in stationary fluid, (c) sphere in a linear flow.

## 3.3 Spheres in Flow

With the general solution in hand we now turn to some of the most important results for particles in flow, the solutions for rigid spheres translating or rotating in an otherwise stationary fluid or suspended in an arbitrary linear flow (Figure 3.1).

### 3.3.1 Translating Sphere

The solution procedure for a sphere with radius $a$ translating with velocity $\boldsymbol{U}$ is closely related to that for a point force. We take the particle to be centered at the origin. The pressure and velocity fields must depend linearly on a vector, which is now $\boldsymbol{U}$ rather than $\boldsymbol{F}$. Carrying over the arguments that led to (3.17) and (3.18), $p$ and $\boldsymbol{v}^{\mathrm{H}}$ have the forms

$$p - p_\infty = \alpha \boldsymbol{U} \cdot \boldsymbol{\nabla} \frac{1}{r} = -\alpha U_i \frac{x_i}{r^3}, \tag{3.22}$$

and

$$v_i^{\mathrm{H}} = \beta \frac{1}{r} U_i + \gamma \left( \frac{\partial}{\partial x_i} \frac{\partial}{\partial x_j} \frac{1}{r} \right) U_j. \tag{3.23}$$

Now, however, $\gamma$ need not vanish since the origin is not part of the flow domain. The overall solution is thus

$$v_i = -\frac{1}{2\eta} \alpha \frac{x_i x_j}{r^3} U_j + \frac{\beta}{r} U_i + \gamma \left( \frac{\partial}{\partial x_i} \frac{\partial}{\partial x_j} \frac{1}{r} \right) U_j. \tag{3.24}$$

The constants $\alpha$, $\beta$, and $\gamma$ are determined from the conditions of incompressibility and no slip.

Incompressibility yields that $\alpha = -2\eta\beta$ so

$$v_i = \frac{\beta}{r} \left( \delta_{ij} + \frac{x_i x_j}{r^2} \right) U_j + \gamma \left( \frac{\partial}{\partial x_i} \frac{\partial}{\partial x_j} \frac{1}{r} \right) U_j. \tag{3.25}$$

The no-slip condition requires that $\boldsymbol{v} = \boldsymbol{U}$ at $r = a$:

$$v_i = U_i = \frac{\beta}{a}\left(\delta_{ij} + \frac{x_i x_j}{a^2}\right)U_j + 3\gamma\left(\frac{x_i x_j}{a^5} - \frac{\delta_{ij}}{3a^3}\right)U_j$$
$$= \left(\frac{\beta}{a} - \frac{\gamma}{a^3}\right)U_i + \left(\frac{\beta}{a^3} + \frac{3\gamma}{a^5}\right)x_i x_j U_j.$$

Since $U_i$ and $x_i x_j U_j$ do not depend in the same way on position, the terms multiplying them must separately cancel out, leaving two equations for the two unknowns $\beta$ and $\gamma$. The solution to these is $\gamma = -a^3/4$, $\beta = 3a/4$, so (3.25) becomes

$$v_i = \frac{3a}{4r}\left(\delta_{ij} + \frac{x_i x_j}{r^2}\right)U_j - \frac{a^3}{4r^3}\left(-\delta_{ij} + 3\frac{x_i x_j}{r^2}\right)U_j. \tag{3.26}$$

The first term in this solution can be recognized as a Stokeslet corresponding to a force

$$\boldsymbol{F} = 6\pi\eta a \boldsymbol{U}. \tag{3.27}$$

This is *Stokes's law*, which we first introduced in Section 2.5. It is often expressed in the form $\boldsymbol{F} = \zeta \boldsymbol{U}$, where $\zeta = 6\pi\eta a$ is called the friction coefficient. Alternatively, we can define the *drag* force $\boldsymbol{F}_{\text{drag}} = -\boldsymbol{F}$ exerted by the fluid on the sphere and write $\boldsymbol{F}_{\text{drag}} = -\zeta \boldsymbol{U}$. Recalling (2.27), the second term in (3.26) can be recognized as a potential dipole. It arises not from the force the sphere exerts on the fluid but rather from the fluid that the sphere displaces as it moves. Furthermore, since it decays as $1/r^3$, it must be part of the quadrupole contribution $\boldsymbol{\nabla\nabla}\mathbf{G}$ of the multipole expansion and in fact is simply $a^2/6\ \nabla^2\mathbf{G}$. Therefore, we can write the velocity field generated by a translating sphere as

$$\boldsymbol{v}(\boldsymbol{x}) = \left(1 + \frac{a^2}{6}\nabla^2\right)\mathbf{G}\cdot\boldsymbol{F}. \tag{3.28}$$

Similarly, noting that $\alpha = -2\eta\beta = -3\eta a/2$, the pressure can be written

$$p(\boldsymbol{x}) - p_\infty = \frac{1}{4\pi r^3}\boldsymbol{x}\cdot\boldsymbol{F}. \tag{3.29}$$

This is identical to the pressure field associated with a Stokeslet. A further important result for this case (Problem 3.1) is that, aside from the normal component due to the uniform background pressure field, the traction on the surface of the sphere is independent of location on the sphere's surface:

$$\hat{\boldsymbol{n}}\cdot\boldsymbol{\sigma}|_{r=a} = -p_\infty\hat{\boldsymbol{n}} - \frac{3\eta\boldsymbol{U}}{2a}, \tag{3.30}$$

where $\hat{\boldsymbol{n}}$ is the outward unit normal on the sphere. The constancy of the velocity and traction on the surface of a translating sphere make this case very useful for applications of the Lorentz reciprocal relations.

### 3.3.2 Rotating Sphere

Consider a sphere with radius $a$ centered at the origin and rotating with an angular velocity $\boldsymbol{\Omega}$ in a fluid that is stationary at infinity. Since $\boldsymbol{\Omega}$ is a pseudovector, the pressure must contain terms like this:

$$\epsilon_{ijk}\Omega_j \frac{\partial}{\partial x_i}\frac{\partial}{\partial x_k}\frac{1}{r}.$$

But since $\epsilon_{ijk}$ is antisymmetric in any pair of its indices and the harmonics are symmetric with respect to the same index exchange, all terms of this form vanish and we must conclude that in this situation

$$p = p_\infty. \tag{3.31}$$

The velocity must have the form

$$v_i = \alpha\epsilon_{ijk}\Omega_j \frac{\partial}{\partial x_k}\frac{1}{r} = -\alpha\epsilon_{ijk}\Omega_j \frac{x_k}{r^3}, \tag{3.32}$$

where $\alpha$ is undetermined. The symmetric traceless property of the harmonics again makes the higher-order terms vanish. The constant $\alpha$ is determined by the no-slip boundary condition at $r = a$:

$$v_i = \epsilon_{ijk}\Omega_j x_k. \tag{3.33}$$

Comparing (3.32) and (3.33) yields that $\alpha = -a^3$, so the velocity field for the rotating rigid sphere is

$$v_i = a^3\epsilon_{ijk}\Omega_j \frac{x_k}{r^3}. \tag{3.34}$$

As it rotates, the sphere is exerting a torque on the fluid and vice versa. In Section 2.3, we saw that the flow associated with a point torque (antisymmetric force dipole) is a rotlet. For a *point* torque $\boldsymbol{T}$ exerted on the fluid, (2.43) gives the velocity field

$$v_i = \frac{1}{8\pi\eta r^3}\epsilon_{ijk}T_j x_k, \tag{3.35}$$

which is identical to (3.34) when

$$\boldsymbol{T} = \zeta_{\mathrm{r}}\boldsymbol{\Omega}, \tag{3.36}$$

with $\zeta_{\mathrm{r}} = 8\pi\eta a^3$. Thus a rotating sphere generates a pure rotlet flow and exerts a torque given by (3.36). This equation is the rotational analogue to Stokes's law, and $\zeta_{\mathrm{r}}$ is called the *rotational friction coefficient*. Alternatively, the *drag* torque $\boldsymbol{T}_{\mathrm{drag}} = -\boldsymbol{T}$ is given by $\boldsymbol{T}_{\mathrm{drag}} = -\zeta_{\mathrm{r}}\boldsymbol{\Omega}$. Observe that $\zeta_{\mathrm{r}} \sim a^3$, in constrast to the $a^1$ scaling of $\zeta$. The two additional factors of $a$ arise from the fact that a torque is a product of moment arm and force: The moment arm scales linearly with $a$, and the force scales with linear velocity on the surface, which is the product of angular velocity and radial position, the latter of which also scales linearly with $a$ (Guazzelli & Morris 2012).

For a sphere that is both translating and rotating, the velocity field is simply a superposition of the two individual fields, and in particular (3.27) and (3.36) apply. Finally, if the fluid velocity $\boldsymbol{v}_\infty$ in the absence of the particle is just a combination of uniform flow with velocity $\boldsymbol{U}_\infty$ and rigid rotation, i.e.,

$$\boldsymbol{v}_\infty = \boldsymbol{U}_\infty + \boldsymbol{\omega}_\infty \times \boldsymbol{x},$$

then these equations become by a change in reference frame

$$\boldsymbol{F} = \zeta\,(\boldsymbol{U} - \boldsymbol{v}_\infty), \tag{3.37}$$

$$\boldsymbol{T} = \zeta_{\mathrm{r}}\,(\boldsymbol{\Omega} - \boldsymbol{\omega}_\infty). \tag{3.38}$$

### 3.3.3 Sphere in a General Linear Flow and the Stress in a Dilute Suspension

A general linear flow field can be written as

$$\boldsymbol{v}_\infty(\boldsymbol{x}) = \boldsymbol{x}\cdot\boldsymbol{\nabla}\boldsymbol{v}_\infty = \mathbf{W}_\infty^{\mathrm{T}}\cdot\boldsymbol{x} + \mathbf{E}_\infty\cdot\boldsymbol{x},$$

where $\boldsymbol{\nabla}\boldsymbol{v}_\infty$ is constant (and thus so are $\mathbf{W}_\infty$ and $\mathbf{E}_\infty$). Equivalently,

$$\boldsymbol{v}_\infty(\boldsymbol{x}) = \boldsymbol{\omega}_\infty\times\boldsymbol{x} + \mathbf{E}_\infty\cdot\boldsymbol{x}.$$

All linear flows satisfy $\nabla^2\boldsymbol{v} = \mathbf{0}$ so their pressure fields must be constant – we set that constant to zero. We determine here the velocity field for a rigid force-and torque-free spherical particle placed at the origin in a flow whose velocity approaches $\boldsymbol{v}_\infty$ as $r\to\infty$.

By superposition, we can write this solution as two pieces, one due to the rotational part $\boldsymbol{\omega}_\infty\times\boldsymbol{x}$ and one due to the straining part $\mathbf{E}_\infty\cdot\boldsymbol{x}$. The piece due to the rotational part is trivial once we realize that the torque-free condition will be satisfied only if the particle is rotating with the same angular velocity as the fluid – this follows from (3.38). There is thus no change in the rotational part of the fluid motion due to the presence of the particle.

Therefore, it remains to compute the velocity field due to a particle in a pure straining field $\mathbf{E}_\infty$. We sketch the solution here; Leal (2007) and Kim & Karrila (1991) provide details.[2] The basic structure of the solution can be ascertained from the expression for the pressure, which must be a scalar function of the symmetric tensor $\mathbf{E}_\infty$. The most general harmonic form for the pressure is

$$p - p_\infty = \alpha\left(\frac{\partial}{\partial x_i}\frac{\partial}{\partial x_j}\frac{1}{r}\right)E_{\infty ij}. \tag{3.39}$$

Similarly, the homogeneous part of the velocity will have the form

$$v_i^{\mathrm{H}} = E_{\infty ij}x_j + \beta\left(\frac{\partial}{\partial x_j}\frac{1}{r}\right)E_{\infty ji} + \gamma\left(\frac{\partial}{\partial x_i}\frac{\partial}{\partial x_j}\frac{\partial}{\partial x_k}\frac{1}{r}\right)E_{\infty jk}. \tag{3.40}$$

The first term on the right-hand side is simply the unperturbed flow, to which the velocity field converges as $r\to\infty$. As discussed earlier, incompressibility and no-slip determine the constants $\alpha$, $\beta$, and $\gamma$. The final result, written in terms of $\mathbf{G}$, is

$$\boldsymbol{v}(\boldsymbol{x}) = \mathbf{E}_\infty\cdot\boldsymbol{x} + \frac{20}{3}\pi\eta a^3\,(\mathbf{E}_\infty\cdot\boldsymbol{\nabla})\cdot\left(1 + \frac{a^2}{10}\nabla^2\right)\mathbf{G}(\boldsymbol{x}). \tag{3.41}$$

A particularly useful feature of this result is that it allows us to readily extract the

[2] Alternatively, one can use a stream function formulation for a sphere in an axisymmetric straining flow (Deen 2012). Any straining motion can be written as a superposition of axisymmetric ones, allowing the solution for any $\mathbf{E}_\infty$ to be determined from the axisymmetric case.

force dipole tensor generated by a sphere in a linear flow. Comparing this expression to the multipole expansion expression, (2.48), yields that

$$\mathbf{D} = \frac{20}{3}\pi\eta a^3 \mathbf{E}_\infty. \tag{3.42}$$

With this result, we can immediately find the viscosity of a suspension of spheres in the dilute limit, where the spheres are very far apart compared to their size, so their force dipoles are given by this isolated particle result. Recall that (2.60) gives the particle contribution to the stress in a suspension of rigid particles with number density $n$ in terms of the dipoles. This equation along with (3.42) yields that

$$\begin{aligned}\mathbf{\Sigma}_\mathrm{P} &= \frac{20}{3} n\pi\eta a^3 \mathbf{E}_\infty, \\ &= 5\phi\eta\mathbf{E}_\infty,\end{aligned}$$

where $\phi = \frac{4}{3}\pi a^3 n$ is the volume fraction of spheres. From this result and the definition of viscosity, we see that the particles contribute a viscosity

$$\eta_\mathrm{P} = \frac{5}{2}\eta\phi \tag{3.43}$$

to the fluid so the total suspension viscosity $\eta_\mathrm{t}$ is given by

$$\eta_\mathrm{t} = \eta + \eta_\mathrm{P} = \eta(1 + \frac{5}{2}\phi). \tag{3.44}$$

This is the classical result of Einstein (1998), although he derived it from an energy dissipation argument rather than a stress computation. The quantity

$$[\eta] = \lim_{\phi\to 0} \frac{\eta_\mathrm{t} - \eta}{\phi\eta} \tag{3.45}$$

is called the *intrinsic viscosity*. Equation (3.44) yields $[\eta] = 5/2$.

Equation (3.43) is a good approximation when $\phi \lesssim 0.03$ (Larson 1999). At higher volume fractions, the hydrodynamic interactions between the spheres cannot be neglected – the stresslet velocity field generated by each particle affects the others. Incorporating this effect rigorously yields an $O(\phi^2)$ correction to (3.43) (Batchelor & Green 1972, Kim & Karrila 1991, Guazzelli & Morris 2012). Rather than presenting this subtle and difficult calculation, we present here a mathematically simpler, albeit heuristic, approach based on the idea that as the volume fraction increases, each particle experiences a fluid with the viscosity not of the background fluid but of the suspension itself (Ball & Richmond 1980, Larson 1999). According to the simplest version of this idea, the differential increment in total viscosity is given by

$$d\eta_\mathrm{t} = [\eta]\eta_\mathrm{t}(\phi)\, d\phi, \tag{3.46}$$

resulting in a viscosity

$$\eta_\mathrm{t} = \eta \exp([\eta]\phi). \tag{3.47}$$

This reduces to the Einstein expression for small $\phi$ and captures the increase in slope of $\eta_\mathrm{t}$ with concentration that is observed in experiment. A refinement of this approach

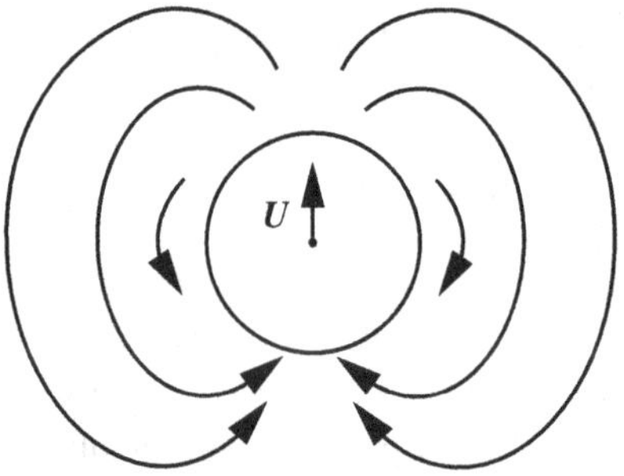


**Figure 3.2** Potential dipole model of a swimming microorganism. The velocity field is sketched in the stationary frame.

incorporates the observation that the viscosity of a suspension diverges as the volume fraction approaches the close-packed limit $\phi_{\text{max}} \approx 0.64$. Modifying (3.46) to account for this effect yields

$$d\eta_{\text{t}} = [\eta]\frac{\eta_{\text{t}}(\phi)}{1 - \frac{\phi}{\phi_{\text{max}}}}\, d\phi. \tag{3.48}$$

This integrates to

$$\eta_{\text{t}} = \eta\left(1 - \frac{\phi}{\phi_{\text{max}}}\right)^{-[\eta]\phi_{\text{max}}}, \tag{3.49}$$

which is known as the *Krieger–Dougherty equation*; it provides a reasonable description of the viscosity of suspensions over the entire range of volume fraction. For a suspension of perfect spheres in the absence of any nonhydrodynamic effects (Brownian motion, particle-particle friction, electrostatic effects, etc.), the linearity of the Stokes equation requires that $\eta_{\text{t}}$ be independent of strain rate. However, such nonidealities are always present; to capture them empirically, one can treat $[\eta]$ as well as the maximum packing fraction $\phi_{\text{max}}$ as adjustable parameters that in general will depend on strain rate, solvent and particle surface properties, and particle size distribution (Hiemenz & Rajagopalan 1997).

## 3.4 A Model Microscale Swimmer

In this section, we turn from passive particles in flow to a model of an *active* particle, one that can propel itself through fluid. We show here that our singularity solutions can be used to construct a special case, illustrated in Figure 3.2, of the so-called *squirmer* model of a spherical self-propelled particle (Lighthill 1952, Blake 1971*b*).

A neutrally buoyant, force- and torque-free particle, self-propelled or not, can have no Stokeslet contribution to its velocity field. At a minimum, if the object has finite size, then it must generate a potential dipole flow as it moves, so we begin by seeking a model of a swimming object that is a pure potential dipole. In a stationary reference frame, with the swimmer instantaneously located at the origin, the flow field will be of the form

$$v_i = c\left(\frac{x_i x_j}{r^5} - \frac{\delta_{ij}}{3r^3}\right)u_j, \tag{3.50}$$

where $c$ is an unknown constant and $\boldsymbol{u}$ the unit vector defining the orientation of the dipole. If the object is swimming with a velocity $\boldsymbol{U} = U\boldsymbol{u}$ with $U > 0$ (the potential dipole flow is axisymmetric, so $\boldsymbol{U}$ must be parallel to $\boldsymbol{u}$), then in the reference frame moving with the swimmer, the velocity field becomes

$$v_i = c\left(\frac{x_i x_j}{r^5} - \frac{\delta_{ij}}{3r^3}\right)u_j - Uu_i. \tag{3.51}$$

As of yet, $c$ and $U$ are arbitrary. We will relate them to one another by requiring that the object be a sphere of radius $a$ and that no fluid can penetrate the sphere. That is, in the moving frame $\boldsymbol{n} \cdot \boldsymbol{v} = 0$ at $r = a$. Since $n_i = x_i/r$ on the surface of a sphere, this condition is equivalent to requiring that $x_i v_i = 0$ at $r = a$. Explicitly,

$$\begin{aligned} x_i v_i = 0 &= c\left(\frac{x_i x_i x_j}{a^5} - \frac{x_i \delta_{ij}}{3a^3}\right)u_j - Ux_i u_i \\ &= c\left(\frac{a^2 x_j}{a^5} - \frac{x_j}{3a^3}\right)u_j - Ux_i u_i \\ &= \left[c\left(\frac{1}{a^3} - \frac{1}{3a^3}\right) - U\right]x_i u_i. \end{aligned} \tag{3.52}$$

This is satisfied if we take

$$c = \frac{3Ua^3}{2}. \tag{3.53}$$

With this choice for $c$, the flow is axisymmetric with respect to $\boldsymbol{u}$ and purely tangential at $r = a$. Some intuition can be gained by recasting the solution in terms of the velocity field that must be imposed on the surface of the sphere to generate the potential dipole flow. To do so, consider the component of $\boldsymbol{v}$ in the $\boldsymbol{u}$ direction at $r = a$ and note that $x_i u_i = r\cos\theta$, where $\theta$ is the angle of $\boldsymbol{x}$ with respect to $\boldsymbol{u}$:

$$\begin{aligned} u_i v_i &= \frac{3Ua^3}{2}\left(\frac{u_i x_i x_j}{a^5} - \frac{u_j}{3a^3}\right)u_j - Uu_i u_i \\ &= \frac{3Ua^3}{2}\left(\frac{\cos^2\theta}{a^3} - \frac{1}{3a^3}\right) - U. \end{aligned} \tag{3.54}$$

This quantity takes on its largest magnitude at the equator,

$$u_i v_i = -\frac{3}{2}U,$$

where it reduces to the tangential velocity. More generally, defining the equatorial speed as $V = 3U/2$, the tangential velocity expressed in spherical coordinates relative to $\boldsymbol{u}$ is easily shown to be

$$v_\theta = V\sin\theta. \tag{3.55}$$

Therefore, generating this velocity distribution on the surface of a sphere leads to net motion of the sphere with velocity $U\boldsymbol{u}$ and a velocity field in the frame moving with the sphere of

$$v_i = Va^3\left(\frac{x_i x_j}{r^5} - \frac{\delta_{ij}}{3r^3}\right)u_j - Uu_i. \tag{3.56}$$

In the stationary frame, the position $\boldsymbol{R}$ of the swimmer is, of course, time-dependent, satisfying

$$\frac{d\boldsymbol{R}}{dt} = \boldsymbol{U},$$

so at time $t$ the velocity field in the stationary frame is

$$v_i(t) = Va^3\left(\frac{(x_i - R_i(t))(x_j - R_j(t))}{|\boldsymbol{x} - \boldsymbol{R}(t)|^5} - \frac{\delta_{ij}}{3\,|\boldsymbol{x} - \boldsymbol{R}(t)|^3}\right)u_j. \tag{3.57}$$

This potential dipole swimmer is a very special case because it generates no stresslet flow and thus has a velocity field that decays as $1/r^3$. The absence of the stresslet arises from the high symmetry of the velocity distribution (3.55): the swimmer is pulling fluid in from ahead and pushing it out behind in equal parts. What happens if we break this symmetry, allowing a stresslet contribution? Previously, it was the combination of uniform flow and the potential dipole that allowed the no-penetration condition at $r = a$ to be satisfied. It turns out that it is not possible (Problem 3.3) to find a pure stresslet solution that satisfies the no-penetration condition at $r = a$. However, the particular combination of a stresslet and an octupole ($\boldsymbol{\nabla\nabla\nabla}\mathbf{G}$ contribution) that corresponds to a surface velocity

$$v_\theta = V' \sin 2\theta \tag{3.58}$$

satisfies this condition. Because this surface velocity is reflection-symmetric across the equator, it makes no contribution to the overall swimming velocity. The octupole contribution to the velocity field decays as $1/r^4$, so when $r \gg a$, it is negligible compared to the stresslet, which decays as $1/r^2$. When $V' > 0$, the far-field stresslet field pulls fluid in from fore and aft as it swims, while when $V' < 0$, the opposite occurs. Swimmers whose stresslet corresponds to the former and latter cases are called "pullers" and "pushers," respectively. For example, a bacterium propelled from behind by a flagellum is a pusher.

## 3.5 Faxén's Laws

The Lorentz reciprocal relation (1.57) relates two Stokes flows that occur in the same geometry:

$$\int_S (\boldsymbol{v}' \cdot \boldsymbol{t}'' - \boldsymbol{v}'' \cdot \boldsymbol{t}')\; dS = 0.$$

This result can be used to determine the force on a moving sphere in an arbitrary unbounded Stokes flow velocity field, using as a starting point the results for a moving sphere in an otherwise stationary fluid (Happel & Brenner 1965). Let $\boldsymbol{v}'(\boldsymbol{x})$ be the solution we found in Section 3.3.1 for a sphere of radius $a$ moving with velocity $\boldsymbol{U}'$ in a fluid that is stationary at infinity, and choose coordinates so the center of the sphere is at the origin. Therefore, boundary conditions on the sphere and at infinity are

$$\boldsymbol{v}'(r = a) = \boldsymbol{U}', \quad \boldsymbol{v}'(\infty) = \boldsymbol{0}, \quad p'(\infty) = p_\infty = 0.$$

The drag force on the sphere in this case is $\boldsymbol{F}'_{\text{drag}} = -6\pi\eta a \boldsymbol{U}'$. What we would like to determine is the force $\boldsymbol{F}_{\text{drag}}$ on a particle moving with velocity $\boldsymbol{U}$ in an unbounded fluid that in the absence of the particle would be moving with velocity $\boldsymbol{v}_\infty(\boldsymbol{x})$. This velocity can be the solution to a Stokes problem with a nonzero force density $\boldsymbol{f}(\boldsymbol{x})$, as long as the force density vanishes when $r \le a$. We denote the velocity in the presence of the particle as $\boldsymbol{v}^*(\boldsymbol{x})$ and let $\boldsymbol{v}''(\boldsymbol{x}) = \boldsymbol{v}^* - \boldsymbol{v}_\infty$ be the perturbation to the velocity field $\boldsymbol{v}_\infty$ induced by the presence of the particle. This perturbation is the solution to an unforced Stokes problem with boundary conditions

$$\boldsymbol{v}''(r = a) = \boldsymbol{U} - \boldsymbol{v}_\infty(r = a), \quad \boldsymbol{v}''(\infty) = 0.$$

The desired drag force $\boldsymbol{F}_{\text{drag}}$ is given by

$$\boldsymbol{F}_{\text{drag}} = -\int_{S_P} \boldsymbol{t}^* \, dS = -\int_{S_P} (\boldsymbol{t}'' + \boldsymbol{t}_\infty) \, dS.$$

However, since $\boldsymbol{v}_\infty$ satisfies the unforced Stokes equation when $r \le a$,

$$\int_{S_P} \boldsymbol{t}_\infty \, dS = \int_{S_P} \boldsymbol{n} \cdot \boldsymbol{\sigma}_\infty \, dS = \int_{V_P} \nabla \cdot \boldsymbol{\sigma}_\infty \, dV = 0.$$

Therefore,

$$\boldsymbol{F}_{\text{drag}} = -\int_{S_P} \boldsymbol{t}'' \, dS. \tag{3.59}$$

Because both $\boldsymbol{v}'$ and $\boldsymbol{v}''$ vanish as $r \to \infty$, only the particle surface $S_P$ contributes to the integrals in the Lorentz reciprocal relation. At every point on $S_P$, $\boldsymbol{v}'(\boldsymbol{x}) = \boldsymbol{U}'$, and furthermore, for a sphere in a stationary fluid, the traction is the same at every point on the sphere (Problem 3.1):

$$\boldsymbol{t}' = \frac{-\boldsymbol{F}'_{\text{drag}}}{4\pi a^2} = \frac{3\eta \boldsymbol{U}'}{2a}.$$

(Recall that $\boldsymbol{t} = \boldsymbol{n} \cdot \boldsymbol{\sigma}$, and since $\boldsymbol{n}$ is the outward unit normal with respect to the fluid, $\boldsymbol{n} \cdot \boldsymbol{\sigma}$ is the traction vector exerted by the sphere on the fluid.) That is, both the velocity $\boldsymbol{v}'$ and the traction $\boldsymbol{t}'$ are constant on the surface of the sphere and thus can be moved outside the integral in the Lorentz relation (1.57). Applying these results and (3.59), the Lorentz reciprocal relation simplifies to

$$\frac{3\eta}{2a} \boldsymbol{U}' \cdot \int_{S_P} \boldsymbol{v}''(\boldsymbol{x}) \, dS = -\boldsymbol{U}' \cdot \boldsymbol{F}_{\text{drag}}.$$

From this, we can infer that

$$\boldsymbol{F}_{\text{drag}} = -\frac{3\eta}{2a} \int_{S_P} \boldsymbol{v}''(\boldsymbol{x}) \, dS = -\frac{3\eta}{2a} \left( 4\pi a^2 \boldsymbol{U} - \int_{S_P} \boldsymbol{v}_\infty(\boldsymbol{x}) \right) dS. \tag{3.60}$$

The integral of $\boldsymbol{v}_\infty$ can be addressed by Taylor expansion around the origin. In Cartesian tensor notation,

$$\int_{S_P} v_{\infty,i}(\boldsymbol{x})\,dS = 4\pi a^2 v_{\infty,i}|_{\boldsymbol{x}=0} + \left.\frac{\partial v_{\infty,i}}{\partial x_j}\right|_{\boldsymbol{x}=0}\int_{S_P} x_j\,dS + \frac{1}{2}\left.\frac{\partial^2 v_{\infty,i}}{\partial x_j \partial x_k}\right|_{\boldsymbol{x}=0}\int_{S_P} x_j x_k\,dS + \dots. \tag{3.61}$$

The integrals over odd powers of $\boldsymbol{x}$ vanish by symmetry, and those over even powers can be evaluated in spherical coordinates by writing $x_i = r(\sin\theta\cos\phi\delta_{i1} + \sin\theta\sin\phi\delta_{i2} + \cos\theta\delta_{i3})$. Specifically,

$$\int_{S_P} x_j x_k\,dS = \frac{4\pi a^4}{3}\delta_{jk}$$

so the last term of (3.61) becomes

$$\frac{2\pi a^4}{3}\nabla^2 \boldsymbol{v}_\infty(0).$$

The higher degree even terms yield multiples of $\nabla^{2n}\boldsymbol{v}_\infty(0)$. These all vanish, because $\boldsymbol{v}_\infty$ satisfies the Stokes equation at the origin, and from (1.37), $\nabla^{2n}\boldsymbol{v}_\infty(0) = 0$ for all $n \geq 2$. Therefore,

$$\int_{S_P} \boldsymbol{v}_\infty(\boldsymbol{x})\,dS = 4\pi a^2 \boldsymbol{v}_\infty(0) + \frac{2\pi a^4}{3}\nabla^2 \boldsymbol{v}_\infty(0) = 4\pi a^2\left(1 + \frac{a^2}{6}\nabla^2\right)\boldsymbol{v}_\infty(0)$$

and (3.60) simplifies to

$$\boldsymbol{F}_{\text{drag}} = 6\pi\eta a\left(\left(1 + \frac{a^2}{6}\nabla^2\right)\boldsymbol{v}_\infty(0) - \boldsymbol{U}\right). \tag{3.62}$$

This result is called *Faxén's first law*. A similar analysis (Problem 3.5) leads to *Faxén's second law*, an expression for the drag torque $\boldsymbol{T}_{\text{drag}}$ exerted by the fluid on the sphere:

$$\boldsymbol{T}_{\text{drag}} = 8\pi\eta a^3\left(\boldsymbol{\omega}_\infty(0) - \boldsymbol{\Omega}\right), \tag{3.63}$$

where $\boldsymbol{\Omega}$ is the angular velocity of the sphere and $\boldsymbol{\omega}_\infty(0)$ the local angular velocity of the fluid in the absence of the sphere. Interestingly, this result is the same as (3.38), which was derived for the simpler case of a sphere rotating in an otherwise stationary fluid. Setting $\boldsymbol{T}_{\text{drag}}$ to zero shows that a torque-free rigid sphere in any unbounded Stokes flow rotates with the local angular velocity that the fluid would have in the sphere's absence. Finally, the results of Section 3.3.3 can be used to find the stresslet for a sphere in an arbitrary unbounded flow:

$$\mathbf{S} = \frac{20}{3}\pi\eta a^3\left(1 + \frac{a^2}{10}\nabla^2\right)\mathbf{E}_\infty(0), \tag{3.64}$$

where $\mathbf{E}_\infty$ is the rate of strain associated with $\boldsymbol{v}_\infty$ (Kim & Karrila 1991).

## 3.6 Mobility of a System of Spheres

In Section 2.5, we derived the mobility for a system of $N$ point particles in an unbounded fluid. Having in hand results for spheres, we revisit this problem here to illustrate the change of the mobility due to the finite size of the particles. We focus on the case of a pair of spheres. Sphere $\alpha$ has radius $a_\alpha$, resides at position $\boldsymbol{R}_\alpha$, and exerts a given force $\boldsymbol{F}_\alpha$ on the fluid. For simplicity, we take the particles to be torque-free, though this assumption is easily relaxed. In the absence of the other particle, particle 1 generates a flow

$$\boldsymbol{v}_1(\boldsymbol{x}) = \left(1 + \frac{a_1^2}{6}\nabla^2\right)\mathbf{G}(\boldsymbol{x} - \boldsymbol{R}_1) \cdot \boldsymbol{F}_1. \tag{3.65}$$

Combining this expression with Faxén's first law, (3.62), for the motion of particle 2 yields

$$\begin{aligned}\boldsymbol{F}_2 &= -6\pi\eta a_2 \left[\left(1 + \frac{a_2^2}{6}\nabla^2\right)\boldsymbol{v}_1(\boldsymbol{x})|_{\boldsymbol{x}=\boldsymbol{R}_2} - \boldsymbol{U}_2\right] \\ &= -6\pi\eta a_2 \left[\left(1 + \frac{a_2^2}{6}\nabla^2\right)\left(1 + \frac{a_1^2}{6}\nabla^2\right)\mathbf{G}(\boldsymbol{x} - \boldsymbol{R}_1)|_{\boldsymbol{x}=\boldsymbol{R}_2} \cdot \boldsymbol{F}_1 - \boldsymbol{U}_2\right].\end{aligned}$$

Solving for $\boldsymbol{U}_2$, we find that

$$\boldsymbol{U}_2 = \frac{1}{\zeta_2}\boldsymbol{F}_2 + \left(1 + \frac{a_2^2}{6}\nabla^2\right)\left(1 + \frac{a_1^2}{6}\nabla^2\right)\mathbf{G}(\boldsymbol{x} - \boldsymbol{R}_1)|_{\boldsymbol{x}=\boldsymbol{R}_2} \cdot \boldsymbol{F}_1. \tag{3.66}$$

Generalizing to a system of $N$ spheres, the mobility tensor becomes

$$\mathbf{M}_{\alpha\beta} = \frac{1}{\zeta_\alpha}\delta_{\alpha\beta}\boldsymbol{\delta} + (1 - \delta_{\alpha\beta})\left(1 + \frac{a_2^2}{6}\nabla^2\right)\left(1 + \frac{a_1^2}{6}\nabla^2\right)\mathbf{G}(\boldsymbol{x} - \boldsymbol{R}_\beta)|_{\boldsymbol{x}=\boldsymbol{R}_\alpha}. \tag{3.67}$$

Evaluation of (3.67) is facilitated by observing that $\nabla^2\nabla^2\mathbf{G} = \mathbf{0}$. This analysis can be generalized to yield expressions for the angular velocity and stresslet each particle induces on the other.

In deriving (3.67) we have neglected the fact that (3.65) does not account for the velocity perturbation at particle 1 due to the flow driven by particle 2 and vice versa. A careful asymptotic analysis of this problem using the "method of reflections" (Kim & Karrila 1991) reveals that the error in (3.67) is $O\left(\left|\boldsymbol{R}_\alpha - \boldsymbol{R}_\beta\right|^{-4}\right)$.

## 3.7 Nonspherical Rigid Particles

Section 3.3 gives us relationships between force, torque, stresslet and ambient fluid velocity, vorticity, and rate of strain for a sphere. For a rigid particle of arbitary shape in a linear flow, the linearity of the Stokes equation allows us to write a general linear relationship between these quantities in the form

$$\begin{bmatrix} \boldsymbol{F}_{\text{drag}} \\ \boldsymbol{T}_{\text{drag}} \\ \mathbf{S} \end{bmatrix} = \mathcal{R} \cdot \begin{bmatrix} \boldsymbol{v}_\infty - \boldsymbol{U} \\ \omega_\infty - \boldsymbol{\Omega} \\ \mathbf{E}_\infty \end{bmatrix}, \tag{3.68}$$

where

$$\mathcal{R} = \begin{bmatrix} \mathbf{R}^{\text{FU}} & \mathbf{R}^{\text{F}\Omega} & \mathbf{R}^{\text{FE}} \\ \mathbf{R}^{\text{TU}} & \mathbf{R}^{\text{T}\Omega} & \mathbf{R}^{\text{TE}} \\ \mathbf{R}^{\text{SU}} & \mathbf{R}^{\text{S}\Omega} & \mathbf{R}^{\text{SE}} \end{bmatrix} \tag{3.69}$$

is called the *grand resistance tensor* (Happel & Brenner 1965, Kim & Karrila 1991, Guazzelli & Morris 2012). As a consequence of the Lorentz reciprocal relations (or equivalently the self-adjointness of the Stokes operator), its elements satisfy the following symmetries:

$$R_{ij}^{\text{FU}} = R_{ji}^{\text{FU}}, \quad R_{ij}^{\text{T}\Omega} = R_{ji}^{\text{T}\Omega},$$
$$R_{ij}^{\text{F}\Omega} = R_{ji}^{\text{TU}},$$
$$R_{ijk}^{\text{FE}} = R_{kij}^{\text{SU}}, \quad R_{ijk}^{\text{TE}} = R_{jki}^{\text{S}\Omega},$$
$$R_{ijkl}^{\text{SE}} = R_{klij}^{\text{SE}}.$$

This tensor must also be positive definite, i.e.,

$$\begin{bmatrix} \boldsymbol{v}_\infty - \boldsymbol{U} \\ \omega_\infty - \boldsymbol{\Omega} \\ \mathbf{E}_\infty \end{bmatrix}^{\text{T}} \cdot \mathcal{R} \cdot \begin{bmatrix} \boldsymbol{v}_\infty - \boldsymbol{U} \\ \omega_\infty - \boldsymbol{\Omega} \\ \mathbf{E}_\infty \end{bmatrix} > 0$$

whenever there is relative motion of the particle and fluid, because the energy dissipation rate in any nontrivial flow must be positive (Kim & Karrila 1991). (These are the same reasons that the $N$-particle mobility matrix $\mathbf{M}$ that we introduced earlier is symmetric and positive definite.) Alternatively, one can write

$$\begin{bmatrix} \boldsymbol{v}_\infty - \boldsymbol{U} \\ \omega_\infty - \boldsymbol{\Omega} \\ \mathbf{E}_\infty \end{bmatrix} = \mathcal{M} \cdot \begin{bmatrix} \boldsymbol{F}_{\text{drag}} \\ \boldsymbol{T}_{\text{drag}} \\ \mathbf{S} \end{bmatrix}, \tag{3.70}$$

where $\mathcal{M} = \mathcal{R}^{-1}$ is called the *grand mobility tensor*.

For spheres, only the diagonal blocks of these matrices are nonzero and their values are implicitly given by the analyses of Section 3.3. At the opposite extreme, for a particle of arbitrary shape, all the elements are nonzero (Happel & Brenner 1965, Kim & Karrila 1991): for example, an imposed torque will lead to translation and vice versa, so $\mathbf{R}^{\text{F}\Omega}$ and $\mathbf{R}^{\text{TU}}$ will be nonzero. (A helical filament is a classic example of a particle with such coupling.) For particles with intermediate degrees of symmetry, constraints on the nature of this coupling can be derived. Consider a particle that is reflection symmetric across some plane – we can choose coordinates such that this is the plane $x_1 = 0$. Symmetry arguments can be used to show that for this particle

$$R_{12}^{\text{FU}} = R_{13}^{\text{FU}} = R_{21}^{\text{FU}} = R_{31}^{\text{FU}} = 0 \tag{3.71}$$

and

$$R^{F\Omega}_{11} = R^{F\Omega}_{22} = R^{F\Omega}_{33} = 0 \tag{3.72}$$

(Guyon et al. 2001). For a particle with three orthogonal planes of symmetry, the same arguments show that there can be no coupling at all between translation and rotation: $\mathbf{R}^{F\Omega} = \mathbf{R}^{TU} = \mathbf{0}$.

We will focus attention on axisymmetric, fore-aft symmetric particles. Specific analytical results can be obtained for the special case of spheroids but all particles with the same symmetries, for example, rigid rods, have qualitatively identical governing equations and dynamics (Kim & Karrila 1991). In this case, the only off-diagonal coupling in the grand resistance is between rotation (or torque) and strain rate.

Consider first the relation between velocity and force. Let $\boldsymbol{u}$ be the unit vector defining the axis of symmetry of the particle. Because of the fore-aft symmetry, all results must be unchanged if we replace $\boldsymbol{u}$ with $-\boldsymbol{u}$. By symmetry, a force in the $\boldsymbol{u}$ direction will lead the particle to move in the same direction, and likewise for a force perpendicular to $\boldsymbol{u}$. Based on these arguments, we can write the analogue of Stokes's law for the particle as

$$\boldsymbol{F}_{\text{drag}} = \boldsymbol{\zeta} \cdot (\boldsymbol{v}_\infty - \boldsymbol{U}), \tag{3.73}$$

$$\boldsymbol{\zeta} = \zeta^{\parallel}\boldsymbol{u}\boldsymbol{u} + \zeta^{\perp}(\boldsymbol{\delta} - \boldsymbol{u}\boldsymbol{u}). \tag{3.74}$$

Here the tensor $\boldsymbol{\zeta}$ (written as $\mathbf{R}^{FU}$ in (3.69)) generalizes the scalar friction coefficient for the sphere; the operators $\boldsymbol{u}\boldsymbol{u}$ and $\boldsymbol{\delta} - \boldsymbol{u}\boldsymbol{u}$ are parallel and perpendicular projection operators[3] with respect to $\boldsymbol{u}$, so the quantities $\zeta^{\parallel}$ and $\zeta^{\perp}$ are the friction coefficients for motion parallel and perpendicular to $\boldsymbol{u}$, respectively. One can easily verify that (3.74) can be inverted to read

$$\boldsymbol{\zeta}^{-1} = \frac{1}{\zeta^{\parallel}}\boldsymbol{u}\boldsymbol{u} + \frac{1}{\zeta^{\perp}}(\boldsymbol{\delta} - \boldsymbol{u}\boldsymbol{u}). \tag{3.75}$$

A simple but important consquence of the anisotropy of $\boldsymbol{\zeta}$ is that a sedimenting anisotropic particle will only move parallel to the direction of the imposed force (such as gravity) if $\boldsymbol{u}$ is parallel or perpendicular to that force. Otherwise, the particle will move at an angle to the applied force.

For the important special case of prolate spheroids, the parallel and perpendicular friction coefficients can be found analytically, either by solving the Stokes equation in spheroidal coordinates (which in turn arises from the solution to the Laplace equation in these coordinates (Lamb 1932)) or by seeking a solution as a distribution of Stokeslets and potential dipoles along the axis of symmetry (Chwang & Wu 1975, Kim & Karrila 1991, Leal 2007) – a form suggested by the solution for a sphere in uniform flow that we found in Section 3.3. For a spheroid with half-length $a$ and equatorial radius (half-width)

[3] For a unit vector $\boldsymbol{u}$, the projection operator $\boldsymbol{u}\boldsymbol{u}$ operating on a vector $\boldsymbol{w}$ yields $\boldsymbol{u}(\boldsymbol{u} \cdot \boldsymbol{w})$, i.e., the component of $\boldsymbol{w}$ parallel to $\boldsymbol{u}$.

$c$, these coefficients are given by

$$\zeta^{\parallel} = 6\pi\eta a\left(\frac{8}{3}e^3\left[-2e + (1+e^2)L\right]^{-1}\right), \tag{3.76}$$

$$\zeta^{\perp} = 6\pi\eta a\left(\frac{16}{3}e^3\left[2e + (3e^2-1)L\right]^{-1}\right). \tag{3.77}$$

(3.78)

Here $\ell = \frac{a}{c}$ is the aspect ratio, $e = \sqrt{1-\ell^{-2}}$ the eccentricity, and $L = \ln\frac{1+e}{1-e}$. (Analogous exact results for *oblate* spheroids, where $\ell < 1$, are given in Kim & Karrila (1991).)

For a needlelike object with aspect ratio $\ell \gg 1$, we let $\epsilon = 1/\ell = \sqrt{1-e^2}$ and can find that

$$\zeta^{\parallel} = 4\pi\eta a\left(\frac{1}{\ln\frac{2}{\epsilon} - \frac{1}{2}} + O(\epsilon^2)\right), \tag{3.79}$$

$$\zeta^{\perp} = 8\pi\eta a\left(\frac{1}{\ln\frac{2}{\epsilon} + \frac{1}{2}} + O(\epsilon^2)\right), \tag{3.80}$$

where $\ln\frac{2}{\epsilon} = \ln 2\ell \gg 1$. Note that $\zeta^{\perp}/\zeta^{\parallel} \to 2$ as $\ell \to \infty$. The difference between perpendicular and parallel friction coefficients is only a factor of two even for an infinitely long thin filament. A needle is not a very streamlined object in Stokes flow! This is because in either case the filament drags along with it a blob of fluid of characteristic size $a$, the length of the filament, *not* $c$, its width. On a related note, because of the slow growth of $\ln\frac{2}{\epsilon}$ as $\ell$ increases (for $\ell = 10$, $\ln\frac{2}{\epsilon} = 3.0$, and for $\ell = 1000$, $\ln\frac{2}{\epsilon} = 7.60$), the drag on a spheroid is not very much less that that on a sphere of radius $a$ unless $\ell$ is very large, despite the enormous difference in volume of the two objects. For example, when $\ell = 10$, $\zeta^{\parallel} \approx 0.26(6\pi\eta a)$, $\zeta^{\perp} \approx 0.4(6\pi\eta a)$. Indeed, this observation is a special case of the general qualitative result that, because of the long-ranged nature of the Green's function, the drag on an object in Stokes flow is primarily determined by its largest dimension. We see this feature again in Sections 3.9 and 8.3.

These results can be used to illustrate the mechanism of bacterial locomotion by means of rotating helical flagella. Consider a segment of the flagellum with local tangent vector $\boldsymbol{u}$ as shown in Figure 3.3 and take the segment to be moving with velocity $\boldsymbol{U} = U\boldsymbol{e}_\theta$ in cylindrical coordinates (i.e. normal to the axis of the flagellum, which is the $z$-direction). To illustrate the basic locomotion mechanism we will neglect the hydrodynamic interactions between this segment and the other parts of the flagellum, and just consider it to have friction coefficients $\zeta^{\parallel}$ and $\zeta^{\perp}$ – this approximate treatment of flagellar hydrodynamics is called *resistive force theory* (Gray & Hancock 1955) and is a special case of the *slender body theory* that is detailed in Section 3.9. An external force $\boldsymbol{F}_{\text{ext}} = 0$ must be exerted on this segment (by the neighboring segments, for example) to move it; the resulting force balance is

$$\boldsymbol{F}_{\text{drag}} + \boldsymbol{F}_{\text{ext}} = 0,$$

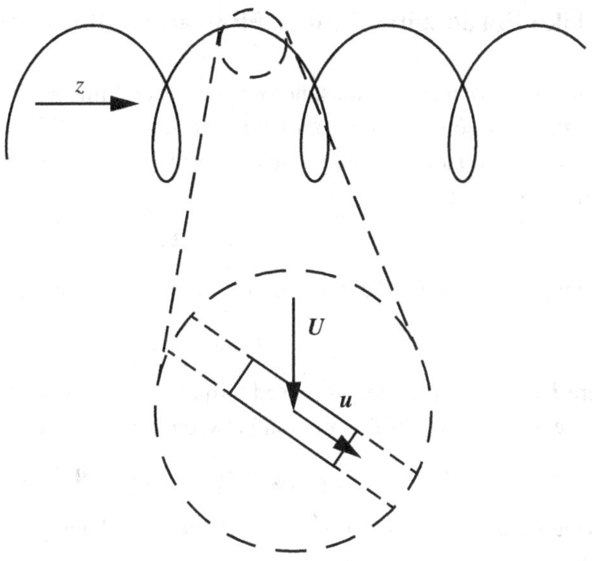


Figure 3.3 A rotating rigid flagellum. In cylindrical coordinates, the highlighted segment is moving with velocity $U\boldsymbol{e}_\theta$.

where $\boldsymbol{F}_{\text{drag}}$ is given by the anisotropic version of Stokes's law, (3.73). Furthermore, the force $\boldsymbol{F}$ that the segment exerts on the fluid is simply $-\boldsymbol{F}_{\text{drag}}$, which by the force balance is $\boldsymbol{F}_{\text{ext}}$. Accordingly, we find that the flagellar segment exerts the following force on the fluid:

$$\boldsymbol{F} = \left(\zeta^{\|}\boldsymbol{u}\boldsymbol{u} + \zeta^{\perp}(\boldsymbol{\delta} - \boldsymbol{u}\boldsymbol{u})\right) \cdot \boldsymbol{U}.$$

In particular, the $z$-component, $F_z$, of the force (i.e., the component of the force along the axis of the flagellum) is given by

$$\begin{aligned} F_x &= \zeta^{\|}(\boldsymbol{e}_z \cdot \boldsymbol{u})\boldsymbol{u} \cdot \boldsymbol{U} + \zeta^{\perp}(\boldsymbol{e}_z \cdot \boldsymbol{U} - (\boldsymbol{e}_z \cdot \boldsymbol{u})(\boldsymbol{u} \cdot \boldsymbol{U})) \\ &= (\zeta^{\|} - \zeta^{\perp})(\boldsymbol{u} \cdot \boldsymbol{e}_z)(\boldsymbol{u} \cdot \boldsymbol{e}_\theta)U. \end{aligned}$$

Because for a slender segment $\zeta^{\|} \neq \zeta^{\perp}$, a motion of the flagellar segment in the $\theta$-direction leads to a force on the fluid in the $z$-direction (thus an equal and opposite force on the flagellum and thus locomotion of the microbe to which the flagellum is attached). Adding up the forces exerted on all segments of the flagellum yields an estimate of the total thrust exerted by the flagellum on the fluid as it rotates. The calculation presented here yields a convenient approximate method for estimating the effect of helix geometry (i.e., $\boldsymbol{u} \cdot \boldsymbol{e}_z$) on thrust. More generally, it provides an important example of a particle shape for which translational and rotational motions are coupled.

## 3.8 Rodlike Particle in a Linear Flow and Jeffery Orbits

We now turn to the case of a spheroid in a general linear flow. Our goal is to derive and understand the equations of motion for the orientation $\boldsymbol{u}$ of the particle – this is equivalent to finding an equation for the particle's angular velocity $\boldsymbol{\Omega}$, since the kinematics of rigid rotation dictate that

$$\dot{\boldsymbol{u}} = \boldsymbol{\Omega} \times \boldsymbol{u}. \tag{3.81}$$

The torque balance for the particle in the absence of inertia is simply

$$\boldsymbol{T}_{\text{drag}} + \boldsymbol{T}_{\text{ext}} = \boldsymbol{0}, \tag{3.82}$$

where $\boldsymbol{T}_{\text{ext}}$ is any externally imposed torque on the particle. For a fore-aft-axisymmetric particle, (3.68) gives us the relation between the drag torque and the imposed flow:

$$\boldsymbol{T}_{\text{drag}} = \mathbf{R}^{\text{T}\Omega} \cdot (\boldsymbol{\omega}_\infty - \boldsymbol{\Omega}) + \mathbf{R}^{\text{TE}} \cdot \mathbf{E}_\infty. \tag{3.83}$$

Solving for $\boldsymbol{\Omega}$, taking the cross product with $\boldsymbol{u}$ and applying (3.81) yields an evolution equation for $\boldsymbol{u}$:

$$\dot{\boldsymbol{u}} = \boldsymbol{\omega}_\infty \times \boldsymbol{u} + \left(\mathbf{R}^{\text{T}\Omega}\right)^{-1} \cdot \mathbf{R}^{\text{TE}} \cdot \mathbf{E}_\infty \times \boldsymbol{u} + \left(\mathbf{R}^{\text{T}\Omega}\right)^{-1} \cdot \boldsymbol{T}_{\text{ext}} \times \boldsymbol{u}. \tag{3.84}$$

For a fore-aft-axisymmetric particle, $\mathbf{R}^{\text{T}\Omega}$ can be viewed as a rotational friction tensor and has the same tensorial form as the translational friction tensor given in (3.74). There are different rotational friction coefficients in the directions parallel and perpendicular to $\boldsymbol{u}$ corresponding to rotations around the polar axis and any axis passing through the equatorial plane, respectively. Thus, we rename $\mathbf{R}^{\text{T}\Omega}$ as $\boldsymbol{\zeta}_{\text{r}}$, where

$$\boldsymbol{\zeta}_{\text{r}} = \zeta_{\text{r}}^{\parallel} \boldsymbol{u}\boldsymbol{u} + \zeta_{\text{r}}^{\perp} (\boldsymbol{\delta} - \boldsymbol{u}\boldsymbol{u}). \tag{3.85}$$

Similarly,

$$\boldsymbol{\zeta}_{\text{r}}^{-1} = \frac{1}{\zeta_{\text{r}}^{\parallel}} \boldsymbol{u}\boldsymbol{u} + \frac{1}{\zeta_{\text{r}}^{\perp}} (\boldsymbol{\delta} - \boldsymbol{u}\boldsymbol{u}). \tag{3.86}$$

(We will not generally be interested in the rotation angle around the axis of symmetry because it plays no role in the dynamics of $\boldsymbol{u}$, so without confusion we can define a scalar rotational friction coefficient $\zeta_{\text{r}} = \zeta_{\text{r}}^{\perp}$.)

The quantity $\mathbf{R}^{\text{TE}}$ can be written

$$R_{kij}^{\text{TE}} = \zeta_{\text{E}}(\epsilon_{ikl} u_j + \epsilon_{jkl} u_i) u_l. \tag{3.87}$$

The form of this expression can be motivated by the fact that it must depend on $u_i$ and $\epsilon_{ijk}$, with the latter dependence arising because $\boldsymbol{T}_{\text{drag}}$ is a pseudovector (Kim & Karrila 1991).

With these results for $\mathbf{R}^{\text{T}\Omega}$ and $\mathbf{R}^{\text{TE}}$ and the double cross product identity (A.8), we can write (3.84) as

$$\dot{\boldsymbol{u}} = \boldsymbol{\Omega} \times \boldsymbol{u} = \boldsymbol{\omega}_\infty \times \boldsymbol{u} + B\,(\boldsymbol{\delta} - \boldsymbol{u}\boldsymbol{u}) \cdot \mathbf{E}_\infty \cdot \boldsymbol{u} + \frac{1}{\zeta_{\text{r}}} \boldsymbol{T}_{\text{ext}} \times \boldsymbol{u}. \tag{3.88}$$

The quantity $B = \frac{\zeta_{\text{E}}}{\zeta_{\text{r}}}$ is called the Bretherton constant. For a spheroid, this quantity can

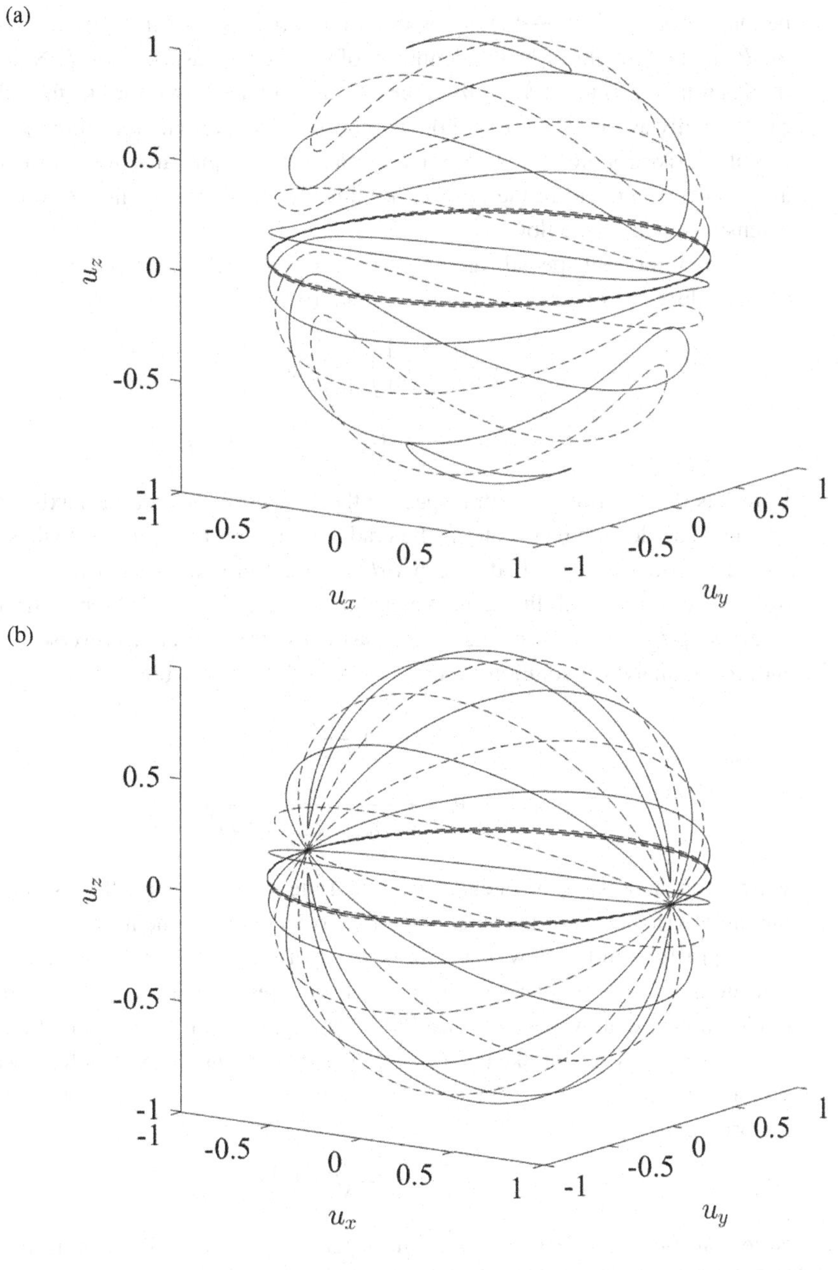


**Figure 3.4** Jeffery orbits of $\boldsymbol{u}$ for (a) $\ell = 5$ and (b) $\ell = 100$, for $C = 0.1, 0.5, 1, 2, 5, 100$, and $\infty$, using alternating solid and dashed lines as $C$ increases. Two orbits for each value are given, one for an initial condition with $u_z > 0$ and one for $u_z < 0$. By symmetry, these are indistinguishable. With the viewpoint shown here, the $x - y$ plane is horizontal.

be found exactly: $B = \frac{\ell^2-1}{\ell^2+1}$. For any other fore-aft-axisymmetric particle, this equation for $B$ can be thought of as the definition of an effective aspect ratio $\ell$. Note that if the fluid is rotating rigidly ($\mathbf{E}_\infty = \mathbf{0}$) in the absence of an external torque, the spheroid will rotate rigidly as well: $\boldsymbol{\Omega} = \boldsymbol{\omega}_\infty$. For any nonspherical particle, the straining motion $\mathbf{E}_\infty$ can also drive rotation. In unixial extension, for example, any initial orientation with a nonzero component in the extensional direction will evolve until $\boldsymbol{u}$ points along the extensional axis of the flow.

For the important special case of a torque-free particle in simple shear, $\boldsymbol{v}_\infty = \dot{\gamma} y \boldsymbol{e}_x$, (3.88) reduces to *Jeffery's equations* (Jeffery 1922):

$$\dot{\theta} = \left(\frac{\ell^2 - 1}{\ell^2 + 1}\right) \frac{\dot{\gamma}}{4} \sin 2\theta \sin 2\phi, \tag{3.89}$$

$$\dot{\phi} = \frac{\dot{\gamma}}{\ell^2 + 1} \left(\ell^2 \cos^2 \phi + \sin^2 \phi\right). \tag{3.90}$$

Here $\theta$ is the polar angle with respect to the $z$ (neutral, or vorticity) axis, and $\phi$ is the azimuthal angle with respect to $y$ (the gradient direction).[4] Solutions to these equations are closed trajectories called *Jeffery orbits*. One can understand why they are closed by considering reversibility, axisymmetry of the particle, and the symmetry of the flow – reversing the direction of the shear must cause the particle to reverse its trajectory. Indeed, an analytical solution for these closed orbits is possible:

$$\tan \phi = \ell \tan \left(\frac{\dot{\gamma} t}{\ell + \ell^{-1}}\right), \tag{3.91}$$

$$\tan \theta = \frac{C\ell}{\left(\ell^2 \cos^2 \phi + \sin^2 \phi\right)^{1/2}}, \tag{3.92}$$

where $C$ is a constant of integration, called the *orbit constant*, that is determined by the initial orientation of the fiber. A particle oriented in the neutral direction ($\boldsymbol{u} = \boldsymbol{e}_z$), which just rolls in the flow, has $C = 0$. By contrast, a particle for which $\boldsymbol{u}$ lies in the flow-gradient ($x - y$) plane has $C = \infty$. It tumbles end-over-end, and $\boldsymbol{u}$ traces out a circle. Intermediate values of $C$ lead to what might be called kayaking motions, as the particle traces out a path like that of a kayak paddle. Figure 3.4 shows Jeffery orbits with various values of $C$ for $\ell = 5$ and $\ell = 100$. From (3.91), it can be seen that the period of the orbit is

$$T = \frac{2\pi}{\dot{\gamma}} \left(\ell + \ell^{-1}\right). \tag{3.93}$$

Note that this is independent of $C$ but depends on $\ell$. In the limit of an infinitely thin rod, $\ell \to \infty$, the period diverges. In this (singular) limit, the particle just rotates, with $\boldsymbol{u}$ following a great circle path on the unit sphere, until it is pointing in the flow direction and stays there without tumbling. (The orbits on Figure 3.4(b), for which $\ell \gg 1$, nearly follow great circles except when the orientation becomes close to the flow direction.) The physical origin of this result can be explained as follows. At any finite aspect ratio, the upper and lower halves of a flow-aligned rod experience tractions in opposite directions,

[4] Observe that this is not the usual convention for spherical coordinates.

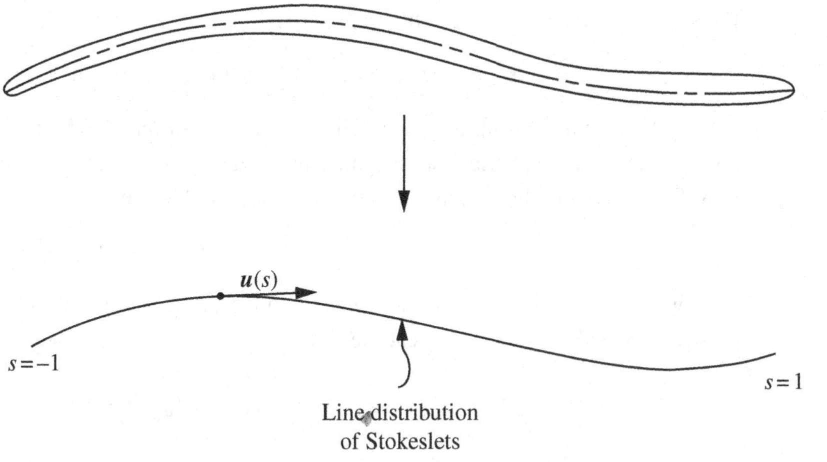


**Figure 3.5** A general slender body.

and since the rod has finite thickness, these lead to a finite moment; this moment is balanced by a hydrodynamic torque associated with rotation of the rod through the shear plane. For an infinitely thin rod aligned with the flow direction, the thickness and thus moment vanish, so the fluid exerts no torque on the rod that would drive its orientation through the shear plane. Even for a rod with infinite aspect ratio, any nonzero amount of Brownian motion can also break the invariance of the flow-aligned state (Section 9.3).

## 3.9 Slender Body Theory

The helical flagellum described earlier is one of many important examples of long, thin filaments in a flow field. We noted previously that for a spheroidal particle, the flow field can be found by placing Stokeslets and potential dipoles along the axis of the particle. The core idea of *slender body theory* is to generalize this approach to long, slender filaments of arbitary shape and cross-section. We will present a very basic form of this theory and show how it can be used to estimate $\zeta^{\|}$ and $\zeta^{\perp}$ for a long, straight axisymmetric filament (Batchelor 1970*a*, Leal 2007).

As in the previous spheroid case, we take the filament to have length $2a$. The filament is aligned with the $x$-axis, and we define a dimensionless arclength variable $s = x/a$ along the filament such that $-1 \leq s \leq 1$. The local radius is given by

$$r(s) = \epsilon a R(s), \tag{3.94}$$

where $R$ is dimensionless and is $O(1)$ so that $\epsilon \ll 1$ measures the relative thickness of the body. Without approximation, we can write that the velocity field around the body is given by

$$v_i(\boldsymbol{x}) - v_{\infty,i}(\boldsymbol{x}) = -\int_{S_P} G_{ij}(\boldsymbol{x} - \boldsymbol{x}')\left(\hat{n}_k \sigma_{kj}(\boldsymbol{x}')\right) dS(\boldsymbol{x}'). \tag{3.95}$$

Points $\boldsymbol{x}'$ on the surface of the filament satisfy

$$\boldsymbol{x}' = as'\boldsymbol{e}_x + a\epsilon R(s')\left(\cos\theta'\boldsymbol{e}_y + \sin\theta'\boldsymbol{e}_z\right),$$

where $\theta'$ is the azimuthal angle measuring position around the local circumference of the filament and the primes indicate dummy variables over which we will integrate. We now Taylor-expand the Green's function around $\epsilon = 0$ to yield

$$G_{ij}(\boldsymbol{x} - \boldsymbol{x}') = G_{ij}(\boldsymbol{x} - as'\boldsymbol{e}_x) + O(\epsilon).$$

Now we insert this expression into (3.95) and integrate around the local circumference $\theta'$ at each $s'$ value to yield (to leading order in $\epsilon$)

$$v_i(\boldsymbol{x}) - v_{\infty,i}(\boldsymbol{x}) = a\int_{-1}^{1} G_{ij}(\boldsymbol{x} - as'\boldsymbol{e}_x)\bar{f}_j(s')\,ds', \tag{3.96}$$

where $\bar{f}_j(s')$ is the traction integrated around the local circumference of the filament at arclength position $s'$ – the net force per unit length exerted by the filament on the fluid. This is just the velocity field driven by a line of Stokeslets with as yet unknown line force density $\bar{\boldsymbol{f}}$. This density is determined by imposing the no-slip boundary condition on the surface of the filament.

As before, but now considering the argument $\boldsymbol{x}$ rather than $\boldsymbol{x}'$ in the Green's function, we note that points $\boldsymbol{x}(s)$ on the surface of the filament satisfy

$$\boldsymbol{x}(s) = as\boldsymbol{e}_x + a\epsilon R(s)\left(\cos\theta\boldsymbol{e}_y + \sin\theta\boldsymbol{e}_z\right).$$

At these points, the no-slip condition requires that $\boldsymbol{v}(\boldsymbol{x}) = \boldsymbol{U}(\boldsymbol{x})$ where $\boldsymbol{U}$ is the velocity of the filament at point $\boldsymbol{x}$. We will assume that the difference between $\boldsymbol{U}$ on the filament surface and on its centerline is negligible and likewise for $\boldsymbol{v}_\infty$. Now (3.96), evaluated for points $\boldsymbol{x}(s)$ on the surface of the filament, can be written

$$U_i((x(s),0,0)^{\mathrm{T}}) - v_{\infty,i}((x(s),0,0)^{\mathrm{T}}) = \frac{a}{8\pi\eta}\int_{-1}^{1}\left(\frac{\delta_{ij}}{\rho(s')} + \frac{a^2(s-s')^2\delta_{i1}\delta_{j1}}{\rho(s')^3}\right)\bar{f}_j(s')ds', \tag{3.97}$$

where

$$\rho(s') = a\sqrt{(s-s')^2 + \epsilon^2R^2(s)),}$$

and again we have neglected terms of $O(\epsilon)$. This is a Fredholm integral equation of the first kind for $\bar{\boldsymbol{f}}$. Equations in this class are ill-posed[5] and thus challenging to solve

[5] In the context of linear differential or integral equations, a problem is mathematically *well-posed* if its solution (1) exists, (2) is unique and (3) depends continuously on the given forcing functions and parameters (the "data") for the problem. Otherwise, it is *ill-posed*. As an illustration of ill-posedness for Fredholm integral equations, consider the equation

$$\int_a^b K(x,y)f(x)\,dx = g(y),$$

where $g(y)$ is known. The solution to this problem is not unique. If $f(x)$ is a solution, then so is $f(x) + h(x)$, where $h(x)$ is any function that satisfies $\int_a^b K(x,y)h(x)\,dx = 0$. Specifically, if $\int_a^b |K(x,y)|\,dx < \infty$, then the *Riemann–Lebesgue lemma* (Stakgold 1998), a basic result in the theory directly. Nevertheless, progress can be made when $\epsilon \ll 1$ by carefully treating the integrand.

Consider the integral containing the first term in parentheses:

$$\int_{-1}^{1} \frac{\bar{f}_i(s')}{\rho}\, ds'.$$

This is nearly singular when $s' \to s$, where $\rho \to O(\epsilon^{-1})$. This integral can be rewritten:

$$\int_{-1}^{1} \frac{\bar{f}_i(s')}{\rho(s')}\, ds' = \bar{f}_i(s) \int_{-1}^{1} \frac{1}{\rho(s')}\, ds' + \int_{-1}^{1} \frac{\bar{f}_i(s') - \bar{f}_i(s)}{\rho(s')}\, ds',$$

which serves to isolate the nearly singular part in the first integral. The numerator in the second integrand changes sign as $s'$ passes through $s$, and that integral can be shown to be $O(1)$ as $\epsilon \to 0$. Now consider a cylindrical filament with $R = 1$, and for the moment allow the lower and upper integration limits to bracket $s$ by an $O(1)$ distance but to be otherwise arbitrary. Now the integral under consideration is

$$\frac{1}{a} \int_{s-\alpha}^{s+\beta} \left((s - s')^2 + \epsilon^2\right)^{-1/2} ds',$$

where $\alpha, \beta = O(1)$. This integrates to

$$\frac{1}{a} \ln \frac{\sqrt{(\alpha^2 + \epsilon^2)} + \alpha}{\sqrt{(\beta^2 + \epsilon^2)} - \beta}.$$

The quantity inside the logarithm Taylor-expands to $-\alpha\beta\epsilon^{-2} + O(1)$ so the final result is

$$\int_{\frac{x}{a}-\alpha}^{\frac{x}{a}+\beta} \left((x - as')^2 + a^2\epsilon^2\right)^{-1/2} ds' = \frac{2}{a} \ln \frac{1}{\epsilon} + O(1)$$

as $\epsilon \to 0$. Observe that at leading order, the upper and lower integration limits do not appear in this expression – it is independent of the overall geometry of the filament as long as the filament is sufficiently slender (though for a filament of constant radius, the error is particularly large near the ends (Johnson 1980)). To summarize, our analysis shows that

$$\int_{-1}^{1} \frac{\bar{f}_i(s')}{\rho}\, ds' = \frac{2}{a} \ln \frac{1}{\epsilon} \bar{f}_i(s) + O(1).$$

Through a similiar analysis, the second term in the parentheses of (3.97) evaluates to a very similar expression:

$$\int_{-1}^{1} \left( \frac{a^2(s - s')^2 \delta_{i1} \delta_{j1}}{\rho(s')^3} \right) \bar{f}_j(s') ds' = \delta_{i1} \frac{2}{a} \ln \frac{1}{\epsilon} \bar{f}_1(s) + O(1).$$

of Fourier transforms, states that

$$\lim_{k \to \infty} \int_a^b K(x, y) \cos(kx)\, dx = \lim_{k \to \infty} \int_a^b K(x, y) \sin(kx)\, dx = 0.$$

Therefore, if $f(x)$ is a solution to the integral equation, then so is $f(x)$ plus any combination of $\sin(kx)$ and $\cos(kx)$ in the limit $k \to \infty$.

Inserting these expressions into (3.97) yields the following relationship between the relative velocity evaluated on the filament centerline $\boldsymbol{x}(s) = as\boldsymbol{e}_x$ and the force density along its length:

$$U_i(\boldsymbol{x}(s)) - v_{\infty,i}(\boldsymbol{x}(s)) = \frac{\ln\frac{1}{\epsilon}}{4\pi\eta}\left(\bar{f}_i(s) + \delta_{i1}\bar{f}_1(s)\right) + O(1). \tag{3.98}$$

Defining friction coefficients per unit length $\bar{\zeta}^{\parallel}$ and $\bar{\zeta}^{\perp}$, we have then that

$$\bar{\zeta}^{\parallel} = \frac{2\pi\eta}{\ln\frac{1}{\epsilon}}, \tag{3.99}$$

$$\bar{\zeta}^{\perp} = \frac{4\pi\eta}{\ln\frac{1}{\epsilon}}. \tag{3.100}$$

The logarithmic dependence of velocity on $\epsilon$ is a reflection of the fact that forces on distant parts of the filament (relative to its cross-sectional dimension $\epsilon a$) contribute to the velocity at each point, due to the slow $1/\rho$ decay of the Stokeslet – qualitatively the logarithmic dependence arises from integration of the $1/\rho$. Nevertheless, because the filament is so long compared to thickness, the friction is local on the scale of the whole filament. Finally, note for a filament moving at uniform velocity $\boldsymbol{U}$ in a stationary fluid we can find the overall friction coefficients:

$$\zeta^{\parallel} = \int_{-a}^{a} \bar{\zeta}^{\parallel}\, dx = \frac{4\pi\eta a}{\ln\frac{1}{\epsilon}}, \tag{3.101}$$

$$\zeta^{\perp} = \int_{-a}^{a} \bar{\zeta}^{\perp}\, dx = \frac{8\pi\eta a}{\ln\frac{1}{\epsilon}}. \tag{3.102}$$

Comparing these results to the asymptotic results for prolate spheroids, (3.79)–(3.80), by observing that $\ln\frac{2}{\epsilon} = \ln\frac{1}{\epsilon} + O(1)$ we see that they agree as $\epsilon \to 0$.

In the preceding presentation, we have specified that the filament be straight. However, we found that to leading order, the integrals in (3.97) are independent of the filament length because the limits of integration drop out. Furthermore, to leading order they are also independent of local curvature as long as it is sufficiently small. Therefore, for a filament of general shape whose local centerline position and orientation at arclength value $s$ are given by $\boldsymbol{x}(s)$ and $\boldsymbol{u}(s)$, respectively, as shown in Figure 3.5, and whose local radius is given by (3.94), (3.98) generalizes to read

$$\boldsymbol{U}(\boldsymbol{x}(s)) - \boldsymbol{v}_{\infty}(\boldsymbol{x}(s)) = \frac{\ln\frac{1}{R(s)\epsilon}}{4\pi\eta}\left(\boldsymbol{\delta} + \boldsymbol{u}(s)\boldsymbol{u}(s)\right)\cdot\bar{\boldsymbol{f}}(s) + O(1). \tag{3.103}$$

This is a local anisotropic Stokes's law expression relating relative velocity to position along the backbone of the filament. It is easily inverted to yield

$$\bar{\boldsymbol{f}}(s) = \frac{4\pi\eta}{\ln\frac{1}{R(s)\epsilon}}\left(\boldsymbol{\delta} - \frac{1}{2}\boldsymbol{u}(s)\boldsymbol{u}(s)\right)\cdot\left(\boldsymbol{U}(\boldsymbol{x}(s)) - \boldsymbol{v}_{\infty}(\boldsymbol{x}(s))\right) + O(1). \tag{3.104}$$

This can be rewritten as

$$\bar{\boldsymbol{f}}(s) = \bar{\zeta}^{\parallel}\left(\boldsymbol{u}(s)\boldsymbol{u}(s) + 2\left(\boldsymbol{\delta} - \boldsymbol{u}(s)\boldsymbol{u}(s)\right)\right)\cdot\left(\boldsymbol{U}(\boldsymbol{x}(s)) - \boldsymbol{v}_{\infty}(\boldsymbol{x}(s))\right) + O(1), \tag{3.105}$$

where

$$\bar{\zeta}^{\parallel} = \frac{2\pi\eta}{\ln \frac{1}{R(s)\epsilon}} \tag{3.106}$$

and the factor of two between the parallel and perpendicular local friction coefficients is built into the expression. This is the result used in the aforementioned resistive force theory as well as more broadly.

The relative error in this local form of slender body theory is rather large: $\ln \frac{1}{\epsilon}$ increases *very* slowly as $\epsilon$ decreases. Improved approximations can be made using matched asymptotic expansions, i.e., by carefully treating the behavior on the scales $\epsilon a$ and $a$ and matching them together (Keller & Rubinow 1976, Johnson 1980, Götz 2000, Tornberg & Shelley 2004). These versions of slender body theory are *nonlocal*: the relative velocity at one point along the filament centerline depends on the force density over the entire filament. Numerical implementations of nonlocal slender body theory require careful treatment (Tornberg & Shelley 2004).

## 3.10 Lubrication Theory

In Section 2.5, we noted that the multipole expansion provides an analytical tool for dealing with hydrodynamic interactions between objects that are very far apart. Here we consider the opposite limit, objects that are very close together, and introduce another analytical tool known as the lubrication approximation. The name comes from the method's origin in analyzing the role of lubricants in reducing friction between moving machine parts. This approximation is valid in very thin domains between surfaces, such that the thickness of the domain $L_z$ is small compared to the length scale $L_{\mathrm{H}}$ over which the thickness varies. This is precisely the case for smooth particles that are very close together.

For definiteness, we will consider a domain like that shown in Figure 3.6. The bottom surface of the domain will be flat and located at $z = 0$ while the top will be at $z = h(x, y, t)$ or $z = h(r, \theta, t)$, depending on whether it is more convenient to work in Cartesian or cylindrical coordinates. The restriction to a flat bottom surface is not necessary but simplifies the presentation. We distingush between the vertical ($z$) (thin) direction and the horizontal ($H$) directions using the following definitions for the horizontal parts of various quantities: $\boldsymbol{v}_{\mathrm{H}} = (\boldsymbol{\delta} - \boldsymbol{e}_z\boldsymbol{e}_z) \cdot \boldsymbol{v}$, $\boldsymbol{\nabla}_{\mathrm{H}} = (\boldsymbol{\delta} - \boldsymbol{e}_z\boldsymbol{e}_z) \cdot \boldsymbol{\nabla}$ and $\nabla_{\mathrm{H}}^2 = \boldsymbol{\nabla}_{\mathrm{H}} \cdot \boldsymbol{\nabla}_{\mathrm{H}}$, noting that $(\boldsymbol{\delta} - \boldsymbol{e}_z\boldsymbol{e}_z)$ is simply the projection operator onto the horizontal plane. Using these, the Stokes and continuity equations become

$$\boldsymbol{0} = -\boldsymbol{\nabla}_{\mathrm{H}} p + \eta \left( \nabla_{\mathrm{H}}^2 + \frac{\partial^2}{\partial z^2} \right) \boldsymbol{v}_{\mathrm{H}}, \tag{3.107}$$

$$0 = -\frac{\partial p}{\partial z} + \eta \left( \nabla_{\mathrm{H}}^2 + \frac{\partial^2}{\partial z^2} \right) v_z, \tag{3.108}$$

$$0 = \boldsymbol{\nabla}_{\mathrm{H}} \cdot \boldsymbol{v} + \frac{\partial v_z}{\partial z}. \tag{3.109}$$

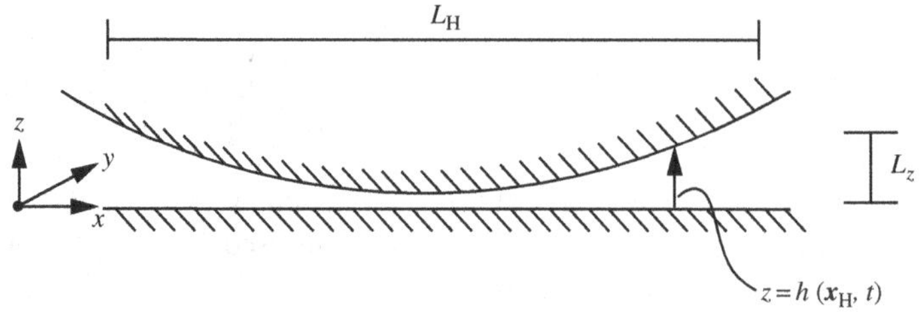


**Figure 3.6** General setup for lubrication theory.

Now we let $U$ and $W$ be characteristic scales for the horizontal and vertical velocities and $\epsilon = L_z/L_H \ll 1$. Since $\nabla_H \sim L_H^{-1}$ and $\frac{\partial}{\partial z} \sim L_z^{-1}$, the continuity equation leads us to conclude that $W/U \sim \epsilon$ – vertical velocities are much smaller than horizontal.

Now consider the horizontal part of the Stokes equation, (3.107). Defining an as yet undetermined pressure scale $\Pi$, we estimate the magnitudes of the terms in these equations as

$$\begin{aligned} \nabla_H p &\sim \frac{\Pi}{L_H}, \\ \eta \nabla_H^2 \boldsymbol{v}_H &\sim \eta \frac{U}{L_H^2}, \\ \eta \frac{\partial^2}{\partial z^2} \boldsymbol{v}_H &\sim \eta \frac{U}{L_z^2}. \end{aligned}$$

Dividing each term by $\eta U/L_z^2$, we find that, relative to the third term, the second term is $O(\epsilon^2)$ and is thus negligible. The term containing pressure cannot vanish from the leading order formulation in $\epsilon$ – otherwise, continuity cannot be satisfied. Therefore, the first and third terms must balance, which leads us to conclude that $\Pi = \eta U/L_z\epsilon$. For a small gap, the magnitude of the pressure can thus be very large. A similar analysis of the vertical momentum equation shows that $\partial p/\partial z = O(\epsilon)$, so to leading order, $p$ does not vary in $z$. Therefore, to leading order in $\epsilon$, the horizontal momentum balance is

$$\mathbf{0} = -\nabla_H p + \eta \frac{\partial^2}{\partial z^2} \boldsymbol{v}_H. \tag{3.110}$$

This equation has very simple solutions. For example, in the planar case, where $v_y = 0$ and $\boldsymbol{v}$ is independent of $y$, (3.110) integrates to

$$v_x = \frac{1}{2\eta}\frac{dp}{dx} z^2 + c_1(x) z + c_2(x). \tag{3.111}$$

Similarly, for axisymmetric flow with no $\theta$ component,

$$v_r = \frac{1}{2\eta}\frac{dp}{dr} z^2 + c_1(r) z + c_2(r). \tag{3.112}$$

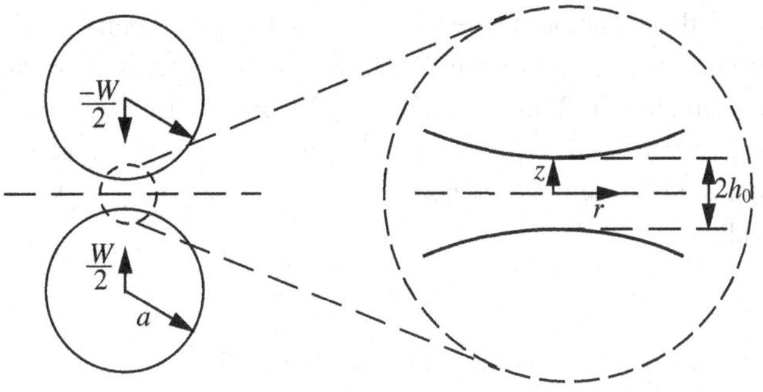


**Figure 3.7** Spheres approaching one another at relative speed $W$ along their line of centers.

More generally, at any horizontal position, solutions to (3.110) are simply a superposition of plane Couette flow (a linear velocity profile) and plane Poiseuille (pressure-driven) flow (a parabolic velocity profile).

The continuity equation (3.109) can be integrated from $z = 0$ to $z = h$:

$$0 = \int_0^h \nabla_{\mathrm{H}} \cdot \boldsymbol{v}_{\mathrm{H}}\, dz + \int_0^h \frac{\partial v_z}{\partial z}\, dz$$
$$= \int_0^h \nabla_{\mathrm{H}} \cdot \boldsymbol{v}_{\mathrm{H}}\, dz + v_z|_{z=h} - v_z|_{z=0}.$$

Now using the fact that

$$v_z|_{z=0} = 0, \qquad v_z|_{z=h} = \frac{\partial h}{\partial t}$$

for surfaces with a no-penetration boundary condition, this can be rewritten:

$$\frac{\partial h}{\partial t} + \nabla_{\mathrm{H}} \cdot \int_0^h \boldsymbol{v}_{\mathrm{H}}\, dz = 0. \tag{3.113}$$

Equations (3.110) and (3.113) along with boundary conditions for $\boldsymbol{v}_{\mathrm{H}}$ at $z = 0$ and $z = h$ form a closed set of equations that can be solved for the problem of interest.

Finally, it is often of interest to compute the stress on the surface $z = h$. Using the scalings found earlier and observing that $\hat{\boldsymbol{n}} = -\boldsymbol{e}_z + O(\epsilon)$ on this surface, it is straightforward to show that to leading order in $\epsilon$, the normal traction $t_z = (\hat{\boldsymbol{n}} \cdot \boldsymbol{\sigma}) \cdot \boldsymbol{e}_z$ and tangential tractions $t_x = (\hat{\boldsymbol{n}} \cdot \boldsymbol{\sigma}) \cdot \boldsymbol{e}_x$ and $t_y = (\hat{\boldsymbol{n}} \cdot \boldsymbol{\sigma}) \cdot \boldsymbol{e}_y$ are

$$t_z = p, \quad t_x = -\eta \frac{\partial v_x}{\partial z}, \quad t_y = -\eta \frac{\partial v_y}{\partial z}. \tag{3.114}$$

These equations and generalizations thereof have been used to address a wide range of problems from lubrication flows in bearings to waves on thin films to coating processes.

We will use this formalism to estimate the force required to move two spheres of radius

$a$ toward one another in the limit where the gap between the spheres is much less than the sphere radius. Let the relative velocity of the spheres be $W$ and the minimum distance between them be $2h_0(t)$, as shown on Figure 3.7. Here $L_z = h_0$ and $L_H = a$ so $\epsilon = h_0/a$. This system is axisymmetric and reflection symmetric across $z = 0$. Because of the latter symmetry, we only need consider the region $z \geq 0$, with a symmetry boundary condition

$$\frac{\partial v_r}{\partial z} = 0$$

at $z = 0$ and a no-penetration boundary condition

$$\frac{\partial h}{\partial t} = v_z = -\frac{W}{2}$$

at $z = h(r,t)$. There is no relative rotational motion of the spheres, so the flow is purely radial. Equation (3.110) becomes

$$0 = -\frac{dp}{dr} + \eta\frac{\partial v_r}{\partial z^2}$$

and applying no-slip on the radial velocity it is straightforward to show that

$$v_r = \frac{1}{2\eta}\frac{dp}{dr}(z^2 - h^2).$$

The radial pressure gradient $dp/dr$ is as yet unknown. We will determine it by application of (3.113), finding that

$$\frac{\partial h}{\partial t} = \frac{1}{3\eta r}\frac{\partial}{\partial r}\left[r\frac{dp}{dr}h^3\right].$$

Recalling that $\partial h/\partial t = -W/2$, this expression can integrated once to yield that

$$h^3\frac{dp}{dr} = -\frac{3\eta W r}{4} + \frac{c}{r}.$$

Boundedness of the pressure gradient at $r = 0$ requires that $c = 0$. In the case $h \ll a$, we can Taylor-expand the shape of the sphere:

$$h(r,t) = a + h_0(t) - \left(a^2 - r^2\right)^{1/2} \approx h_0(t) + \frac{a}{2}\left(\frac{r^2}{a^2}\right) + O\left(\frac{r^4}{a^4}\right)$$

and the equation for pressure gradient becomes

$$\frac{dp}{dr} = -\frac{3\eta W r}{4\left(h_0(t) + \frac{a}{2}\left(\frac{r^2}{a^2}\right)\right)^3}. \tag{3.115}$$

Now we need to consider the boundary condition on pressure for $r \gg h_0$ (which is infinitely far from the origin as $\epsilon \to 0$). Sufficiently far from the gap between the particles, the pressure must approach the background value, which we will take to be zero. Integrating (3.115) from $r$ to $\infty$ (because for any fixed $r$, $r/h_0 \to \infty$ as $\epsilon \to 0$), and setting $p(\infty) = 0$, we find that

$$p(r) = \frac{3\eta W a^3}{2}\left(2ah_0 + r^2\right)^{-2} = \frac{3\eta W}{8\epsilon h_0}\left(1 + \frac{\epsilon}{2}\left(\frac{r}{h_0}\right)^2\right)^{-2}. \tag{3.116}$$

Thus the so-called *lubrication pressure* is large ($O(\epsilon^{-1})$) right between the spheres and decays as $(r/h_0)^{-4}$ once $r/h_0 \gg 1/\sqrt{\epsilon}$. The total upward force $F_z$ exerted by the fluid on the sphere is determined by this pressure field:

$$F_z = \int_0^\infty p(r)\, 2\pi r\, dr \tag{3.117}$$

$$= \frac{3\pi\eta W a^2}{4h_0}. \tag{3.118}$$

Normalizing this result with the Stokes drag and using the total gap $H = 2h_0$, we find that

$$\frac{F_z}{6\pi\eta a W} = \frac{1}{4}\frac{a}{H}. \tag{3.119}$$

Thus at constant velocity, the upward force exerted by the fluid on the particle diverges as $H^{-1}$. If we consider instead the case where the particles are driven together at a constant force $F_{\text{ext},z} = -F_z$, we find that

$$\frac{dH}{dt} = -\frac{2F_{\text{ext},z}}{3\pi\eta a^2}H. \tag{3.120}$$

Thus at constant force, the distance between the particles will decay exponentially in time, but they will never touch. In reality, roughness effects and short-range forces (such as van der Waals interactions, which become important at distances on the order of 1–10 nm) will allow the particles to come into contact (Russel, Saville & Schowalter 1989). Furthermore, porous particles can come into contact because pressure between them is relieved by the motion of fluid through the particles, which cannot happen for rigid particles. Nevertheless, the point remains that a large force (relative to normal Stokes drag) is required to bring impenetrable particles into close proximity. For details of the asymptotic analysis of this flow as well as results for spheres in relative tangential motion or rotation, see Kim & Karrila (1991).

## 3.11 Stokesian Dynamics

In many applications, analytical solutions to the Stokes equation will not be available and numerical methods must be applied. Of course, many approaches are available for solving partial differential equations, and these can be implemented for the Stokes equation (Prosperetti & Tryggvason 2007). On the other hand, some specialized approaches that take special advantage of the linearity of the Stokes equation and analytical solutions that are already available for it can be developed. We present here the outline of one such approach, known as *Stokesian Dynamics*. In Section 4.6, we describe another, called the *boundary integral method*.

Stokesian Dynamics was developed to allow simulations of suspensions of rigid spherical particles in Stokes flow (Durlofsky, Brady & Bossis 1987, Phillips, Brady & Bossis 1988, Sierou & Brady 2001, Banchio & Brady 2003). The main idea is that when the particles are far apart, the multipole expansion can be used to approximate the many-body hydrodynamic interactions between the particles, while for nearby particles pairwise lubrication interactions are dominant. In Section 2.5, we described the relationship between the velocities and forces on a system of $N$ point particles using mobility and resistance tensors, while in Section 3.7 we did the same for a single finite-size particle, including the torques and stresslets present in this case. Here we combine the ideas of these earlier sections to generalize (3.68) to a system of $N$ spherical particles in an unbounded or periodic domain. Carrying over the notation of Section 2.5, we let $\mathsf{F}_{\mathrm{drag}} = -\mathsf{F}$ be the $3N$-vector of hydrodynamic (drag) forces exerted on the particles, $\mathsf{T}_{\mathrm{drag}}$ be the analogous quantity for the hydrodynamic torques, and $\mathsf{S}$ be the same for the stresslets. The spheres are subjected to a linear flow $\boldsymbol{v}_\infty = \boldsymbol{U}_\infty + \boldsymbol{\omega}_\infty \times \boldsymbol{x} + \mathbf{E}_\infty \cdot \boldsymbol{x}$, where $\boldsymbol{U}_\infty$ is a uniform background velocity. Finally, we define $\mathsf{O}$ and $\mathsf{o}_\infty$ as $3N$-vectors containing the angular velocities of the particles and $N$ copies of the position-independent fluid angular velocity, respectively.

The linearity of Stokes flow allows the drag force on particle $\alpha$ in this system to be written in terms of the relative linear and angular velocities and strain rate, as follows:

$$\boldsymbol{F}_{\mathrm{drag},\alpha} = \sum_{\beta=1}^{N} \left[ \mathsf{R}^{\mathrm{FU}}_{\alpha\beta} \cdot \left( \boldsymbol{v}_\infty(\boldsymbol{R}_\beta) - \boldsymbol{U}_\beta \right) + \mathsf{R}^{\mathrm{FW}}_{\alpha\beta} \cdot \left( \boldsymbol{\omega}_\infty(\boldsymbol{R}_\beta) - \boldsymbol{\Omega}_\beta \right) \right] + \mathsf{R}^{\mathrm{FE}}_{\alpha} : \mathbf{E}_\infty \quad (3.121)$$

and similarly for the torque and stresslet. The elements $\mathsf{R}_{\mathrm{FU}}$, etc., that relate these quantities to the relative linear and angular velocities and rate of strain can be combined succinctly into a grand resistance matrix

$$\mathscr{R} = \begin{bmatrix} \mathsf{R}^{\mathrm{FU}} & \mathsf{R}^{\mathrm{F\Omega}} & \mathsf{R}^{\mathrm{FE}} \\ \mathsf{R}^{\mathrm{TU}} & \mathsf{R}^{\mathrm{T\Omega}} & \mathsf{R}^{\mathrm{TE}} \\ \mathsf{R}^{\mathrm{SU}} & \mathsf{R}^{\mathrm{S\Omega}} & \mathsf{R}^{\mathrm{SE}} \end{bmatrix} \quad (3.122)$$

and we can write

$$\begin{bmatrix} \mathsf{F}_{\mathrm{drag}} \\ \mathsf{T}_{\mathrm{drag}} \\ \mathsf{S} \end{bmatrix} = \mathscr{R} \cdot \begin{bmatrix} \mathsf{V}_\infty - \mathsf{U} \\ \mathsf{o}_\infty - \mathsf{O} \\ \mathbf{E}_\infty \end{bmatrix}. \quad (3.123)$$

If we take the particles to be torque-free, then in the absence of particle inertia $\mathsf{T}_{\mathrm{drag}} = \mathsf{0}$. If the particles are subjected to external or interparticle nonhydrodynamic forces contained in $3N$-vector $\mathsf{F}_{\mathrm{ext}}$, then in the absence of inertia the force balance reads

$$\mathsf{F}_{\mathrm{drag}} + \mathsf{F}_{\mathrm{ext}} = \mathsf{0}. \quad (3.124)$$

Finally, given $\mathsf{U}$ at a given instant, the particle positions can be updated according to

$$\frac{d\mathsf{R}}{dt} = \mathsf{U}. \quad (3.125)$$

If the nonhydrodynamic forces are spherically symmetric (i.e., independent of particle orientations), then we do not need to track the particle orientations.

Up to this point, no simplifications have been made to the hydrodynamics – in the absence of inertia, the preceding description is exact. To yield a tractable computational scheme that captures the dominant physics of the system, the Stokesian Dynamics method makes a number of judicious approximations. The long-range interactions are computed using the leading terms of the multipole expansion for a sphere, force – torque, stresslet, and potential dipole – combined with Fáxen's laws, to form an approximation $\mathscr{M}^{\infty}$ of the grand mobility matrix $\mathscr{M} = \mathscr{R}^{-1}$. This expression captures the fact that each sphere interacts hydrodynamically with all other spheres in the system. However, it cannot capture the short-range lubrication interactions between neighboring particles. To account for this effect, exact resistance results for pairs of spheres (Onishi & Jeffrey 1984, Kim & Karrila 1991) are used, yielding the two-body resistance matrix $\mathscr{R}_{2B}$. The final approximation to the grand resistance is given by adding these two results and subtracting off the long-range component $\mathscr{R}_{2B}^{\infty}$ of the exact pair resistance to avoid double-counting of the long-range effects. The final result is

$$\mathscr{R} = (\mathscr{M}^{\infty})^{-1} + \mathscr{R}_{2B} - \mathscr{R}_{2B}^{\infty}. \tag{3.126}$$

Given $\mathscr{R}$, (3.123)–(3.125) can be integrated forward in time to simulate the dynamics of the system.

For $N$ particles in an unbounded domain, direct construction of $\mathscr{M}^{\infty}$ requires $O(N^2)$ operations and its inversion $O(N^3)$ operations. However, to study large systems, periodic boundary conditions can be imposed and the Ewald sum solution derived in Section 2.7 applied along with fast Fourier transforms and iterative linear system solvers to improve the scaling to $O(N \log N)$ (Sierou & Brady 2001). This methodology has been widely used to study a variety of problems in suspension dynamics and rheology, not least of which is the viscosity of a dense suspension, allowing understanding of the problem well beyond the simple results we described in Section 3.3.3.

## Problems

**3.1** Consider a uniform distribution of stresses on the surface $S$ of a sphere with radius $a$. That is, the traction vector $\boldsymbol{t} = \boldsymbol{n} \cdot \boldsymbol{\sigma}$ is a constant vector independent of position. Evaluate the multipole expansion, keeping the first three terms, and show that the resulting velocity field is the same as that of translating sphere in a stationary fluid. Use the identity

$$\int_S n_i n_j \, dS = \frac{4\pi a^2}{3} \delta_{ij}, \tag{3.127}$$

which can be easily derived using spherical coordinates.

**3.2** Derive the velocity field and friction coefficient for a spherical bubble with radius $a$ translating with velocity $\boldsymbol{U}$ through a stationary fluid. On the surface of the bubble, the no penetration condition $\boldsymbol{n} \cdot \boldsymbol{v} = 0$ holds, but the no-slip condition on a rigid sphere is replaced by the no-shear-stress condition $(\boldsymbol{\delta} - \boldsymbol{nn}) \cdot (\boldsymbol{n} \cdot \boldsymbol{\sigma}) = \mathbf{0}$. In Cartesian coordinates, $n_i = -x_i/a$.

**3.3** The most general symmetric force dipole that will generate an axisymmetric flow with respect to an axis $\boldsymbol{u}$ is

$$\mathbf{D} = \kappa_1 \boldsymbol{uu} + \kappa_2(\boldsymbol{\delta} - \boldsymbol{uu}).$$

If we attempt to supplement the squirmer model with a stresslet, we would thus have the velocity field

$$v_i = T^{\text{STR}}_{ijk} S_{jk} + c\left(\frac{x_i x_j}{r^5} - \frac{\delta_{ij}}{3r^3}\right) u_j - U u_i, \tag{3.128}$$

where $\mathbf{S}$ is the traceless part of $\mathbf{D}$. Show that no nonzero choices of $\kappa_1$ and $\kappa_2$ lead to a squirmer flow that satisfies the no-penetration condition. Hint: Collect terms of like dependence on $\theta$ where $a\cos\theta = \boldsymbol{x}\cdot\boldsymbol{u}$.

**3.4** Use (3.62) to compute the velocity $\boldsymbol{U}$ of a force-free sphere with radius $a$ positioned at the origin due to the flow generated by a point force $\boldsymbol{F}$ at position $\boldsymbol{x}_0 = \ell\boldsymbol{e}_x$ (where $\ell > a$).

**3.5** The velocity field generated by a sphere of radius $a$ rotating around its center with angular velocity $\boldsymbol{\Omega}$ in a stationary fluid is simply a rotlet:

$$v_i = -\frac{\epsilon_{ijk} x_k T_{\text{drag},l}}{8\pi\eta r^3},$$

where $\boldsymbol{T}_{\text{drag}} = -8\pi\eta a^3\boldsymbol{\Omega}$ is the rotational version of Stokes's law. Use this result as $\boldsymbol{v}'$ in the Lorentz reciprocal relation to derive Faxén's relation between torque and angular velocity in an arbitrary Stokes flow, (3.63). Hint: The net torque exerted by a fluid on a sphere is $\int_{S_P} \boldsymbol{r}\times(\hat{\boldsymbol{n}}\cdot\boldsymbol{\sigma})\,dS$, where $\boldsymbol{r} = a\hat{\boldsymbol{n}}$, and for a rigid body the velocity at any point $\boldsymbol{r}$ (measured from the center of rotation) is $\boldsymbol{v} = \boldsymbol{\Omega}\times\boldsymbol{r}$. As before, $\hat{\boldsymbol{n}} = -\boldsymbol{n}$ is the unit normal pointing into the fluid.

**3.6** Consider a general spherical swimmer with radius $a$. If the velocity on its surface relative to the overall velocity $\boldsymbol{U}$ is given by $\boldsymbol{v}_s$ then in a stationary frame the velocity on the surface is $\boldsymbol{U}+\boldsymbol{v}_s$. Given $\boldsymbol{v}_s$, use the Lorentz reciprocal relation to show that for a force-free swimmer,

$$\boldsymbol{U} = -\frac{1}{4\pi a^2}\int \boldsymbol{v}_s\,dS$$

(Stone & Samuel 1996). Verify that the results of Section 3.4 are consistent with this expression.

**3.7** Evaluate (3.67) for the case $\boldsymbol{R}_\alpha - \boldsymbol{R}_\beta = \ell\boldsymbol{d}$, where $\boldsymbol{d}$ is a unit vector.

**3.8** Find $\zeta^{\parallel}$ and $\zeta^{\perp}$ for a particle comprised of two spheres of radius $a$ connected by a rod of length $\ell$, in the limit $a/\ell \ll 1$. The rod is assumed not to affect the fluid. Use the mobility tensor analysis presented in Section 3.6.

**3.9** (a) Use slender body theory to estimate the drag torque $\boldsymbol{T}_{\text{drag}}$ on a rod with half-length $a$ and aspect ratio $1/\epsilon$ whose long axis is rotating with angular velocity $\boldsymbol{\Omega}$ through a stationary fluid (i.e., $\boldsymbol{v}_\infty = \boldsymbol{0}$). This result gives the rotational friction coefficient for a rod. For simplicity, let the rod be instantaneously aligned with the $x$ direction and rotating around the $z$ axis, so $\boldsymbol{\Omega} = \omega\boldsymbol{e}_z$.

(b) Now find the result that would be obtained by considering the rod as a line of $N$ osculating spheres each of radius $c = a/N$ arranged along a line (where $N$ is even), and neglecting the hydrodynamic interactions between the spheres. Is the friction coefficient based on this (incorrect) estimate bigger or smaller than the slender body theory estimate?

(c) Now repeat part (b) for $N = 100$, but including (in an approximate way) the hydrodynamic interactions between the spheres – you will have to do this numerically. Specifically, solve the problem

$$\mathbf{U} = \mathbf{M} \cdot \mathbf{F},$$

using (3.67) as an estimate for the mobility. Use the result you get for the vector $\mathbf{F}$ to determine the torque. Hint: For a filament aligned in $x$ and rotated around $\boldsymbol{e}_z$, the resulting instantaneous velocities and forces will be only in the $y$-direction.

**3.10** Consider one period of a helix with radius $R$ made from a rod of radius $a \ll R$. The pitch angle – the angle of the helix with respect to its central axis – is given by $\psi$ (if $\psi = 0$, then the helix reduces to a ring). If the local orientation of the rod is $\boldsymbol{u}$ and the helical axis is oriented along $\boldsymbol{e}_z$ then $\boldsymbol{u} \cdot \boldsymbol{e}_z = \sin\psi$. Use slender body theory to estimate the $z$-component of the force and torque exerted on the fluid if (a) the helix is rotated around its axis with angular velocity $\Omega$ without translating and (b) the helix is translated along its axis with speed $U$ without rotating. These quantities determine the $zz$ components of the matrices $\mathbf{R}^{\mathrm{FU}}$, $\mathbf{R}^{\mathrm{TU}}$, $\mathbf{R}^{\mathrm{F\Omega}}$, and $\mathbf{R}^{\mathrm{T\Omega}}$ of (3.69).

**3.11** Repeat the derivation leading to (3.120) but for a pair of drops with negligible interior viscosity, so that the no-slip boundary condition at $z = h(r)$ is replaced by a free-slip condition $\partial v_r/\partial z = 0$. Compare to the no-slip result.

# 4 Fundamental Solutions for Bounded Geometries

Up to this point, we have only considered flows in unbounded geometries. Unboundedness is a good approximation for particulate flows if the particles under consideration are small compared to the flow domain and far from its walls. In many applications, however, we will be interested in precisely the opposite case, where the effects of the boundaries may be dominant. Therefore, we now turn to the case of bounded geometries.

In bounded geometries, analytical solutions are sometimes still possible, although they become complex. Therefore, our focus here will not be on exact analytical results but rather on general principles.

## 4.1 General Reciprocity Result for the Green's Function

We begin with an important general result for the Green's function in an arbitrary geometry. Consider the flow field $\boldsymbol{v}'$ that arises from a point force $\boldsymbol{F}'$ exerted at an arbitrary position $\boldsymbol{x}'$ in a bounded domain $V_{\mathrm{D}}$ as shown in Figure 4.1. For definiteness, we impose a no-slip boundary condition. Since $\boldsymbol{v}'$ must be a linear function of $\boldsymbol{F}'$, we can write the velocity field that solves this problem as

$$\boldsymbol{v}'(\boldsymbol{x}) = \mathbf{G}^{\mathrm{B}}(\boldsymbol{x}, \boldsymbol{x}') \cdot \boldsymbol{F}', \tag{4.1}$$

where $\mathbf{G}^{\mathrm{B}}(\boldsymbol{x}, \boldsymbol{x}')$ is the as yet unknown Green's function for the bounded geometry under consideration. The associated stress tensor $\boldsymbol{\sigma}'$ satisfies

$$\boldsymbol{\nabla} \cdot \boldsymbol{\sigma}' + \boldsymbol{F}'\delta(\boldsymbol{x} - \boldsymbol{x}') = \mathbf{0}. \tag{4.2}$$

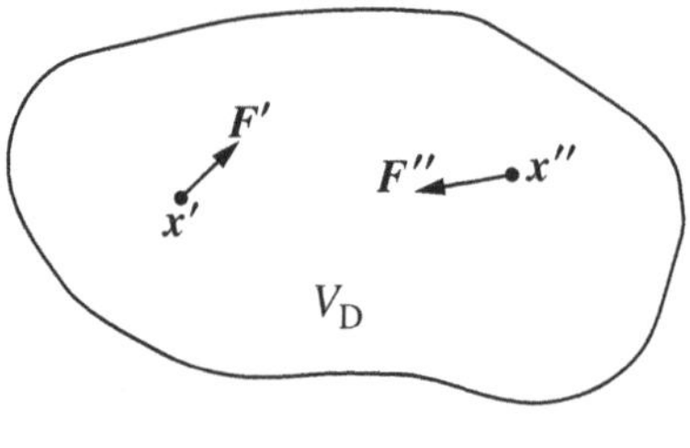


**Figure 4.1** Point forces at positions $\boldsymbol{x}'$ and $\boldsymbol{x}''$ in a bounded domain $V_{\mathrm{D}}$.

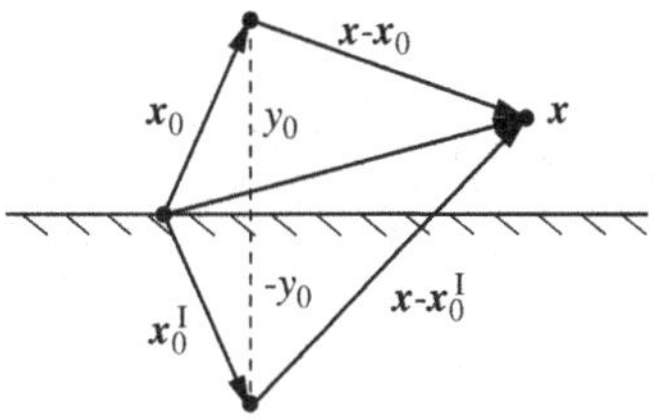


**Figure 4.2** Geometry for singularity solutions with a plane wall.

Similarly, the solution for a point force $\boldsymbol{F}''$ located at position $\boldsymbol{x}''$ yields the velocity field

$$\boldsymbol{v}''(\boldsymbol{x}) = \mathbf{G}^{\mathrm{B}}(\boldsymbol{x}, \boldsymbol{x}'') \cdot \boldsymbol{F}''$$

and a result analogous to (4.2) for the stress tensor $\boldsymbol{\sigma}''$.

Now a natural question arises: what is the relationship between $\mathbf{G}^{\mathrm{B}}(\boldsymbol{x}', \boldsymbol{x}'')$ and $\mathbf{G}^{\mathrm{B}}(\boldsymbol{x}'', \boldsymbol{x}')$, or equivalently, between the velocity at point $\boldsymbol{x}'$ generated by a point force at position $\boldsymbol{x}''$ and the velocity at point $\boldsymbol{x}''$ generated by a point force at position $\boldsymbol{x}'$? To address this question, we return to the Lorentz reciprocal relation (1.55):

$$\int_{V_{\mathrm{D}}} (\boldsymbol{v}' \cdot \nabla \cdot \boldsymbol{\sigma}'' - \boldsymbol{v}'' \cdot \nabla \cdot \boldsymbol{\sigma}')\, dV = \int_{S_{\mathrm{D}}} (\boldsymbol{v}' \cdot (\boldsymbol{n} \cdot \boldsymbol{\sigma}'') - \boldsymbol{v}'' \cdot (\boldsymbol{n} \cdot \boldsymbol{\sigma}'))\, dS.$$

Inserting (4.1) and (4.2) and their analogues for $\boldsymbol{v}''$ and noting that $\boldsymbol{v}'$ and $\boldsymbol{v}''$ both obey the no-slip boundary condition, the right-hand side of this expression vanishes and the left-hand side reduces to

$$\boldsymbol{F}'' \cdot \mathbf{G}^{\mathrm{B}}(\boldsymbol{x}'', \boldsymbol{x}') \cdot \boldsymbol{F}' - \boldsymbol{F}' \cdot \mathbf{G}^{\mathrm{B}}(\boldsymbol{x}', \boldsymbol{x}'') \cdot \boldsymbol{F}'' = 0.$$

Since $\boldsymbol{F}'$ and $\boldsymbol{F}''$ are arbitrary, we see that

$$\mathbf{G}^{\mathrm{B}}(\boldsymbol{x}', \boldsymbol{x}'') = (\mathbf{G}^{\mathrm{B}}(\boldsymbol{x}'', \boldsymbol{x}'))^{\mathrm{T}}. \tag{4.3}$$

This result holds regardless of the boundary conditions chosen – the integral on the right-hand side will alway vanish because $\boldsymbol{v}'$ and $\boldsymbol{v}''$ satisfy the same boundary conditions. Mathematically speaking, this reciprocity property of the Green's function is a general consequence of the self-adjointness of the Stokes equation. Referring back to the development of the mobility tensor in Sections 2.5 and 3.6, the symmetry of this tensor is a general consequence of (4.3), regardless of the domain containing the particles.

## 4.2 Point Force above a Plane Wall

To set the stage for the solution for a point force in the semi-infinite domain above a plane wall, we first describe the simpler situation of a point-source solution to the

Poisson equation in the presence of a wall:

$$\nabla^2\phi + Q\delta(\boldsymbol{x} - \boldsymbol{x}_0) = 0,$$

$$\frac{\partial \phi}{\partial y} = 0 \text{ at } y = 0, \quad \phi \to 0 \text{ as } \boldsymbol{x} \to \infty.$$

Here we have imposed a homogeneous Neumann boundary condition; the homogeneous Dirichlet boundary condition will be considered shortly. In the absence of the wall, the solution would simply be

$$\phi_+ = \frac{Q}{4\pi|\boldsymbol{x} - \boldsymbol{x}_0|}.$$

Now consider the case of a point source "below" the wall, at position $\boldsymbol{x}_0^I = (x_0, -y_0, z_0)^{\mathrm{T}}$; see Figure 4.2. With this source, the solution is

$$\phi_- = \frac{Q}{4\pi|\boldsymbol{x} - \boldsymbol{x}_0^{\mathrm{I}}|}.$$

Both $\phi_+$ and $\phi_-$ are spherically symmetric about theirs source points, and it is straightforward to show by symmetry that the superposition solution

$$\phi(\boldsymbol{x}) = \phi_+ + \phi_- = \frac{Q}{4\pi}\left(\frac{1}{|\boldsymbol{x} - \boldsymbol{x}_0|} + \frac{1}{|\boldsymbol{x} - \boldsymbol{x}_0^{\mathrm{I}}|}\right) \tag{4.4}$$

satisfies the boundary condition at $y = 0$. Specifically, the two individual solutions have equal and opposite $y$-derivatives at the wall, so the sum of them satisfies the Neumann boundary condition. The point source below the wall is called an "image" of the point source above the wall, and this approach to construction of solutions is called the *method of images*. Similarly, if a Dirichlet boundary condition $\phi = 0$ is applied at the wall, the solution is

$$\phi(\boldsymbol{x}) = \phi_+ - \phi_- = \frac{Q}{4\pi}\left(\frac{1}{|\boldsymbol{x} - \boldsymbol{x}_0|} - \frac{1}{|\boldsymbol{x} - \boldsymbol{x}_0^{\mathrm{I}}|}\right). \tag{4.5}$$

Finally, although in the case of unbounded flow we constructed a potential flow velocity field $\boldsymbol{v} = -\boldsymbol{\nabla}\phi$, in the bounded flow case the velocities corresponding to solutions (4.4) and (4.5) cannot be made to satisfy a no-slip boundary condition at $y = 0$. One can either satisfy no penetration (with (4.4)) or no tangential velocity (with (4.5)) but not both. So the point source flow above a plane no-slip wall is not a potential flow. Nevertheless, the image idea introduced here is a useful one.

Turning to the solution for a point force above a plane wall, the construction of an analogous solution in terms of images is relatively straightforward in the case of a free-slip boundary condition at $y = 0$, so we consider that case first. Specifically, we aim to solve the following:

$$-\boldsymbol{\nabla}p + \eta\nabla^2\boldsymbol{v} + \boldsymbol{F}\delta(\boldsymbol{x} - \boldsymbol{x}_0) = \boldsymbol{0}$$

$$\boldsymbol{\nabla}\cdot\boldsymbol{v} = 0$$

subject to the no-penetration condition

$$v_y = 0$$

and the no-shear stress condition

$$(\boldsymbol{\delta} - \boldsymbol{nn}) \cdot (\boldsymbol{n} \cdot \boldsymbol{\sigma}) = \mathbf{0}$$

at $y = 0$. Using the no-penetration condition, this equation reduces to

$$\frac{\partial v_x}{\partial y} = \frac{\partial v_z}{\partial y} = 0.$$

Symmetry helps us in this case. Let

$$\boldsymbol{F}^{\mathrm{I}} = (F_x, -F_y, F_z)^{\mathrm{T}}$$

be the"image force," i.e., the reflection of the original force across the plane $y = 0$. Now from symmetry we see that a solution that satisfies the boundary conditions is just the following superposition of Stokeslets:

$$\boldsymbol{v}(\boldsymbol{x}) = \mathbf{G}(\boldsymbol{x} - \boldsymbol{x}_0) \cdot \boldsymbol{F} + \mathbf{G}(\boldsymbol{x} - \boldsymbol{x}_0^{\mathrm{I}}) \cdot \boldsymbol{F}^{\mathrm{I}}. \tag{4.6}$$

By changing the sign of the $y$-component to construct $\boldsymbol{F}^{\mathrm{I}}$, we satisfy the no-penetration condition and no-shear-stress condition by symmetry. Noting that we can write $\boldsymbol{F}^{\mathrm{I}} = \left(\boldsymbol{\delta} - 2\boldsymbol{e}_y\boldsymbol{e}_y\right) \cdot \boldsymbol{F}$, where $\left(\boldsymbol{\delta} - 2\boldsymbol{e}_y\boldsymbol{e}_y\right)$ is a reflection operator, (4.6) can be written

$$\boldsymbol{v}(\boldsymbol{x}) = \left[\mathbf{G}(\boldsymbol{x} - \boldsymbol{x}_0) + \mathbf{G}(\boldsymbol{x} - \boldsymbol{x}_0^{\mathrm{I}}) \cdot \left(\boldsymbol{\delta} - 2\boldsymbol{e}_y\boldsymbol{e}_y\right)\right] \cdot \boldsymbol{F}. \tag{4.7}$$

The quantity within the angle brackets is the Green's function for the Stokes equation with the boundary conditions imposed: i.e.,

$$\boldsymbol{v}(\boldsymbol{x}) = \mathbf{G}^{\mathrm{B}}(\boldsymbol{x}, \boldsymbol{x}_0) \cdot \boldsymbol{F},$$

where

$$\mathbf{G}^{\mathrm{B}}(\boldsymbol{x}, \boldsymbol{x}_0) = \mathbf{G}(\boldsymbol{x} - \boldsymbol{x}_0) + \mathbf{G}(\boldsymbol{x} - \boldsymbol{x}_0^{\mathrm{I}}) \cdot \left(\boldsymbol{\delta} - 2\boldsymbol{e}_y\boldsymbol{e}_y\right). \tag{4.8}$$

Note that in this case the velocity field decays as $1/r$ and that the velocity is substantially larger than it would have been in free space: no momentum is absorbed by the boundary – it is "reflected" back into the domain, leading to a larger velocity than if the boundary had not been there.

In the case of a no-slip boundary condition, the construction of an image solution is much more involved (Blake 1971*a*, Pozrikidis 1997) and will not be described here. (In fact, in Blake (1971*a*), the solution is first constructed using Fourier transform methods, and only afterward is it found to be comprised of a finite set of singularity solutions.) The end result, sometimes called a "Blakelet," is

$$\begin{aligned} G_{ij}^{\mathrm{B}}(\boldsymbol{x}, \boldsymbol{x}_0) = {} & G_{ij}(\boldsymbol{x} - \boldsymbol{x}_0) - \left[G_{ij}(\boldsymbol{x} - \boldsymbol{x}_0^{\mathrm{I}})\right] \\ & + \frac{y_0^2}{\eta}\left(1 - 2\delta_{j2}\right)\frac{\partial}{\partial x_j}\left[\frac{1}{4\pi}\left(\frac{x_i - x_{0i}^{\mathrm{I}}}{|\boldsymbol{x} - \boldsymbol{x}_0^{\mathrm{I}}|^3}\right)\right] \\ & - 2y_0\left(1 - 2\delta_{j2}\right)\left[\frac{\partial}{\partial x_j}G_{i2}(\boldsymbol{x} - \boldsymbol{x}_0^{\mathrm{I}})\right]. \end{aligned} \tag{4.9}$$

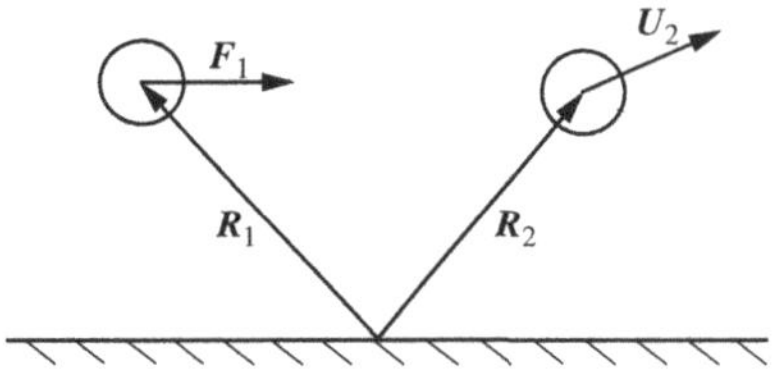


**Figure 4.3** Pair of particles moving near a wall.

(See Gimbutas et al. (2015) for an alternate singularity representation.) The three terms in brackets are singularity solutions centered at the image point $\boldsymbol{x}_0^{\mathrm{I}}$ and are, respectively, the Stokeslet corresponding to a point force $-\boldsymbol{F}$, a potential dipole, and a Stokeslet doublet. An important observation about this solution is that all the momentum imparted to the fluid by the force is absorbed by the wall (the image force has equal magnitude and opposite sign), and as a result, far from the wall, the velocity field decays as $1/r^2$ rather than $1/r$ as would be the case for unbounded or free-slip domain. Furthermore, the presence of the wall breaks the fore-aft symmetry found in the unbounded Stokeslet flow. This result has important implications for particle dynamics near walls.

## 4.3 Mobility of Point Particles in a Confined Geometry

Without loss of generality, we can write the Green's function for any geometry in the form

$$\mathbf{G}^{\mathrm{B}}(\boldsymbol{x}, \boldsymbol{x}_0) = \mathbf{G}(\boldsymbol{x} - \boldsymbol{x}_0) + \mathbf{G}^{\mathrm{W}}(\boldsymbol{x}, \boldsymbol{x}_0),$$

where $\mathbf{G}^{\mathrm{W}}(\boldsymbol{x}, \boldsymbol{x}_0)$ is the "wall correction" to the free-space Green's function $\mathbf{G}(\boldsymbol{x} - \boldsymbol{x}_0)$. It depends on the boundary conditions imposed and is not singular at $\boldsymbol{x} = \boldsymbol{x}_0$. By linearity, this splitting can always be achieved, although in general $\mathbf{G}^{\mathrm{W}}(\boldsymbol{x}, \boldsymbol{x}_0)$ will not be known analytically. In any case, this splitting will help us to understand how to write and interpret the mobility for a system of particles in a confined geometry. Generalizing the approach of Section 2.5, the mobility tensor for a system of point particles in a bounded domain can be written

$$\mathbf{M}_{\alpha\beta} = \frac{1}{\zeta}\boldsymbol{\delta}\delta_{\alpha\beta} + (1 - \delta_{\alpha\beta})\mathbf{G}(\boldsymbol{R}_\alpha - \boldsymbol{R}_\beta) + \mathbf{G}^{\mathrm{W}}(\boldsymbol{R}_\alpha, \boldsymbol{R}_\beta). \tag{4.10}$$

The contribution of $\mathbf{G}^{\mathrm{W}}$ to the diagonal terms ($\alpha = \beta$) reflects the change in mobility of an individual particle due to its proximity to boundaries, while the contribution to the off-diagonal terms ($\alpha \neq \beta$) represents how the confinement alters the hydrodynamic interaction between particles.

To understand the qualitative effect of confinement, consider $\mathbf{M}$ for particles above a plane wall, beginning with the change in the single-particle mobility $\mathbf{M}_{\alpha\alpha}$. The off-diagonal components of $\mathbf{G}^{\mathrm{W}}(\boldsymbol{R}_\alpha, \boldsymbol{R}_\alpha)$ are zero by symmetry, and $G^{\mathrm{W}}_{xx}(\boldsymbol{R}_\alpha, \boldsymbol{R}_\alpha)$,

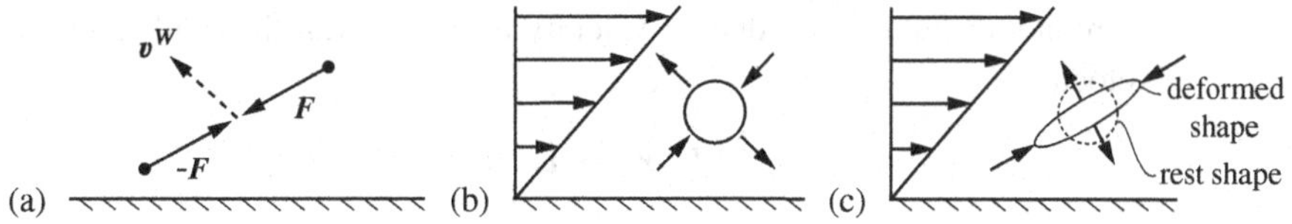


**Figure 4.4** (a) A simple force dipole near a wall and the velocity $\boldsymbol{v}^{\mathrm{W}}$ induced by it. (b–c) The dipole due to a rigid sphere (b) or deformable particle (c) in shear flow near a wall. The arrows indicate the directions of the forces exerted by the particle on the fluid.

$G^{\mathrm{W}}_{yy}(\boldsymbol{R}_\alpha, \boldsymbol{R}_\alpha)$ and $G^{\mathrm{W}}_{zz}(\boldsymbol{R}_\alpha, \boldsymbol{R}_\alpha)$ are all negative, indicating that the mobility of a particle is decreased due to the presence of a wall. With regard to the pair mobility, consider the special case of a pair of particles at the same distance $y_0$ from the wall. If particle 1 is moving so as to exert a horizontal force, the vertical motion of particle 2 will be determined by $G^{\mathrm{W}}_{21}(\boldsymbol{R}_2, \boldsymbol{R}_1)$. If particle 2 is "in front of" particle 1 as shown in Figure 4.3, then it will move upward in response: i.e., $G^{\mathrm{W}}_{21} > 0$. In other words, vertical forces and horizontal displacements are coupled and vice versa: this coupling complicates understanding of situations such as electrostatically interacting colloidal particles near a surface (Dufresne et al. 2000, Anekal & Bevan 2005) and leads to an important phenomenon in the dynamics of deformable particles during flow in confined geometries, which we now describe.

## 4.4 Hydrodynamic Migration of Particles in a Confined Geometry

As we know, a force- and torque-free particle exerts a force dipole on the surrounding fluid. To understand how the presence of a wall affects the motion of such a particle, we begin by considering the simple force dipole shown in Figure 4.4(a). Take the positions of the two point forces comprising the dipole to be $\boldsymbol{x}_0 \pm \boldsymbol{u}\frac{h}{2}$ and the corresponding forces to be $\pm\boldsymbol{F}$, with $\boldsymbol{F} \to \boldsymbol{u}$. At position $\boldsymbol{x}_0$, the part of the velocity field at position $\boldsymbol{x}_0$ that is due to the presence of the boundaries, $\boldsymbol{v}^{\mathrm{W}}$, is

$$v_i^{\mathrm{W}}(\boldsymbol{x}_0) = \left( G^{\mathrm{W}}_{ij}(\boldsymbol{x}_0, \boldsymbol{x}_0 + \boldsymbol{u}\frac{h}{2}) - G^{\mathrm{W}}_{ij}(\boldsymbol{x}_0, \boldsymbol{x}_0 - \boldsymbol{u}\frac{h}{2}) \right) F_j.$$

Taylor-expanding the Green's functions with respect to their second argument and taking the limit as $h \to 0$ as in the previous example yields that

$$v_i^{\mathrm{W}}(\boldsymbol{x}_0) = \left.\frac{\partial G^{\mathrm{W}}_{ij}(\boldsymbol{x}_0, \boldsymbol{q})}{\partial q_k}\right|_{\boldsymbol{q}=\boldsymbol{x}_0} u_k F_{dj} = -\frac{\partial G^{\mathrm{W}}_{ij}}{\partial q_k} D_{jk}, \tag{4.11}$$

where $\frac{\partial}{\partial q_k}$ is the derivative with respect to the second argument. (This expression actually does not require the particle to be torque-free, so it would be valid for a particle driven to rotate by an externally-imposed torque.)

This result implies that a force-free point particle that exerts a force dipole on the surrounding fluid will induce a velocity at its position that will lead it to move with velocity

$$\boldsymbol{U}(\boldsymbol{x}_0) = \boldsymbol{v}_\infty(\boldsymbol{x}_0) + \boldsymbol{U}^{\mathrm{W}}(\boldsymbol{x}_0), \tag{4.12}$$

where $\boldsymbol{U}^{\mathrm{W}} = \boldsymbol{v}^{\mathrm{W}}$. The wall-normal component of $\boldsymbol{U}^{\mathrm{W}}$ is called the *migration velocity* and the tangential component the *slip velocity*. For a force- and torque-free particle, a distance $y_0$ above a plane wall in a semi-infinite domain, the Blake solution can be used to determine $\boldsymbol{v}^{\mathrm{W}}$ in closed form. In particular, the migration velocity is given by

$$\begin{aligned} U_y^{\mathrm{W}} &= \frac{3}{64\pi\eta y_0^2}\left(D_{xx} + D_{zz} - 2D_{yy}\right) \\ &= -\frac{9S_{yy}}{64\pi\eta y_0^2}, \end{aligned} \tag{4.13}$$

where we have used (2.35) and the fact that $\mathrm{Tr}\,\mathbf{S} = 0$. The $1/y_0^2$ dependence of this force arises from the $1/r^2$ decay of the stresslet – the migration is simply due to the component of the stresslet flow induced by the presence of the wall.

In Section 1.3.1, we learned that a rigid sphere in Stokes flow above a wall does not migrate the wall-normal direction. To put this result in the context of the present discussion, we observe that the dipole tensor for a sphere of radius $a$ in unbounded simple shear flow with shear rate $\dot{\gamma}$ is given (see Section 3.3.3) by

$$\mathbf{D} = \frac{10}{3}\pi\eta a^3 \begin{bmatrix} 0 & \dot{\gamma} & 0 \\ \dot{\gamma} & 0 & 0 \\ 0 & 0 & 0 \end{bmatrix}. \tag{4.14}$$

This can be rewritten

$$\mathbf{D} = \frac{10}{3}\pi\eta a^3\dot{\gamma}\left(\boldsymbol{p}_1\boldsymbol{p}_1 - \boldsymbol{p}_2\boldsymbol{p}_2\right), \tag{4.15}$$

where $\boldsymbol{p}_1 = (\boldsymbol{e}_x + \boldsymbol{e}_y)/\sqrt{2}$ and $\boldsymbol{p}_2 = (\boldsymbol{e}_x - \boldsymbol{e}_y)/\sqrt{2}$ are unit vectors that point along the extensional and compressional directions of the deformation rate tensor, respectively (Figure 4.4(b)). By symmetry, the tensorial structure of **D** remains unchanged even if we move beyond the point particle approximation and find the exact solution in the presence of the wall. The wall-normal contributions to $\boldsymbol{v}^{\mathrm{W}}$ corresponding to the dipoles $\boldsymbol{p}_1\boldsymbol{p}_1$ and $\boldsymbol{p}_2\boldsymbol{p}_2$ cancel out due to the left-right symmetry of the problem. On the other hand, there is no up-down symmetry and the slip velocity is nonzero (Problem 4.3).

For deformable particles, the symmetry arguments of the previous paragraph do not hold; indeed, suspended liquid drops (Smart & Leighton 1991), red blood cells (Kim et al. 2009), and polymer molecules (Graham 2011) are all experimentally observed to migrate away from walls during flow. The direction of migration can be understood in terms of the components of **D**; deformable objects tend to stretch and align with flow, leading $D_{xx}$ to be positive and to be the dominant term in (4.13) as illustrated in Figure 4.4(c).

## 4.5 Point Force in Slit and Tube Geometries: Key Features

For geometries more complex than the semi-infinite domain above a plane wall, simple analytical solutions for $\mathbf{G}^{\mathrm{B}}(\boldsymbol{x}, \boldsymbol{x}_0)$ are not available. Nevertheless, for slit and duct geometries, some important results, which exemplify the behavior in complex domains, can be found. Consider first a slit geometry, a channel that is unbounded in $x$ and $z$, but bounded by plane no-slip walls at $y = 0$ and $y = 2H$. For a point force exerted at position $\boldsymbol{x}_0 = [0, y_0, 0]$, i.e., a distance $y_0$ above the bottom wall, there is an exact solution, derived using Fourier transforms in $x$ and $z$, but it is very cumbersome (Liron & Mochon 1976, Mucha et al. 2004, Hernández-Ortiz et al. 2006), so we will not describe it here. There are, however, two approximate solutions for the Green's function that provide insights into the nature of the flow in this geometry.

The first approximation simply superimposes Blake single-wall solutions individually for each wall. This is very crude: the image singularities for each wall violate the no-slip boundary condition on the opposite wall. Nevertheless, this approximation yields a simple and useful expression for the migration velocity:

$$U_y^W = -\frac{9S_{yy}}{64\pi\eta}\left(\frac{1}{y^2} - \frac{1}{(2H-y)^2}\right). \tag{4.16}$$

A slightly more refined approximation, derived for the specific case of deformable drops, is given by Chan & Leal (1979).

The second approximation is the far-field limit of the exact solution of Liron & Mochon (1976) valid in the limit $\rho^2 = (x^2 + z^2) \gg H^2$. For a force exerted in the wall-normal ($y$) direction, all components of the velocity are exponentially small as $\rho \to \infty$. Furthermore, regardless of the direction of the imposed force, the wall-normal velocity is exponentially small. Thus the wall-normal component of the hydrodynamic interactions in a slit is strongly screened by the boundary. For a force exerted parallel to the wall, however, the behavior of the wall-parallel velocity is very different, with the Green's function behaving approximately as follows:

$$G_{ij}^{\mathrm{B}}(\boldsymbol{x}, \boldsymbol{x}_0) = -\frac{3H}{\pi\eta}\frac{y}{H}\left(1 - \frac{y}{H}\right)\frac{y_0}{H}\left(1 - \frac{y_0}{H}\right)\frac{1}{\rho^2}\left[\frac{1}{2}\delta_{ij} - \frac{x_i x_j}{\rho^2}\right](1-\delta_{i2})(1-\delta_{j2}). \tag{4.17}$$

This is a flow in the $x$- and $z$-directions with a parabolic profile in the $y$-direction, so it appears locally like plane Poiseuille flow. (The last two factors set $\boldsymbol{v}$ to zero if the force or velocity component are in the wall-normal direction.) In the $x - z$ plane, the flow structure is that of a *two-dimensional source dipole*, for which $\boldsymbol{v} \sim \boldsymbol{\nabla\nabla} \ln \rho \sim 1/\rho^2$, leading to the perhaps counterintuitive feature that a force in, say, the positive $x$-direction will drive a velocity field that at some positions is in the negative $x$-direction. This phenomenon has been dubbed "antidrag" (Cui et al. 2004). This far-field result can be understood by noting that the average velocity in a slit on scales much larger than the slit height is well approximated by the two-dimensional version of Darcy's law, (2.105), with the permeability

$$k = \frac{(2H)^2}{12}$$

determined by the relation between mean velocity and pressure drop for the parabolic flow profile across the slit. Thus flow in a slit and flow in a porous medium are closely analogous. In particular, hydrodynamic interactions between distant particles in a slit are screened just as they are in a porous medium as described in Section 2.8 (Alvarez & Soto 2005, Tlusty 2006, Balducci et al. 2006). In the slit, the walls play the momentum-absorbing role played in the porous medium case by the stationary obstacles.

Finally, as with the slit geometry, exact point force solutions can be obtained for a tube or rectangular duct via eigenfunction expansions (see, e.g., Happel & Brenner 1965, Harden & Doi 1992). The main general result is that the velocity decays exponentially on the scale of the tube radius $R$ – there are no long-range interactions due to point forces. The momentum introduced to the fluid by a point force is all removed within a small distance away from the particle by shear stress at the walls. For finite-sized particles, the situation is more complex. In a tube of length $L$ with open ends at equal pressure, an axially moving particle will drive a flow that reduces to a simple Poiseuille flow far upstream and downstream of the particle. (Depending on the particle shape and size, some fluid will also squeeze around the particle from front to back.) As $L \to \infty$, the net axial flow must vanish: a nonvanishing Poiseuille flow in this limit would imply an infinite total force exerted by the fluid on the walls (finite shear stress times infinite area), which would require an infinite force to be exerted by the particle on the fluid. At finite $L$, however, the magnitude of the Poiseuille flow in a tube with open ends decays as $R/L$, resulting in long-ranged hydrodynamic interactions between confined particles (Misiunas et al. 2015). In a closed tube, no net axial flow can arise, and hydrodynamic interactions are negligible for particles an axial distance more than about a tube radius apart.

## 4.6 Integral Representation of Stokes Flow and the Boundary Integral Method

Given appropriate boundary conditions, the velocity field in Stokes flow is determined uniquely by solution of the Stokes equation. This observation implies that there might be an explicit formula that yields the velocity at a point in the domain in terms of quantities on the boundary. To derive such an expression, we consider for definiteness the domain shown in Figure 4.5 and recall the Lorentz reciprocal relation (1.55):

$$\int_{V_\mathrm{D}} (\boldsymbol{v}' \cdot (\boldsymbol{\nabla} \cdot \boldsymbol{\sigma}'') - \boldsymbol{v}'' \cdot (\boldsymbol{\nabla} \cdot \boldsymbol{\sigma}'))\, dV = \int_{S_\mathrm{D}} (\boldsymbol{v}' \cdot (\boldsymbol{n} \cdot \boldsymbol{\sigma}'') - \boldsymbol{v}'' \cdot (\boldsymbol{n} \cdot \boldsymbol{\sigma}'))\, dS.$$

We will take $\boldsymbol{v}'$ and $\boldsymbol{\sigma}'$ to be a Stokeslet solution: i.e.,

$$\boldsymbol{\nabla} \cdot \boldsymbol{\sigma}' = -\boldsymbol{F}'\delta(\boldsymbol{x} - \boldsymbol{x}_0) \tag{4.18}$$

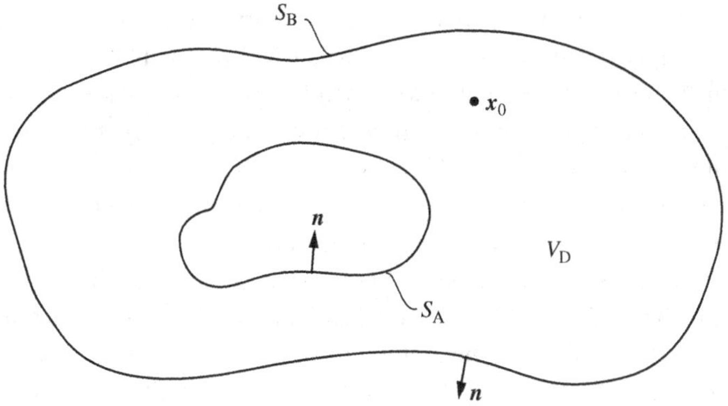


**Figure 4.5** Domain for the boundary integral representation of Stokes flow.

in an *unbounded* domain, so

$$v_i'(\boldsymbol{x}) = G_{ij}(\boldsymbol{x}-\boldsymbol{x}_0)F_j', \quad \sigma_{ik}'(\boldsymbol{x}_0) = T_{ikj}(\boldsymbol{x}-\boldsymbol{x}_0)F_j'. \tag{4.19}$$

Recall that $\mathbf{T}(\boldsymbol{x}-\boldsymbol{x}_0)$ is given by (2.15). Here $\boldsymbol{x}_0$ can be any point in the interior of $V_D$. The double-primed problem is the one of interest: it solves the Stokes equation with as yet unspecified boundary conditions on $S_D = S_A + S_B$. Therefore, $\boldsymbol{\nabla}\cdot\boldsymbol{\sigma}'' = \mathbf{0}$ everywhere in the domain. Applying this condition as well as (4.18) and (4.19) in the Lorentz relation yields that

$$v_j''(\boldsymbol{x}_0)F_j' = \int_{S_D} \Big( (G_{ij}(\boldsymbol{x}-\boldsymbol{x}_0)F_j')(n_l\sigma_{li}''(\boldsymbol{x})) - v_k''(\boldsymbol{x})(n_i T_{ikj}(\boldsymbol{x}-\boldsymbol{x}_0)F_j')\Big)\, dS(\boldsymbol{x}), \tag{4.20}$$

where here we write the differential area element $dS(\boldsymbol{x})$ to explicitly indicate that points $\boldsymbol{x}$ are on the boundary. Dropping the primes and noting that $F_j$ is arbitrary, we can rewrite this expression as

$$\begin{aligned} v_j(\boldsymbol{x}_0) &= \int_{S_D} \Big( (G_{ij}(\boldsymbol{x}-\boldsymbol{x}_0))(n_l\sigma_{li}(\boldsymbol{x})) - v_k(\boldsymbol{x})(n_i T_{ikj}(\boldsymbol{x}-\boldsymbol{x}_0))\Big)\, dS(\boldsymbol{x}) \\ &= \int_{S_D} n_l \Big( \sigma_{li}(\boldsymbol{x})G_{ij}(\boldsymbol{x}-\boldsymbol{x}_0) - v_k(\boldsymbol{x})T_{lkj}(\boldsymbol{x}-\boldsymbol{x}_0)\Big)\, dS(\boldsymbol{x}). \end{aligned} \tag{4.21}$$

This is the fundamental integral relation for Stokes flow. The two terms under the integral are called *single-layer* and *double-layer* terms, respectively, by analogy with the related results from electrostatics, where the field in the interior can be viewed as arising from a layer of point charges and of point dipoles (doublets) at the boundaries. Here the velocity arises from a layer of point forces and one of point force dipoles.

Given $\boldsymbol{v}$ and $\boldsymbol{\sigma}$ on the boundary, (4.21) determines the velocity at any point in the interior. However, both these quantities are not generally given in the statement of a specific problem. For example, if we are given the linear and angular velocities $\boldsymbol{U}$ and

$\boldsymbol{\Omega}$ of a rigid particle, we do not know a priori what $\boldsymbol{\sigma}$ is on its surface. Alternatively, we might be given $\boldsymbol{\sigma} = -p\boldsymbol{\delta}$ on the surface of a bubble, in which case the velocity on the bubble surface is unknown. Therefore, (4.21) does not yet completely determine the solution to a given problem. If we knew how to modify it for points $\boldsymbol{x}_0$ *on* the boundary, then we would have a self-consistent *integral equation* representation of Stokes flow: i.e., an equation that relates unknown quantities on the boundary to known quantities, without reference to the interior of the domain.

Two issues arise in modifying (4.21) for points on the boundary, which we denote $\boldsymbol{x}_S$. First, the left-hand side of the equation arises from evaluating an integral containing a delta function, and some ambiguity arises for such integrals when the delta function is centered on a boundary. Second, $\mathbf{G}(\boldsymbol{x} - \boldsymbol{x}_0)$ and $\mathbf{T}(\boldsymbol{x} - \boldsymbol{x}_0)$ are singular as $\boldsymbol{x} - \boldsymbol{x}_0 \to 0$ (Ladyzhenskaya 1969, Pozrikidis 1992, Stakgold 1998, Blawzdziewicz 2007), so the integrals containing them must be treated carefully. The single layer term is not problematic: its value is the same whether $\boldsymbol{x}_0 \to \boldsymbol{x}_S$ from the interior or exterior. Additionally, although $\mathbf{G} \sim 1/r$, where $r = |\boldsymbol{x}_0 - \boldsymbol{x}|$, the area element in local polar coordinates is $dS = r d\theta\, dr$, so the single-layer integral is well behaved. The double-layer term is trickier: its sign is different depending on whether $\boldsymbol{x}_0 \to \boldsymbol{x}_S$ from the interior or the exterior, leading to strongly singular behavior as $\boldsymbol{x}_0 \to \boldsymbol{x}_S$. Furthermore, since $\mathbf{T} \sim 1/r^2$, a naive attempt to integrate this term directly will fail.

Both of these issues can be treated at once (Blawzdziewicz 2007). First we define an indicator function $\chi(\boldsymbol{x}_0)$ with the property that

$$\int_{V_D} f(\boldsymbol{x})\delta(\boldsymbol{x} - \boldsymbol{x}_0)\, dV(\boldsymbol{x}) = \chi(\boldsymbol{x}_0) f(\boldsymbol{x}_0). \tag{4.22}$$

This function changes discontinuously from unity to zero as $\boldsymbol{x}_0$ crosses from the interior of $V_D$ to the exterior. For the moment, we leave aside the issue of what value $\chi(\boldsymbol{x}_0)$ takes when $\boldsymbol{x}_0$ is on $S_D$. Using this definition, we derive from (1.55) a generalization of (4.21):

$$v_j(\boldsymbol{x}_0)\chi(\boldsymbol{x}_0) = \int_{S_D} n_l \left(\sigma_{li}(\boldsymbol{x}) G_{ij}(\boldsymbol{x} - \boldsymbol{x}_0) - v_k(\boldsymbol{x}) T_{lkj}(\boldsymbol{x} - \boldsymbol{x}_0)\right)\, dS(\boldsymbol{x}), \tag{4.23}$$

which is valid both on the interior and the boundary. Next, we rewrite the double-layer integral by subtraction and addition of its singular part:

$$\int_{S_D} v_k(\boldsymbol{x}) n_l T_{lkj}(\boldsymbol{x} - \boldsymbol{x}_0)\, dS(\boldsymbol{x}) = \int_{S_D} (v_k(\boldsymbol{x}) - v_k(\boldsymbol{x}_0))\, n_l T_{lkj}(\boldsymbol{x} - \boldsymbol{x}_0)\, dS(\boldsymbol{x}) \tag{4.24}$$

$$+\, v_k(\boldsymbol{x}_0) \int_{S_D} n_l T_{lkj}(\boldsymbol{x} - \boldsymbol{x}_0)\, dS(\boldsymbol{x}). \tag{4.25}$$

(We used a similar trick in Section 3.9.) As long as $\boldsymbol{v}$ varies smoothly on the boundary, $\boldsymbol{v}(\boldsymbol{x}) - \boldsymbol{v}(\boldsymbol{x}_0) \sim r$, rendering the first integral nonsingular – we have isolated the singularity in the second one. To compute this singular integral, it is useful to recall (2.18):

$$\boldsymbol{\nabla} \cdot \mathbf{T}(\boldsymbol{x} - \boldsymbol{x}_0) + \boldsymbol{\delta}\delta(\boldsymbol{x} - \boldsymbol{x}_0) = \mathbf{0}.$$

Using this result along with the divergence theorem and (4.22), the singular double-layer integral can be determined:

$$\begin{aligned} v_k(\boldsymbol{x}_0) \int_{S_\mathrm{D}} n_l T_{lkj}(\boldsymbol{x}-\boldsymbol{x}_0)\, dS(\boldsymbol{x}) &= v_k(\boldsymbol{x}_0) \int_{V_\mathrm{D}} \frac{\partial}{\partial x_l} T_{lkj}(\boldsymbol{x}-\boldsymbol{x}_0)\, dV(\boldsymbol{x}) \\ &= v_k(\boldsymbol{x}_0) \int_{V_\mathrm{D}} -\delta_{kj}\delta(\boldsymbol{x}-\boldsymbol{x}_0)\, dV(\boldsymbol{x}) \\ &= -v_j(\boldsymbol{x}_0)\chi(\boldsymbol{x}_0). \end{aligned} \tag{4.26}$$

Finally, combining (4.26), (4.25), and (4.23) yields that

$$\begin{aligned} v_j(\boldsymbol{x}_0)\chi(\boldsymbol{x}_0) = &\int_{S_\mathrm{D}} G_{ij}(\boldsymbol{x}-\boldsymbol{x}_0) n_l \sigma_{li}(\boldsymbol{x})\, dS(\boldsymbol{x}) \\ &- \int_{S_\mathrm{D}} (v_k(\boldsymbol{x}) - v_k(\boldsymbol{x}_0))\, n_l T_{lkj}(\boldsymbol{x}-\boldsymbol{x}_0)\, dS(\boldsymbol{x}) + v_j(\boldsymbol{x}_0)\chi(\boldsymbol{x}_0). \end{aligned}$$

The terms containing $\chi(\boldsymbol{x}_0)$ cancel (so its value when $\boldsymbol{x}_0$ is on $S_\mathrm{D}$ need not be specified), yielding an expression that is valid for any point $\boldsymbol{x}_0$ in the interior or on the boundary of $V_\mathrm{D}$:

$$\begin{aligned} 0 = &\int_{S_\mathrm{D}} G_{ij}(\boldsymbol{x}-\boldsymbol{x}_0) n_l \sigma_{li}(\boldsymbol{x})\, dS(\boldsymbol{x}) \\ &- \int_{S_\mathrm{D}} (v_k(\boldsymbol{x}) - v_k(\boldsymbol{x}_0))\, n_l T_{lkj}(\boldsymbol{x}-\boldsymbol{x}_0)\, dS(\boldsymbol{x}). \end{aligned} \tag{4.27}$$

In particular, this is a self-consistent *boundary integral equation* for the values of $\boldsymbol{v}$ and $\boldsymbol{n}\cdot\boldsymbol{\sigma}$ for points $\boldsymbol{x}_0$ on $S_\mathrm{D}$. If $\boldsymbol{v}$ is given on the boundary (e.g., if $S_\mathrm{A}$ on Figure 4.5 is the surface of a rigid particle moving with velocity $\boldsymbol{U}$ and $S_\mathrm{B}$ is a stationary wall where $\boldsymbol{v} = \boldsymbol{0}$), then the traction vector $\boldsymbol{n}\cdot\boldsymbol{\sigma}$ is the unknown for which (4.27) must be solved. Alternatively, if the tractions are given, then $\boldsymbol{v}$ is the unknown. In many problems, the traction will be given on parts of the domain and the velocity on other parts, in which case the unknowns will be a combination of velocities and tractions.

For a smooth boundary, an alternate form of the boundary integral equation (4.27), which is more commonly presented (e.g., in Pozrikidis 1992), is

$$\frac{1}{2} v_j(\boldsymbol{x}_0) = \int_{S_\mathrm{D}} n_l \sigma_{li}(\boldsymbol{x}) G_{ij}(\boldsymbol{x}-\boldsymbol{x}_0)\, dS(\boldsymbol{x}) - ⨍_{S_\mathrm{D}} n_l v_k(\boldsymbol{x}) T_{lkj}(\boldsymbol{x}-\boldsymbol{x}_0)\, dS(\boldsymbol{x}), \tag{4.28}$$

where $⨍$ denotes a Cauchy principal value integral.[1] While this equation more closely parallels the form of (4.21), evaluation of the principal value integral necessitates performing a singularity subtraction operation that transforms it to (4.27). Thus the latter equation, which contains no singular integrals and is valid both in the interior and on the boundary, is more directly useful.

[1] For an integrand with a singularity at position $\boldsymbol{x}_0$, the Cauchy principal value integral excludes an infinitesimal ball of vanishing radius $\epsilon$ centered on $\boldsymbol{x}_0$ from the domain of integration. For example, in one dimension with $a < x_0 < b$,

$$⨍_a^b \frac{f(x)}{x - x_0}\, dx = \lim_{\epsilon\to 0} \left( \int_a^{x_0-\epsilon} \frac{f(x)}{x-x_0}\, dx + \int_{x_0+\epsilon}^b \frac{f(x)}{x-x_0}\, dx \right).$$

Discretizing the integrals in (4.27) results in a computational approach called the *boundary integral method*. In contrast to computational methods that directly solve the Stokes equation by discretizing the domain $V_D$, here only the boundaries need to be discretized. This feature makes the boundary integral method very appealing for problems involving deforming or moving boundaries such as are found in studies of flowing suspensions, and it is widely used in this context. If $N_E$ is the number of unknowns in the discretized equations, computation time for solution of the system of discretized equations scales with the time to solve a linear system, which is $O(N_E^3)$ in the case of direct solution method and $O(N_E^2)$ in the case of an efficient iterative solver.

Here we have only described the most basic form of the boundary integral equation. Many generalizations are possible. Flow around a particle in an unbounded domain with a prescribed background velocity $\boldsymbol{v}_\infty$ can be described, and drops or capsules with different interior and exterior viscosities can be studied (Pozrikidis 1992, Blawzdziewicz 2007). Suspensions flowing in periodic domains can be studied by replacing the Green's function for an unbounded domain with (2.101) (Metsi 2000, Saintillan et al. 2005) and using fast Fourier transform–based methods that can reduce the computation time to $O(N_E \log N_E)$. For a suspension flowing in a bounded domain, the Green's function for that domain can be used instead of the free-space Green's function, obviating discretization of the domain boundary (Kumar & Graham 2012). Pozrikidis (1992) provides an excellent starting point for further information on this important method.

## Problems

**4.1** Find the mobility to leading order in $a/y_0$ of a rigid sphere near a free-slip and a no-slip boundary.

**4.2** Use (4.9) and (4.11) to derive (4.13).

**4.3** Find the tangential (slip) component of $\boldsymbol{U}^{\mathrm{W}}$ for a rigid sphere in simple shear flow (to leading order in $a/y_0$). Does the sphere move slower or faster than the fluid at its position would move in the sphere's absence?

**4.4** What is $\boldsymbol{U}^{\mathrm{W}}$ for a *rotlet* $R_{ij} = \frac{1}{2}\epsilon_{ijk}M_k$ near a wall? In addition to a stresslet, a bacterium propelled by a single rotating flagellum also generates a rotlet doublet, an equal and opposite pair of rotlets – to conserve angular momentum, the body rotates in the direction opposite to that of the flagellum. Explain why a bacterium swimming due to a single rotating flagellum near and parallel to a wall moves in a curved path.

**4.5** Consider the solution to the Laplace equation $\nabla^2\phi = 0$ in a simple interior domain $V_D$ as shown in Figure 4.6, and recall that the Green's function for Laplace's equation (solution to $\nabla^2 G(\boldsymbol{x}-\boldsymbol{x}_0) + \delta(\boldsymbol{x}-\boldsymbol{x}_0) = 0$) is $G(\boldsymbol{x}-\boldsymbol{x}_0) = 1/4\pi\,|\boldsymbol{x}-\boldsymbol{x}_0|$. Find a boundary integral equation for $\phi(\boldsymbol{x}_0)$ as follows. Use Green's second identity and split

If $f'(x_0) \neq 0$, then this can be rewritten

$$\fint_a^b \frac{f(x)}{x-x_0}\,dx = \int_{-\infty}^{\infty} \frac{f(x)-f(x_0)}{x-x_0}\,dx + f(x_0)\fint_a^b \frac{1}{x-x_0}\,dx$$
$$= \int_a^b \frac{f(x)-f(x_0)}{x-x_0}\,dx + f(x_0)\ln\frac{b-x_0}{x_0-a}$$

and the remaining integral is not singular at $x_0$.

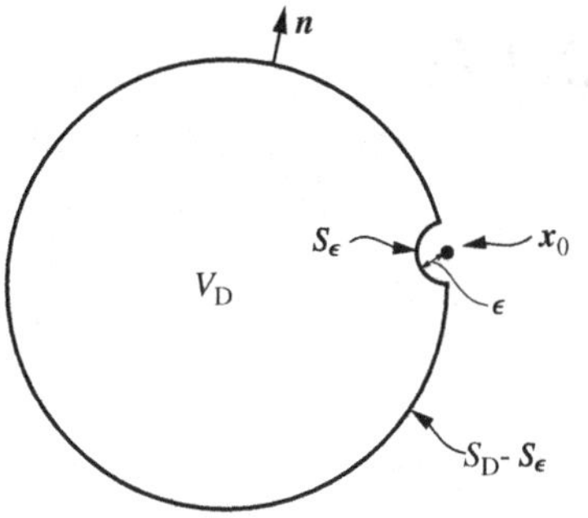


**Figure 4.6** Domain for derivation of boundary integral equations by means of principal value integrals.

the boundary $S_D$ into domains $S_\epsilon$ and $S_D - S_\epsilon$ as shown in the figure, where $S_\epsilon$ is a hemisphere of radius $\epsilon$ centered on the boundary point $\boldsymbol{x}_0$. As $\epsilon \to 0$, you will obtain the analogue of (4.28) for the Laplace equation:

$$\frac{1}{2}\phi(\boldsymbol{x}_0) = \int_{S_D} \frac{\partial \phi}{\partial n}(\boldsymbol{x}) G(\boldsymbol{x} - \boldsymbol{x}_0)\, dS(\boldsymbol{x}) - \fint_{S_D} \phi(\boldsymbol{x}) \frac{\partial G}{\partial n}(\boldsymbol{x} - \boldsymbol{x}_0)\, dS(\boldsymbol{x}),$$

where $\partial/\partial n = \boldsymbol{n} \cdot \boldsymbol{\nabla}$.

# 5 First Effects of Inertia

In the parameter regime where the Stokes equation is a valid approximation, we have powerful tools to analyze and understand flow phenomena. Nevertheless, it is crucially important to understand in what situations the complete neglect of inertia becomes invalid, and why. Recall that in the Navier–Stokes equation, inertia is manifested in two terms that have different physical interpretations. The first is the time-derivative term

$$\rho \frac{\partial \boldsymbol{v}}{\partial t},$$

which has the straightforward interpretation as the acceleration of the fluid. The second is the term

$$\rho \boldsymbol{v} \cdot \boldsymbol{\nabla} \boldsymbol{v},$$

which corresponds to the convection of momentum by the flow itself and arises even if the flow is steady. The inertialess approximation can break down due to either of these terms becoming important. Referring back to Section 1.2.2, the acceleration term becomes significant once $\mathrm{ReSr}^{-1} = L^2/\nu T$ becomes appreciable: this quantity measures the time scale $L^2/\nu$ required for momentum to diffuse over the length scale $L$ of the system versus the externally imposed time scale $T$. The convection term becomes significant as Re, the ratio of diffusive and convective time scales for momentum, increases. By example, we address a number of important cases where $\mathrm{ReSr}^{-1}$ or Re are nonnegligible.

## 5.1 Unbounded Uniform Flow around a Sphere at Small but Nonzero Reynolds Number

We begin by revisiting the flow driven by motion of a sphere in a stationary fluid that we studied in Section 3.3.1. To illustrate the role of inertia, we will compare the term $\rho \boldsymbol{v} \cdot \boldsymbol{\nabla} \boldsymbol{v}$ that we neglected in the Stokes flow approximation with the viscous term $\eta \nabla^2 \boldsymbol{v}$ that we kept. From the Stokes flow solution, (3.26), we can extract the following scalings

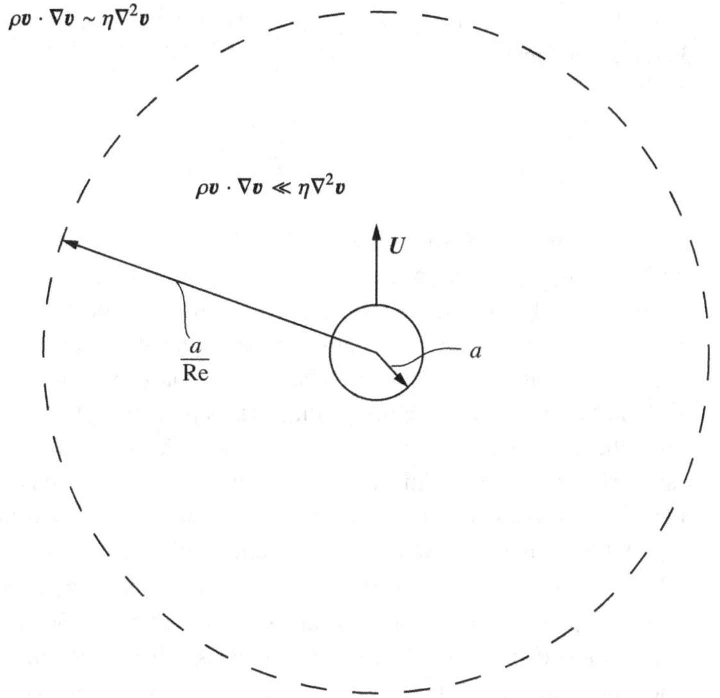


**Figure 5.1** A sphere moving in stationary fluid: illustration of regimes where Stokes flow is valid or breaks down.

for $r \gg a$:

$$v \sim U\frac{a}{r},$$
$$\nabla v \sim U\frac{a}{r^2},$$
$$\rho v \cdot \nabla v \sim \frac{\rho U^2 a}{r^2},$$
$$\eta \nabla^2 v \sim \eta \frac{Ua}{r^3},$$

where $U = |U|$. The ratio between the last two of these terms (inertial and viscous) is

$$\frac{\frac{\rho U^2 a}{r^2}}{\frac{Ua}{r^3}} = \frac{\rho U r}{\eta}.$$

Using the definition of the Reynolds number based on the size of the sphere, $\mathrm{Re} = \rho U a/\eta$, this result becomes

$$\frac{\frac{\rho U^2 a}{r^2}}{\frac{U a}{r^3}} = \mathrm{Re}\frac{r}{a}.$$

Thus even when the Reynolds number based on the size of the sphere is small, once we are far enough from the sphere, $r = O(a/\mathrm{Re})$, the convective term is not negligible in comparison with the viscous terms and the inertialess approximation fails, as illustrated in Figure 5.1. Physically, this failure is attributable to the implicit assumption that viscosity will carry momentum from the sphere surface to infinity in an infinitesimal time, while in fact the viscous diffusion time scales quadratically with distance. Therefore, while the viscous diffusion over the *particle* scale $a$ is fast compared to convective transport when $Re \ll 1$, diffusion over a scale $a/\mathrm{Re}$ is not. In other words, in an infinite domain the Stokes flow approximation must fail at some point because momentum cannot diffuse an infinite distance in an infinitesimal time.

More broadly, when considering hydrodynamic interactions between particles (or between a particle and a wall), this analysis shows that the Stokes flow approximation is only valid if the distance $L$ between objects satisfies $\mathrm{Re}L/a \ll 1$, or equivalently (and probably unsurprisingly) $\mathrm{Re}_L \equiv \rho U L/\eta \ll 1$. Note that this complication does not affect the accuracy of the flow field in the vicinity of the sphere – in this region, the transport of momentum by diffusion is indeed fast relative to convective transport.

Mathematically, these arguments imply that a full solution of the flow around a sphere is not simply given by the Stokes solution even in the limit $\mathrm{Re} \to 0$. This result is called "Whitehead's paradox" and is resolved by performing a matched asymptotic expansion in which the Stokes flow solution, which is valid close to the sphere, is combined with a solution far from the sphere in which convection is not neglected (Proudman & Pearson 1957, Deen 2012). This analysis leads to a modification of Stokes's law of the following form:

$$\boldsymbol{F}_{\mathrm{drag}} = -6\pi\eta a\left(1 + \frac{3}{8}\mathrm{Re} + O(\mathrm{Re}^2 \ln \mathrm{Re})\right)\boldsymbol{U}. \tag{5.1}$$

In two dimensions, the problem with Stokes flow in an unbounded domain is even worse. If we followed the procedure of Section 3.1 for steady Stokes around a cylinder in an unbounded domain, we would find a velocity that behaves as $\ln r$ rather than $1/r$ – at zero Reynolds number, we cannot satisfy the boundary condition that the velocity vanishes as $r \to \infty$. This result is clearly unphysical, a fact known as "Stokes's paradox," and is the reason that there is no Stokes drag expression for flow around a cylinder at zero Reynolds number. A hint of this problem appears in the friction coefficients we found in Sections 3.7 and 3.9 for slender bodies, which contain the term $\ln \frac{1}{\epsilon}$. The singular case $\epsilon = 0$ would represent a cylinder, for which we cannot evaluate this expression.

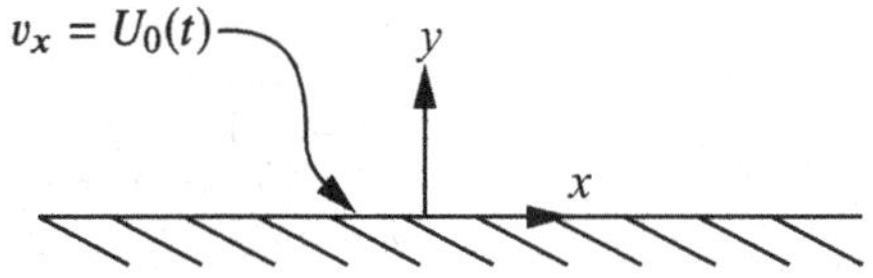


**Figure 5.2** Flow above a horizontally moving wall.

## 5.2 Flow Near a Moving Wall: Transient Acceleration and Viscous Diffusion

The preceding example concerns steady flow and the impact of the convective contribution to the inertia term. When we constructed the nondimensional Navier–Stokes equation in Section 1.2.2, the Strouhal number Sr arose as a ratio between an imposed time scale $T$ and the time scale $L/U$ for flow at speed $U$ over a distance $L$. When $\mathrm{ReSr}^{-1} = O(1)$, but $\mathrm{Re} \ll 1$, it is appropriate to consider the *transient Stokes equation*, (1.38), which we repeat here:

$$\rho \frac{\partial \boldsymbol{v}}{\partial t} = -\boldsymbol{\nabla} p + \eta \nabla^2 \boldsymbol{v}.$$

A very simple example of transient flow is the case of a semi-infinite domain bounded by a flat wall at $y = 0$ that is impulsively set in motion at $t = t_0$ in the $x$-direction with speed $U_0$ (Figure 5.2). In this case, we can take the velocity to be only in the $x$-direction and take the pressure gradient to vanish, and the steady Stokes equation would simplify to

$$0 = \frac{d^2 v_x}{dy^2}.$$

Solutions to this that satisfy the boundary condition at $y = 0$ are simply linear profiles $v_x = U_0 + cy$ and no choice of $c$ can satisfy the condition that $v_x$ vanish as $y \to \infty$. Therefore, we must consider the transient Stokes equation, which in this case reduces to a transient diffusion equation for $v_x$:

$$\frac{\partial v_x}{\partial t} = \nu \frac{\partial^2 v_x}{\partial y^2}. \tag{5.2}$$

This problem has the classical similarity solution

$$\frac{v_x}{U_0} = 1 - \mathrm{erf}\left(\frac{y}{2\sqrt{\nu(t - t_0)}}\right).$$

A boundary layer with time-dependent thickness $\delta_t = \sqrt{\nu(t - t_0)}$ develops as momentum diffuses away from the moving surface. Physically, this solution reflects the fact that the time scale for diffusion over a distance $\ell$ scales as $\ell^2/\nu$ – in Chapter 6, we will address the physical origin of this scaling in depth in the context of diffusion of a Brownian particle. The stress exerted by the fluid on the wall (i.e., the drag) during this flow is

$$\tau_{yx}|_{y=0} = -\frac{\eta U_0}{\sqrt{\pi\nu(t - t')}} = -\frac{\eta U_0}{\sqrt{\pi}\delta_t}, \tag{5.3}$$

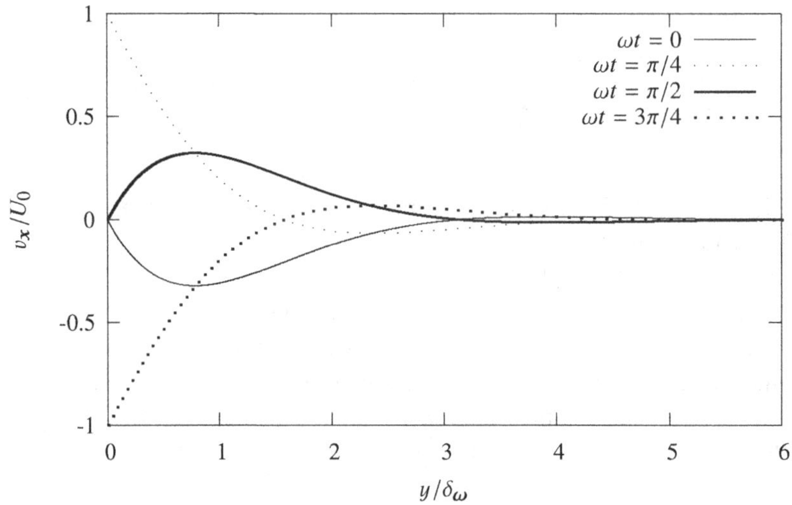


**Figure 5.3** Velocity versus position in a Stokes layer; solution to (5.4) over half a period of oscillation.

consistent with the $1/\delta_t$ scaling of the velocity gradient at the wall.

Now consider instead the case where the wall oscillates with velocity $v_x(0, t) = U_0 \sin(\omega t)$. The velocity field again obeys (5.2) and we will seek a time-periodic solution of the form

$$v_x(y, t) = f(y)e^{i\omega t} + \text{c.c.}$$

Inserting this into (5.2) and picking out the terms multiplying $e^{i\omega t}$ yields that

$$i\omega f = \nu f'',$$

where $f'' = d^2 f/dy^2$. This has solutions

$$f(y) = f_0 \exp\left(\pm\sqrt{\frac{i\omega}{\nu}}y\right) = f_0 \exp\left(\pm\sqrt{\frac{\omega}{2\nu}}(1+i)y\right),$$

where the result $\sqrt{i} = e^{i\pi/4} = (1+i)/\sqrt{2}$ has been applied. To satisfy the boundary condition $v_x \to 0$ as $y \to \infty$ requires that the minus sign be chosen, so

$$v_x = f_0 \exp\left(-\sqrt{\frac{\omega}{2\nu}}y\right) \exp\left(i\left(\omega t - \sqrt{\frac{\omega}{2\nu}}y\right)\right) + \text{c.c.}$$

Applying the boundary condition $v_x(0, t) = U_0 \sin(\omega t)$ and recalling that $\sin x = (e^{ix} - e^{-ix})/2i$, we find that $f_0 = U_0/2i$, so the solution becomes

$$v_x(y, t) = U_0 \exp\left(-\sqrt{\frac{\omega}{2\nu}}y\right) \sin\left(\omega t - \sqrt{\frac{\omega}{2\nu}}y\right).$$

Defining a frequency-dependent length scale $\delta_\omega = \sqrt{2\nu/\omega}$ we can rewrite this expression as

$$v_x(y,t) = U_0 \exp\left(-\frac{y}{\delta_\omega}\right) \sin\left(\omega t - \frac{y}{\delta_\omega}\right). \tag{5.4}$$

This velocity field is shown in Figure 5.3. The region near the wall over which the velocity decays is called the "Stokes layer"; its thickness $\delta_\omega$ corresponds to the distance momentum can diffuse in a time scale $\omega^{-1}$, i.e., one period of the boundary oscillation. Similarly, the phase lag $y/\delta_\omega$ in the time dependence of the velocity at a given distance $y$ from the wall is proportional to that distance scaled with the Stokes layer thickness $\delta_\omega$. Similarly to the preceding example, the force the fluid exerts on the wall during this flow scales inversely with $\delta_\omega$:

$$\tau_{yx}|_{y=0} = -\frac{\eta U_0}{\delta_\omega}(\sin\omega t + \cos\omega t). \tag{5.5}$$

## 5.3 Corrections to Stokes Drag for an Accelerating Sphere

Nontrivial inertial effects also arise even at negligible Reynolds number for a sphere moving with a time-dependent velocity $\boldsymbol{U}(t)$ in a stationary fluid. Rather than going through the details of solving the transient Stokes equation in this case (Landau & Lifschitz 1959, Kim & Karrila 1991), we simply report here the resulting drag force $\boldsymbol{F}_{\text{drag}}(t)$. This has three parts:

$$\boldsymbol{F}_{\text{drag}}(t) = \boldsymbol{F}^{\text{s}}_{\text{drag}}(t) + \boldsymbol{F}^{\text{h}}_{\text{drag}}(t) + \boldsymbol{F}^{\text{a}}_{\text{drag}}(t), \tag{5.6}$$

where

$$\boldsymbol{F}^{\text{s}}_{\text{drag}}(t) = -6\pi\eta a \boldsymbol{U}(t), \tag{5.7}$$

$$\boldsymbol{F}^{\text{h}}_{\text{drag}}(t) = -6\eta a^2\sqrt{\pi}\int_{-\infty}^{t}\frac{\dot{\boldsymbol{U}}(t')\,dt'}{\sqrt{\nu(t-t')}}, \tag{5.8}$$

$$\boldsymbol{F}^{\text{a}}_{\text{drag}}(t) = -\frac{2}{3}\pi\rho a^3\dot{\boldsymbol{U}}(t). \tag{5.9}$$

The first term, $\boldsymbol{F}^{\text{s}}_{\text{drag}}$, is simply a quasisteady Stokes drag term – this is the result we would obtain from naively using the steady Stokes equation in an unsteady situation.

The second term, $\boldsymbol{F}^{\text{h}}_{\text{drag}}$, results from the past history of the particle velocity and is often called the *Basset force*. This term is an example of what is called a "memory integral," as it relates the force at the present time to the particle velocities at past times. It arises from the drag on the sphere at the present time associated with the momentum imparted to the fluid at a past time $t'$, which has diffused a distance $\delta_t = \sqrt{\nu(t-t')}$ from the sphere surface, thus producing a stress that scales as $\delta_t^{-1} = 1/\sqrt{\nu(t-t')}$. This is the same behavior found earlier for flow near a moving plate – note the similarity with (5.3). Further insight into the nature of this term can be gained from considering the special case where the sphere simply oscillates back and forth along the $x$-axis so $\boldsymbol{U} = U_0 \boldsymbol{e}_x \sin \omega t$. Here the integral in (5.8) can be evaluated analytically, yielding that

$$\boldsymbol{F}^{\mathrm{h}}_{\mathrm{drag}}(t) = -6\pi\eta a U_0 \frac{a}{\delta_\omega} (\sin \omega t + \cos \omega t)\, \boldsymbol{e}_x, \tag{5.10}$$

where again $\delta_\omega = \sqrt{2\nu/\omega}$, corresponding to a Stokes layer thickness around the sphere. The ratio $\delta_\omega / a$ is the distance relative to the sphere radius $a$ that momentum is transported by viscosity in one period of oscillation – as expected the stress exerted on the sphere is inversely proportional to this distance.

The third term, $\boldsymbol{F}^{\mathrm{a}}_{\mathrm{drag}}$, is independent of viscosity and results from the force required to accelerate the fluid surrounding the sphere (Problem 5.2): it has the form $m_a \boldsymbol{a}$, where $\boldsymbol{a} = \dot{\boldsymbol{U}}$ and $m_a = 2\pi\rho a^3/3$, i.e., half the mass of fluid displaced by the sphere. Thus this term is called the "added mass" (also called the *virtual mass* or *hydrodynamic mass*) contribution to the force. Observe that $\boldsymbol{F}^{\mathrm{s}}_{\mathrm{drag}} \sim a$, $\boldsymbol{F}^{\mathrm{h}}_{\mathrm{drag}} \sim a^2$, and $\boldsymbol{F}^{\mathrm{a}}_{\mathrm{drag}} \sim a^3$ so that with decreasing particle size, the quasisteady Stokes drag term becomes dominant except for very large particle accelerations.

The discussion in this section focused on the solution to the transient Stokes equation, which is valid when Re is negligible but $\mathrm{ReSr}^{-1}$ is not. By contrast, Section 5.1 addressed the opposite case. The situation where both transient and convective effects are nonnegligible is addressed in Lovalenti & Brady (1993).

## 5.4 Inertial Migration of a Sphere in Confined Flow

By symmetry and Stokes flow reversibility (Section 1.3.1), a neutrally buoyant sphere in plane Couette or Poiseuille flow does not migrate in the wall-normal direction during flow. At any nonzero Reynolds number, the reversibility argument fails to hold. Experiments show that in plane Couette flow a sphere migrates to the centerline and in pressure-driven flow to a position intermediate between the wall and centerline, indicating the presence of an inertial lift force $\boldsymbol{F}_{\mathrm{lift}}$ on the sphere. This phenomenon is often called the *Segré–Silberberg effect* (Segre & Silberberg 1962, Di Carlo 2009).

Ho and Leal (1974) performed an analysis for a rigid sphere of radius $a$ in an arbitrary combination of Couette and Poiseuille flow in a planar channel of height $2H$ at a small but nonzero Reynolds number. Letting $V$ be some characteristic velocity for the bulk flow, such as the relative velocity of the walls in Couette flow or the centerline velocity in Poiseuille flow, a characteristic shear rate $\dot{\gamma}_c = V/H$ and a particle Reynolds number $\mathrm{Re} = \rho\dot{\gamma}_c a^2/\eta$ can be defined. Using a regular perturbation analysis for $\mathrm{Re} \ll a^2/H^2 \ll 1$ and the Lorentz reciprocal relations, they found that to leading order, the nondimensional wall-normal lift force on the sphere scales as $\mathrm{Re}\, a^2/H^2$. In dimensional form, with $y$ being the wall-normal distance from one wall, this becomes

$$F_{\mathrm{lift},y} = \frac{\rho V^2 a^4}{H^2} G(y/H), \tag{5.11}$$

where $G(y/H)$ depends on the specific velocity profile chosen and is odd with respect to the channel centerline $y = H$. Observe that this force is independent of the fluid viscosity $\eta$ and scales with $\rho V^2$, indicating its origin in the fluid inertia. For a force-free sphere, this lift force is balanced by viscous drag, leading to a wall-normal migration velocity

$$U_y^{\mathrm{W}} = \frac{1}{\zeta} F_{\mathrm{lift},\, y}. \tag{5.12}$$

For Couette flow, $G > 0$ for $y < H$ and vice versa, so the lift force drives the particle to the centerline in agreement with experimental observations. For Poiseuille flow, there is a position $y_{\mathrm{e}} \approx 0.4H$ such that $G > 0$ for $y < y_{\mathrm{e}}$ and $G < 0$ for $y_{\mathrm{e}} < y < H$. Thus the lift force drives the particle to a steady-state position $y = y_{\mathrm{e}}$. These results agree with experimental observations, and despite the fact that the latter calculation is for plane Poiseuille flow, it also agrees with experimental observations for a Poiseuille flow in a tube of radius $R$ if we replace $H$ by $R$ and $y$ by $R - r$.

The migration toward the centerline in the Couette flow case is driven by the hydrodynamic interaction between the particle stresslet and the wall. In the inertia-less case, the stresslet for a sphere in simple shear is a superposition of dipoles with opposite sign oriented at angles of $\pm\pi/4$ with respect to the flow direction as described in Section 4.4, and the flows generated by the images of these two dipoles precisely cancel out. At finite Reynolds number, this perfect symmetry is broken due to the nonlinearity of the convective term, resulting in migration away from walls. In Poiseuille flow, this effect dominates near the walls, but is countered near the centerline by a migration *toward* the walls that drives the particle to regions of higher velocity gradient. The two effects cancel at $y = y_{\mathrm{e}}$. Both of these mechanisms are distinct from classical aerodynamic lift, which requires neither a nearby wall nor a velocity gradient in the undisturbed flow.

## 5.5 Steady Streaming Due to Oscillatory Boundary Motion

In many situations, it is observed that everting a purely oscillatory pressure drop or boundary motion in a fluid can generate a fluid motion with a steady component, a phenomenon called *steady streaming*.[1] The Stokes and transient Stokes equations are linear with time-independent coefficients, so any oscillatory forcing of the these equations can only lead to a response with the same frequencies of the forcing. Thus the steady-streaming effect must have its origin in the nonlinear convective term in the Navier–Stokes equations.

Consider the Navier–Stokes equation in dimensionless form

$$\beta \frac{\partial \boldsymbol{v}}{\partial t} + \mathrm{Re}\boldsymbol{v} \cdot \boldsymbol{\nabla} \boldsymbol{v} = -\boldsymbol{\nabla} p + \nabla^2 \boldsymbol{v}, \tag{5.13}$$

where $\beta = \mathrm{Re}\mathrm{Sr}^{-1} = O(1)$, $\mathrm{Re} \ll 1$, and the flow is driven by some purely sinusoidal boundary motion with period $T = 2\pi/\omega$ and characteristic velocity $U$. (As always in

[1] The phenomenon is sometimes also called *acoustic streaming*, but this term implies the presence of sound waves and thus compressible flow. Sound waves can indeed generate steady streaming, but are not a necessary condition for it.

this book, incompressibility is implied.) We will perform a regular perturbation analysis of the problem, seeking a solution

$$\boldsymbol{v}(\boldsymbol{x},t) = \boldsymbol{v}_0(\boldsymbol{x},t) + \mathrm{Re}\boldsymbol{v}_1(\boldsymbol{x},t) + O(\mathrm{Re}^2). \tag{5.14}$$

The leading order solution $\boldsymbol{v}_0$ will be a solution to the transient Stokes equation, i.e.,

$$\beta\frac{\partial \boldsymbol{v}_0}{\partial t} + \boldsymbol{\nabla} p_0 - \nabla^2\boldsymbol{v}_0 = \boldsymbol{0} \tag{5.15}$$

and will have the form

$$\boldsymbol{v}_0(\boldsymbol{x},t) = \boldsymbol{v}_\omega(\boldsymbol{x})e^{i\omega t} + \bar{\boldsymbol{v}}_\omega(\boldsymbol{x})e^{-i\omega t}. \tag{5.16}$$

Here $\bar{}$ indicates complex conjugate. The first correction at finite Reynolds number is the solution to

$$\beta\frac{\partial \boldsymbol{v}_1}{\partial t} + \boldsymbol{\nabla} p_1 - \nabla^2\boldsymbol{v}_1 = -\boldsymbol{v}_0\cdot\boldsymbol{\nabla}\boldsymbol{v}_0. \tag{5.17}$$

This is a transient Stokes problem as well, but now with a *forcing* term $-\boldsymbol{v}_0\cdot\boldsymbol{\nabla}\boldsymbol{v}_0$. If $\boldsymbol{v}_0$ is a unidirectional flow, then this term is zero and $\boldsymbol{v}_0$ is an exact solution to the problem. This was the case, for example, in the flow generated by a tangentially translating plane wall that was described in Section 5.2. In any nonunidirectional flow, however, this term is nonzero, and inserting (5.16) into it yields that

$$-\boldsymbol{v}_0\cdot\boldsymbol{\nabla}\boldsymbol{v}_0 = -\left(\boldsymbol{v}_\omega\cdot\boldsymbol{\nabla}\bar{\boldsymbol{v}}_\omega + \bar{\boldsymbol{v}}_\omega\cdot\boldsymbol{\nabla}\boldsymbol{v}_\omega\right) - \left(\boldsymbol{v}_\omega\cdot\boldsymbol{\nabla}\boldsymbol{v}_\omega e^{2i\omega t} + \bar{\boldsymbol{v}}_\omega\cdot\boldsymbol{\nabla}\bar{\boldsymbol{v}}_\omega e^{-2i\omega t}\right).$$

The second term on the right-hand side has frequency $2\omega$ while the first is time independent. This forcing will lead to a solution $\boldsymbol{v}_1$ that has an oscillatory component but also a time-independent one – this is the steady-streaming flow. Because it appears at $O(\mathrm{Re})$ in the perturbation expansion, the steady-streaming flow is proportional to Re for small but nonzero Reynolds numbers.

## Problems

**5.1** As a way of gaining understanding of the Basset force that arises in the force on a sphere in transient Stokes flow, consider the analogous problem for transient heat conduction, the temperature field due to a sphere whose surface displays a time-dependent temperature $T_a(t)$. The governing equation is the transient diffusion equation in spherical coordinates:

$$\frac{\partial T}{\partial t} = \alpha\frac{1}{r^2}\frac{\partial}{\partial r}r^2\frac{\partial T}{\partial r},$$

where $T(r,0) = 0$, $T(r = a, t > 0) = T_a(t)$, $T(r,t) \to 0$ as $r \to \infty$. Solve this problem using Laplace transforms (Appendix A.3) in time. Let $T = \psi/r$ (this is a common trick in spherical coordinates), and use the following result:

$$L^{-1}\left\{s^{-1}e^{-2b\sqrt{s}}\right\} = \mathrm{erfc}\left(\frac{b}{\sqrt{t}}\right).$$

**5.2** The kinetic energy content $K$ in a flowing fluid is

$$K = \int_V \frac{1}{2}\rho\boldsymbol{v}\cdot\boldsymbol{v}\,dV.$$

The potential dipole flow field

$$v_i = \frac{3a^3}{2}\left(\frac{x_i x_j}{r^5} - \frac{\delta_{ij}}{3r^3}\right)U_j$$

corresponds to a potential flow solution for a sphere moving with velocity $\boldsymbol{U}$ through a fluid. Show that the kinetic energy contained in this flow is given by

$$K = \frac{1}{2}m_a U^2,$$

where

$$m_a = \frac{2}{3}\pi a^3 \rho.$$

This quantity is the *added mass* associated with the motion of the fluid. That is, the kinetic energy of the fluid in this case is equivalent to that of a sphere with mass $m_a$. Thus a sphere moving through fluid and generating this flow field behaves as if it were more massive by an amount $m_a$. As noted in Section 5.3, to accelerate the sphere one must also accelerate the surrounding fluid.

# 6 Thermal Fluctuations and Brownian Motion

The incessant random motions of the molecules of a fluid lead to fluctuations in the motion of suspended particles. We neglected these fluctuations in the previous chapters: for macroscopic particles, this is appropriate, but as particles reach the micron scale or smaller, the fluctuations become an important aspect of their dynamics. The first careful observations of such motions were made in 1827 by Robert Brown, who observed that intracellular granules within grains of pollen in water under his microscope jiggled around randomly and eventually left the field of view (Perrin 1916). This *Brownian motion* is a manifestation of these random fluctuations and a good starting point for understanding the dynamics of complex fluids at the level of individual particles or molecules. Figure 6.1 shows an experimentally observed trajectory of a microscopic particle in liquid from the classic work of Perrin (1916) as well as a trajectory simulated using an algorithm presented in Section 6.7. This chapter introduces Brownian motion from several physical points of view. Chapter 7 provides further mathematical background on modeling Brownian motion and other stochastic processes.

## 6.1 Brownian Motion of a Particle in Fluid: The Langevin Equation

Consider a neutrally buoyant rigid spherical particle with mass $m$, position $\boldsymbol{R}$, and velocity $\boldsymbol{U}$, suspended in a fluid at constant temperature $T$. We will only consider situations in which $T$ is constant. If this particle is in thermal equilibrium with the surrounding fluid, then at any position $\boldsymbol{R}$, its velocity distribution (or more precisely, the probability distribution function for velocity) will be Maxwellian (McQuarrie 2000): i.e., the probability density function $p(\boldsymbol{U})$ will satisfy

$$p_{\text{Max}}(\boldsymbol{U}) = \left(\frac{m}{2\pi k_B T}\right)^{3/2} \exp\left(-\boldsymbol{U} \cdot \boldsymbol{U}/2k_B T\right), \tag{6.1}$$

where $k_B$ is Boltzmann's constant. From this distribution, we can find the mean kinetic energy of the particle at equilibrium:

$$\frac{1}{2}m\langle \boldsymbol{U} \cdot \boldsymbol{U}\rangle = \int \boldsymbol{U} \cdot \boldsymbol{U} p_{\text{Max}}(\boldsymbol{U})\, d\boldsymbol{U} = \frac{3}{2}k_B T, \tag{6.2}$$

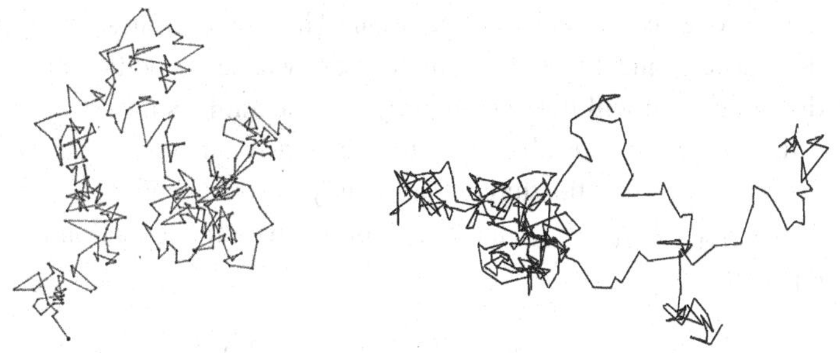

**Figure 6.1** Left: horizontal positions of a 0.53 $\mu$m diameter particle suspended in liquid at equal time intervals as observed experimentally by Perrin. Right: simulated trajectory of a Brownian particle using a Brownian dynamics algorithm.

where the angle brackets indicate ensemble averages (see Appendix A.5). More generally, in an isotropic fluid, the velocity components will be uncorrelated, so

$$\frac{1}{2}m\langle \boldsymbol{U}\boldsymbol{U}\rangle = \frac{1}{2}k_BT\boldsymbol{\delta}. \tag{6.3}$$

Therefore, even at equilibrium the particle must be moving around with a characteristic velocity of $\sqrt{k_BT/m}$, called the *thermal velocity*. This result is a specific case of the equipartition principle of classical equilibrium statistical mechanics (McQuarrie 2000): each translational and rotational degree of freedom of a system at equilibrium contains on average $\frac{1}{2}k_BT$ of energy.

We will now build a model of the particle dynamics, requiring that the proper equilibrium behavior be satisfied. The equations of motion for the particle are simply

$$m\frac{d\boldsymbol{U}}{dt} = \boldsymbol{F}, \tag{6.4}$$

$$\frac{d\boldsymbol{R}}{dt} = \boldsymbol{U}, \tag{6.5}$$

and the key question to be addressed is the nature of the forces $\boldsymbol{F}$ exerted on the particle by the surrounding fluid. In general, there may be conservative external forces $\boldsymbol{F}_{\text{ext}}(\boldsymbol{R}(t), t)$ on the particle: i.e., forces such as gravitational or electrostatic that can be written as the gradient of a potential function. There will also be viscous drag; as a simple model, we will consider Stokes's law to be valid on the scale of the particle: $\boldsymbol{F}_{\text{drag}} = -\zeta\,(\boldsymbol{U} - \boldsymbol{v}(\boldsymbol{R}))$. That is, at the scale of the particle the fluid appears to be a continuum that satisfies the Stokes equation. (A more detailed model of the drag force would incorporate the transient contributions using (5.6)–(5.9).) If the only force acting on the particle were Stokes drag, however, the particle would quickly come to a stop, so additionally, there must be another force that keeps the particle in motion and satisfying the Maxwellian velocity distribution. This force comes from the collisions between the particle and the solvent molecules, whose velocities also satisfy a Maxwellian distribution. It will fluctuate rapidly and randomly, so we denote it $\boldsymbol{F}_{\text{fluc}}$. General properties of random variables are summarized in Appendix A.5. We will determine later the properties that this force must satisfy for the particle to obey the Maxwellian. Now $\boldsymbol{F} = \boldsymbol{F}_{\text{drag}} + \boldsymbol{F}_{\text{ext}} + \boldsymbol{F}_{\text{fluc}}$, and the resulting evolution equation for $\boldsymbol{U}$ is called a *Langevin* equation:

$$m\frac{d\boldsymbol{U}}{dt} = -\zeta\left(\boldsymbol{U} - \boldsymbol{v}(\boldsymbol{R})\right) + \boldsymbol{F}_{\text{ext}} + \boldsymbol{F}_{\text{fluc}}. \tag{6.6}$$

Because of the random forcing, $\boldsymbol{U}$ is now also a random variable. This equation is a special case of a general class of dynamical models called *stochastic differential equations*; Chapter 7 will further develop the mathematical background for such models.

Note that $m/\zeta \equiv 1/\lambda_{\text{v}}$ is a time scale; it is known as the *inertial-viscous* relaxation time and is the time over which the velocity distribution for the particle equilibrates to the Maxwellian distribution, (6.1). For a spherical particle with radius $a$, this time varies as $a^2$ and is on the order of microseconds or smaller for colloidal particles. Many processes of interest take place on time scales much slower than $m/\zeta$, in which case the velocity distribution remains very nearly Maxwellian. Section 7.4.2 contains further discussion and analysis of this point.

In the Langevin equation, the fluctuating force is determined by the solvent dynamics, not the particle dynamics, and will be assumed to be independent of the velocity of the particle. This will be an excellent approximation if the particle is much larger (and thus moves much more slowly) than the solvent molecules. In this situation, any perturbation to the solvent motion arising from the motion of the particle will relax to equilibrium on a time scale much smaller that $1/\lambda_v$ (Snook 2007). For the moment, we also take this fluctuating force to be independent of position, an assumption that will be revisited in Section 6.7.

With these assumptions, we consider now the dynamics of a particle that is freely moving in a stationary fluid: $\boldsymbol{F}_{\text{ext}} = \boldsymbol{v} = \boldsymbol{0}$. In this case, (6.6) has a simple analytical solution. Taking its Laplace transform with initial condition $\boldsymbol{U}(0)$ yields

$$ms\tilde{\boldsymbol{U}}(s) - m\boldsymbol{U}(0) = -\zeta\tilde{\boldsymbol{U}}(s) + \tilde{\boldsymbol{F}}_{\text{fluc}}(s), \tag{6.7}$$

which we can rearrange to

$$\tilde{\boldsymbol{U}}(s) = \frac{1}{ms+\zeta}\left(m\boldsymbol{U}(0) + \tilde{\boldsymbol{F}}_{\text{fluc}}(s)\right). \tag{6.8}$$

Inverting yields

$$\boldsymbol{U}(t) = \boldsymbol{U}(0)e^{-\lambda_{\text{v}}t} + \frac{1}{m}e^{-\lambda_{\text{v}}t}\int_0^t e^{\lambda_{\text{v}}t'}\boldsymbol{F}_{\text{fluc}}(t')\,dt'. \tag{6.9}$$

This solution is not directly useful yet, as we do not know the properties of $\boldsymbol{F}_{\text{fluc}}$. Nevertheless, we can use this solution to work toward determining these.

## 6.2 Velocity Autocorrelation Function and Properties of the Fluctuating Force

We cannot possibly give an exact analytical description of the fluctuating force – it arises from the motions of an enormous number of solvent molecules, but our goal is to understand phenomena whose time and length scales are much larger than those of a solvent molecule. What we can describe are statistics, and we will see that we can develop a model of the time dependence of this force. We will assume that the fluctuating force is stationary, i.e., that its statistics are independent of the origin we choose for time, as well as isotropic and homogeneous, so that

$$\langle \boldsymbol{F}_{\text{fluc}}(t)\rangle = \boldsymbol{0}, \tag{6.10}$$

$$\langle \boldsymbol{F}_{\text{fluc}}(t)\boldsymbol{F}_{\text{fluc}}(t')\rangle = g(t-t')\boldsymbol{\delta}. \tag{6.11}$$

The second quantity is the *time correlation function* or *autocorrelation* of the fluctuating force; here $g(t-t')$ expresses the stationarity requirement in that it is even (we derive shortly the connection between stationarity and evenness) and depends only on the difference between the time instants $t$ and $t'$. Otherwise, its functional form is as yet unknown; we will determine it later in this section. Appendix A.5.4 contains general information about time correlation functions.

Our starting point for gaining further understanding of $\boldsymbol{F}_{\text{fluc}}$ is the *velocity autocorrelation function* for the particle motion (Landau & Lifschitz 1980),

$$\phi_{\text{v},ij}(t,\tau) = \left\langle U_i(t)U_j(t+\tau)\right\rangle, \tag{6.12}$$

which relates the particle velocity at two different times $\tau$ time units apart. We will also assume that the velocity autocorrelation function is stationary – physically, this means that we have let the system relax to thermal equilibrium. Thus the origin of time in (6.12) is arbitrary, so we can take it to be zero and write

$$\phi_{\text{v},ij}(\tau) = \left\langle U_i(0)U_j(\tau)\right\rangle. \tag{6.13}$$

Because the fluctuating force is isotropic, so will be $\phi_{v,ij}(\tau)$. For simplicity, we will work with its $xx$-component and generalize as appropriate.

Multiplying the $x$-component of (6.9) by $U_x(0)$ and averaging yields

$$\langle U_x(0)U_x(\tau)\rangle = \left\langle U_x^2(0)\right\rangle e^{-\lambda_\text{v}\tau} + \left\langle \frac{1}{m}U_x(0)e^{-\lambda_\text{v}\tau}\int_0^\tau e^{\lambda_\text{v}t'}F_{\text{fluc},x}(t')\,dt'\right\rangle.$$

For $\tau \geq 0$, the second term in this equation vanishes, because $U_x(0)$ is uncorrelated with $F_{\text{fluc},x}(t')$ for $t' \geq 0$. This is a result of the *causality* constraint – the velocity at the present time cannot be affected by values of $\boldsymbol{F}_{\text{fluc}}$ in the future. Therefore,

$$\phi_{\text{v},xx}(\tau) = \left\langle U_x^2(0)\right\rangle e^{-\lambda_\text{v}\tau}.$$

By the stationarity assumption, the system is at equilibrium at $t = 0$, so from (6.2) we know that $\left\langle U_x^2(0)\right\rangle_{\text{eq}} = k_BT/m$ and thus that

$$\phi_{\text{v},xx}(\tau) = \frac{k_BT}{m}e^{-\lambda_\text{v}\tau}.$$

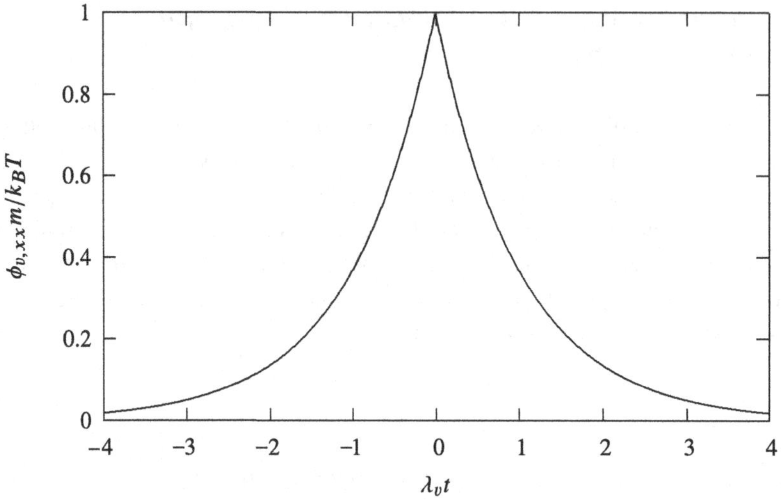


**Figure 6.2** The $xx$-component of the velocity autocorrelation function for a Brownian particle, (6.15). The off-diagonal components are zero.

More generally,

$$\phi_{\mathrm{v},ij}(\tau) = \frac{k_B T}{m} e^{-\lambda_{\mathrm{v}}\tau} \delta_{ij}. \tag{6.14}$$

The preceding analysis fails if $\tau < 0$, because $\boldsymbol{U}$ is *not* uncorrelated with past values of $\boldsymbol{F}_{\mathrm{fluc}}$, so one further condition is necessary to establish the form of $\phi_{\mathrm{v},ij}(\tau)$ for arbitrary values of $\tau$. This condition is stationarity, which allows us to shift the origin of time from 0 to $-\tau$ and write that

$$\langle U_x(0)U_x(\tau)\rangle = \langle U_x(-\tau)U_x(0)\rangle\,.$$

But

$$\langle U_x(0)U_x(\tau)\rangle = \langle U_x(\tau)U_x(0)\rangle$$

as well, so we must conclude that

$$\langle U_x(0)U_x(\tau)\rangle = \langle U_x(0)U_x(-\tau)\rangle\,;$$

the time correlation function of a stationary process is even. Therefore, we can generalize (6.14) for arbitrary values of $\tau$:

$$\phi_{\mathrm{v},ij}(\tau) = \frac{k_B T}{m} e^{-\lambda_{\mathrm{v}}|\tau|} \delta_{ij}. \tag{6.15}$$

The $xx$ component of this equation is shown in Figure 6.2.

The expression we just found for the velocity autocorrelation function relies only on causality, stationarity, and the nature of the friction force in the model. Problem 9.7 broadens these results to more general drag models. We have not needed to specify anything about the function $g(t - t')$ in the definition of the time correlation function of the fluctuating force, (6.11), nor have we learned anything about it. To do so, we will begin with an important result, the *Wiener–Khinchin theorem*, which states that the time correlation function for a stationary ergodic stochastic process $u(t)$ has this property:

$$F\{\langle u(t)u(t+\tau)\rangle\} = \langle \hat{u}(\omega)\hat{u}(-\omega)\rangle, \tag{6.16}$$

where $F\{f(t)\} = \hat{f}(\omega)$ is the Fourier transform in time of the function $f(t)$. This result is derived in Appendix A.5.4. The Fourier transform of the Langevin equation, (6.6), is

$$i\omega m\hat{\boldsymbol{U}}(\omega) = -\zeta\hat{\boldsymbol{U}}(\omega) + \hat{\boldsymbol{F}}_{\text{fluc}}(\omega), \tag{6.17}$$

which is easily solved for $\hat{\boldsymbol{U}}(\omega)$:

$$\hat{\boldsymbol{U}}(\omega) = \frac{\hat{\boldsymbol{F}}_{\text{fluc}}(\omega)}{i\omega m + \zeta}. \tag{6.18}$$

Therefore,

$$\left\langle \hat{\boldsymbol{U}}(\omega)\hat{\boldsymbol{U}}(-\omega)\right\rangle = \frac{1}{m^2\omega^2 + \zeta^2}\left\langle \hat{\boldsymbol{F}}_{\text{fluc}}(\omega)\hat{\boldsymbol{F}}_{\text{fluc}}(-\omega)\right\rangle. \tag{6.19}$$

From the Wiener–Khinchin theorem, we see that the left-hand side of this equation is simply the Fourier transform of the velocity autocorrelation function $\boldsymbol{\phi}_v(\tau)$, which we determined earlier. Therefore,

$$\left\langle \hat{\boldsymbol{U}}(\omega)\hat{\boldsymbol{U}}(-\omega)\right\rangle = F\{\boldsymbol{\phi}_v(\tau)\} = F\left\{\frac{k_BT}{m}e^{-\lambda_v|\tau|}\boldsymbol{\delta}\right\} = \frac{k_BT}{m}\frac{2\lambda_v}{\lambda_\text{v}^2+\omega^2}\boldsymbol{\delta}. \tag{6.20}$$

Combining (6.19) and (6.20), we find that

$$\left\langle \hat{\boldsymbol{F}}_{\text{fluc}}(\omega)\hat{\boldsymbol{F}}_{\text{fluc}}(-\omega)\right\rangle = 2\zeta k_BT\boldsymbol{\delta}. \tag{6.21}$$

Reverting to the time domain and again using Wiener–Khinchin yields

$$\langle \boldsymbol{F}_{\text{fluc}}(t)\boldsymbol{F}_{\text{fluc}}(t+\tau)\rangle = 2\zeta kT\delta(\tau)\boldsymbol{\delta}. \tag{6.22}$$

Thus the function $g(t-t')$ introduced in (6.11) is $2\zeta k_BT\delta(t-t')$. Using (6.3), (6.22) can be recast as

$$\langle \boldsymbol{F}_{\text{fluc}}(t)\boldsymbol{F}_{\text{fluc}}(t+\tau)\rangle = 2\lambda_\text{v}m^2\langle \boldsymbol{U}\boldsymbol{U}\rangle_{\text{eq}}\,\delta(\tau) = 2\lambda_\text{v}m^2\left\langle U^2\right\rangle_{\text{eq}}\delta(\tau)\boldsymbol{\delta}, \tag{6.23}$$

where the subscript eq indicates that the ensemble average is to be taken at thermal equilibrium.

For a system of $N$ particles, (e.g., a bead–spring chain model of a polymer, as further described in Chapter 8), with $\mathbf{F}_{\text{fluc}} = [\boldsymbol{F}_{\text{fluc},1}, \boldsymbol{F}_{\text{fluc},2}, \ldots \boldsymbol{F}_{\text{fluc},N}]^\text{T}$, this analysis generalizes to yield that

$$\langle \mathbf{F}_{\text{fluc}}(t)\mathbf{F}_{\text{fluc}}(t+\tau)\rangle = 2k_BT\mathbf{M}^{-1}\delta(\tau). \tag{6.24}$$

Equations (6.21), (6.22), and (6.24) are special cases of the *fluctuation-dissipation theorem* (Kubo 1966, Reichl 1998, McQuarrie 2000). They indicate the intimate connection between the fluctuating forces and the friction on the particle – both effects arise from its interactions with the solvent. In particular, the fluctuating forces must restore to the particle the kinetic energy dissipated due to viscous drag. Otherwise, the temperature would not approach its proper value.

On physical grounds, the delta-function nature of the time correlation function for $\boldsymbol{F}_{\text{fluc}}$ makes sense because we expect the fluctuations to vary on an extremely short time scale relative to that of the particle motion – they arise due to the very rapid, and rapidly fluctuating, motions of the solvent molecules. Nevertheless, it is crucially important to realize that nowhere have we explicitly assumed this to be the case. Rather, the preceding derivation shows that the structure of $\langle \boldsymbol{F}_{\text{fluc}}(t)\boldsymbol{F}_{\text{fluc}}(t+\tau)\rangle$ follows inevitably from the model of the particle dynamics. The choice of Stokes's law friction for the particle implies the exponential decay of the velocity autocorrelation function and the delta-function structure of $\langle \boldsymbol{F}_{\text{fluc}}(t)\boldsymbol{F}_{\text{fluc}}(t+\tau)\rangle$. If we had chosen to model the drag force on the particle with the more detailed model given by (5.6)–(5.9), which arises from using the transient Stokes equation, we would have obtained somewhat different results for the velocity and force correlations (Alder & Wainwright 1970, Zwanzig & Bixon 1970, Hinch 1975, Kubo, Toda & Hashitsume 1991, Franosch et al. 2011, Lesnicki et al. 2016). In fact, at long times the velocity autocorrelation decays not exponentially as found in the preceding example, but as $t^{-3/2}$ (Problem 9.7). We briefly revisit the relationship between dynamic models and noise models in a broader context in Section 6.8.

## 6.3 Thermal Fluctuations and the Navier–Stokes Equation

Until now, we have focused on the motion of particles in a fluid, incorporating the effects of thermal fluctuations of the particle motion by a fluctuating force on the particle. But of course the origin of this fluctuating force is the fluid itself, so the fluid must be undergoing fluctuating motions even in the absence of any particles (Landau & Lifschitz 1959, 1980, Reichl 1998). Consider a fluid of constant density $\rho$ that is macroscopically at rest and subdivide the fluid domain $V$ into small cubes of volume $\Delta V$ and mass $m = \rho\Delta V$. Assume for the moment that the velocity $\boldsymbol{v}(\boldsymbol{x})$ is constant within each volume. In each of these volumes, the fluctuating fluid velocity must satisfy the equilibrium relation between kinetic energy and temperature $\frac{1}{2}m\langle \boldsymbol{v}\boldsymbol{v}\rangle = \frac{1}{2}k_BT\boldsymbol{\delta}$. Additionally, because the velocity fluctuations originate in atomistic-scale motions, we expect that they will only be spatially correlated over atomistic scales. Thus, from the continuum point of view, we can take the velocity fluctuations in distinct cubes to be independent. Based on these considerations, the spatial correlation function between velocity fluctuations at points $\boldsymbol{x}$ and $\boldsymbol{x}'$ has the form

$$\left\langle v_i(\boldsymbol{x})v_j(\boldsymbol{x}')\right\rangle = \frac{k_BT}{\rho\Delta V}I(\boldsymbol{x},\boldsymbol{x}')\delta_{ij}, \tag{6.25}$$

where

$$I(\boldsymbol{x},\boldsymbol{x}') = \begin{cases} 1 & \boldsymbol{x} \text{ and } \boldsymbol{x}' \text{ are in the same cube} \\ 0 & \text{otherwise}\ . \end{cases}$$

In the continuum limit $\Delta V \to 0$, the quantity $I(\boldsymbol{x}, \boldsymbol{x}')/\Delta V \to \delta(\boldsymbol{x} - \boldsymbol{x}')$, so (6.25) becomes

$$\left\langle v_i(\boldsymbol{x}) v_j(\boldsymbol{x}') \right\rangle = \frac{k_B T}{\rho} \delta(\boldsymbol{x} - \boldsymbol{x}') \delta_{ij}. \tag{6.26}$$

Observe that here we have given each cube a thermal energy of $3k_BT/2$. In shrinking the cubes to zero size while continuing to give them finite thermal energy, we have made the total thermal energy of the system infinite. This artificial infinity arises because any given volume of fluid is not a continuum with an infinite number of degrees of freedom but rather is composed of a finite number of molecules. If the number density of molecules is $n_m$, then the total thermal energy is $n_m V(3k_BT/2)$. Similarly, in (6.26), the delta function must in reality be a finitely narrow and finitely tall function – otherwise, the velocity fluctuations at a point would be infinite. To construct a simple estimate of the actual width of the spatial velocity correlation, observe that volumes $\Delta V$ such that $\Delta V < n_m^{-1}$ contain less than one molecule on average, so we must regard the width of the correlation function to be no less than $n^{-1/3}$, the average spacing between solvent molecules. For our purposes, this limitation of the continuum approach to fluctuations in a fluid does not present any practical problems, but it should nevertheless be understood to exist.

The classical Navier–Stokes equation will not satisfy (6.26) – in the absence of external forces, the velocity will decay to zero. In the case of a single Brownian particle in a fluid, we addressed this issue by including in the momentum balance the fluctuating force that the fluid exerts on the particle. Similarly, now we supplement the momentum balance for the fluid with a fluctuating force density $\boldsymbol{f}_{\text{fluc}}$ that will maintain the fluctuating velocity at its thermal equilibrium value, yielding the *fluctuating Navier–Stokes equation*, sometimes called the *Landau–Lifschitz–Navier–Stokes equation* (Landau & Lifschitz 1959, Reichl 1998):

$$\rho \frac{D\boldsymbol{v}}{Dt} = -\boldsymbol{\nabla} p + \eta \nabla^2 \boldsymbol{v} + \boldsymbol{f}_{\text{fluc}}. \tag{6.27}$$

At the molecular level, $\boldsymbol{f}_{\text{fluc}}(\boldsymbol{x}, t)$ arises through collisions between molecules of the fluid. Since these collisions conserve momentum, they cannot impart any net momentum or angular momentum to the fluid. This fact implies that the force density must in fact be a symmetric force *dipole* (stresslet) density, which can be written as the divergence of a fluctuating symmetric stress tensor:

$$\boldsymbol{f}_{\text{fluc}}(\boldsymbol{x}, t) = \boldsymbol{\nabla} \cdot \boldsymbol{\sigma}_{\text{fluc}}(\boldsymbol{x}, t). \tag{6.28}$$

By isotropy,

$$\langle \boldsymbol{f}_{\text{fluc}} \rangle = \boldsymbol{0}, \quad \langle \boldsymbol{\sigma}_{\text{fluc}} \rangle = \boldsymbol{0}. \tag{6.29}$$

As with the Langevin equation, (6.6), for motion of a single particle, the nature of the fluctuating stress tensor $\boldsymbol{\sigma}_{\text{fluc}}$ is determined from the velocity autocorrelation function, which now depends on both position and time. To address the position dependence, it will be useful to work with spatial Fourier transforms of the velocity field. Neglecting the nonlinear terms in (6.27) on the assumption that they are small for a fluid at thermal equilibrium and taking the spatial Fourier transform of the momentum and continuity equations yields

$$\rho \frac{d\hat{\boldsymbol{v}}(\boldsymbol{k})}{dt} = -i\boldsymbol{k}\hat{p}(\boldsymbol{k}) - \eta k^2 \hat{\boldsymbol{v}}(\boldsymbol{k}) + i\boldsymbol{k} \cdot \hat{\boldsymbol{\sigma}}_{\text{fluc}}(\boldsymbol{k}), \tag{6.30}$$

$$i\boldsymbol{k} \cdot \hat{\boldsymbol{v}}(\boldsymbol{k}) = 0. \tag{6.31}$$

Here $k = \sqrt{\boldsymbol{k} \cdot \boldsymbol{k}}$. The continuity equation requires that at each wavevector $\boldsymbol{k}$, the velocity is constrained to lie in the plane orthogonal to $\boldsymbol{k}$.

Using the Wiener–Khinchin theorem in the spatial variable $\boldsymbol{x}$, the spatial correlation function of the velocity, (6.26), can be written

$$\langle \hat{\boldsymbol{v}}(\boldsymbol{k})\hat{\boldsymbol{v}}(-\boldsymbol{k})\rangle = \frac{k_B T}{\rho}\boldsymbol{\delta}. \tag{6.32}$$

This must be modified in the case of incompressible flow, as we will see shortly.

A general flow field in Fourier space, $\hat{\boldsymbol{v}}(\boldsymbol{k})$, can be expressed using orthogonal basis vectors parallel and perpendicular to $\boldsymbol{k}$. Letting $\boldsymbol{u}^{\parallel}(\boldsymbol{k}) = \boldsymbol{k}/k$ be a unit vector parallel to $\boldsymbol{k}$, and $\boldsymbol{u}^{\perp}(\boldsymbol{k})$ and $\boldsymbol{u}^{\perp\perp}(\boldsymbol{k})$ be unit vectors perpendicular to $\boldsymbol{k}$ and to each other, we can represent $\hat{\boldsymbol{v}}(\boldsymbol{k})$ as follows (Reichl 1998):

$$\hat{\boldsymbol{v}}(\boldsymbol{k}) = v^{\parallel}(\boldsymbol{k})\boldsymbol{u}^{\parallel}(\boldsymbol{k}) + v^{\perp}(\boldsymbol{k})\boldsymbol{u}^{\perp}(\boldsymbol{k}) + v^{\perp\perp}(\boldsymbol{k})\boldsymbol{u}^{\perp\perp}(\boldsymbol{k}). \tag{6.33}$$

It will be convenient to use the same basis vectors for $\hat{\boldsymbol{v}}(-\boldsymbol{k})$ as for $\hat{\boldsymbol{v}}(\boldsymbol{k})$. Using this notation, (6.32) for the spatial correlation function of the velocity becomes

$$\left\langle v^{\parallel}(\boldsymbol{k})v^{\parallel}(-\boldsymbol{k})\boldsymbol{u}^{\parallel}\boldsymbol{u}^{\parallel} + v^{\perp}(\boldsymbol{k})v^{\perp}(-\boldsymbol{k})\boldsymbol{u}^{\perp}\boldsymbol{u}^{\perp} + v^{\perp\perp}(\boldsymbol{k})v^{\perp\perp}(-\boldsymbol{k})\boldsymbol{u}^{\perp\perp}\boldsymbol{u}^{\perp\perp}\right\rangle = \frac{k_B T}{\rho}\boldsymbol{\delta}. \tag{6.34}$$

Since the right-hand side is isotropic, the left-hand side must be as well, implying that

$$\left\langle v^{\parallel}(\boldsymbol{k})v^{\parallel}(-\boldsymbol{k})\right\rangle = \frac{k_B T}{\rho}, \tag{6.35}$$

$$\left\langle v^{\perp}(\boldsymbol{k})v^{\perp}(-\boldsymbol{k})\right\rangle = \left\langle v^{\perp\perp}(\boldsymbol{k})v^{\perp\perp}(-\boldsymbol{k})\right\rangle = \frac{k_B T}{\rho}, \tag{6.36}$$

$$\left\langle v^{\parallel}(\boldsymbol{k})v^{\perp}(-\boldsymbol{k})\right\rangle = \left\langle v^{\parallel}(\boldsymbol{k})v^{\perp\perp}(-\boldsymbol{k})\right\rangle = \left\langle v^{\perp}(\boldsymbol{k})v^{\perp\perp}(-\boldsymbol{k})\right\rangle = 0. \tag{6.37}$$

As elsewhere in this book, we focus here on incompressible flow. In this case, $\boldsymbol{k} \cdot \hat{\boldsymbol{v}}(\boldsymbol{k}) = 0$ and thus $v^{\parallel}(\boldsymbol{k}) = 0$. Therefore, (6.35) cannot be satisfied. The resolution to this seeming contradiction is that for a perfectly incompressible fluid, there is simply not a degree of freedom associated with $v^{\parallel}(\boldsymbol{k})$, so the equipartition principle does not apply and (6.35) need not be enforced. Thus in the incompressible case (6.32) is replaced by

$$\langle \hat{\boldsymbol{v}}(\boldsymbol{k})\hat{\boldsymbol{v}}(-\boldsymbol{k})\rangle = \frac{k_B T}{\rho}\left(\boldsymbol{\delta} - \frac{1}{k^2}\boldsymbol{k}\boldsymbol{k}\right). \tag{6.38}$$

We return now to consider the momentum equation, (6.30), in the incompressible case where $v^{\parallel}(\boldsymbol{k}) = 0$. The fluctuating force $i\boldsymbol{k} \cdot \hat{\boldsymbol{\sigma}}_{\text{fluc}}(\boldsymbol{k})$ can be written

$$i\boldsymbol{k} \cdot \hat{\boldsymbol{\sigma}}_{\text{fluc}}(\boldsymbol{k}) = ik\left(\sigma^{\perp}_{\text{fluc}}(\boldsymbol{k})\boldsymbol{u}^{\perp} + \sigma^{\perp\perp}_{\text{fluc}}(\boldsymbol{k})\boldsymbol{u}^{\perp\perp}\right), \tag{6.39}$$

where $\sigma^{\perp}(\boldsymbol{k}) = \boldsymbol{u}^{\parallel}\cdot\hat{\boldsymbol{\sigma}}_{\text{fluc}}(\boldsymbol{k})\cdot\boldsymbol{u}^{\perp}$ and $\sigma^{\perp\perp}(\boldsymbol{k}) = \boldsymbol{u}^{\parallel}\cdot\hat{\boldsymbol{\sigma}}_{\text{fluc}}(\boldsymbol{k})\cdot\boldsymbol{u}^{\perp\perp}$. We can neglect any component of the fluctuating force parallel to $\boldsymbol{k}$ as it can be absorbed into the pressure. Inserting (6.33) and (6.39) into (6.30) yields

$$\rho\frac{dv^{\perp}(\boldsymbol{k})}{dt} = -k^2\eta v^{\perp}(\boldsymbol{k}) + ik\sigma^{\perp}_{\text{fluc}} \tag{6.40}$$

and an identical equation with $^{\perp}$ replaced with $^{\perp\perp}$. This equation for $v^{\perp}(\boldsymbol{k})$ has precisely the same structure as the Langevin equation for the velocity $\boldsymbol{U}$ of a single Brownian particle, (6.6), with $m$, $\zeta$, $\lambda_v$, and $\boldsymbol{F}_{\text{fluc}}$ replaced with $\rho$, $k^2\eta$, $k^2\nu$, and $ik\sigma^{\perp}_{\text{fluc}}(\boldsymbol{k})$, respectively. Therefore, derivation of the proper form of the fluctuating stress parallels what we did for that situation. The solution for $v^{\perp}(\boldsymbol{k})$ has the same form as (6.9):

$$v^{\perp}(\boldsymbol{k},t) = v^{\perp}(\boldsymbol{k},0)e^{-k^2\nu t} + \frac{1}{\rho}e^{-k^2\nu t}\int_0^t e^{k^2\nu t'}ik\sigma^{\perp}_{\text{fluc}}(\boldsymbol{k},t')\,dt'. \tag{6.41}$$

Multiplying by $v^{\perp}(-\boldsymbol{k},0)$, ensemble averaging, applying (6.36), and repeating the steps that led to (6.15) leads to an expression for the velocity autocorrelation function:

$$\left\langle v^{\perp}(\boldsymbol{k},\tau)v^{\perp}(-\boldsymbol{k},0)\right\rangle = \frac{k_BT}{\rho}e^{-k^2\nu|\tau|}. \tag{6.42}$$

Taking the Fourier transform in time and applying the Wiener–Khinchin theorem, (6.16), yields that

$$\left\langle v^{\perp}(\boldsymbol{k},\omega)v^{\perp}(-\boldsymbol{k},-\omega)\right\rangle = \frac{k_BT}{\rho}\frac{2k^2\nu}{\omega^2+(k^2\nu)^2}, \tag{6.43}$$

where here we allow the arguments to indicate that the variables are in the Fourier domain in both space and time.

Continuing to parallel the discussion of the single-particle Langevin equation, we solve (6.40) via temporal Fourier transform to yield

$$v^{\perp}(\boldsymbol{k},\omega) = \frac{1}{\rho}\frac{ik\sigma^{\perp}_{\text{fluc}}(\boldsymbol{k},\omega)}{i\omega+k^2\nu}. \tag{6.44}$$

Likewise,

$$v^{\perp}(-\boldsymbol{k},-\omega) = \frac{1}{\rho}\frac{-ik\sigma^{\perp}_{\text{fluc}}(-\boldsymbol{k},-\omega)}{-i\omega+k^2\nu}, \tag{6.45}$$

so

$$\left\langle v^{\perp}(\boldsymbol{k},\omega)v^{\perp}(-\boldsymbol{k},-\omega)\right\rangle = \frac{1}{\rho^2}\frac{k^2\left\langle\sigma^{\perp}_{\text{fluc}}(\boldsymbol{k},\omega)\sigma^{\perp}_{\text{fluc}}(-\boldsymbol{k},-\omega)\right\rangle}{\omega^2+(k^2\nu)^2}. \tag{6.46}$$

Combining (6.43) and (6.46), we find that

$$\left\langle\sigma^{\perp}_{\text{fluc}}(\boldsymbol{k},\omega)\sigma^{\perp}_{\text{fluc}}(-\boldsymbol{k},-\omega)\right\rangle = 2k_BT\eta. \tag{6.47}$$

An identical equation holds for $\sigma^{\perp\perp}_{\text{fluc}}$.

To express (6.47) in Cartesian coordinates, we first write it as

$$u_i^{\parallel}u_j^{\perp}u_l^{\parallel}u_m^{\perp}\left\langle\sigma_{\text{fluc},ij}(\boldsymbol{k},\omega)\sigma_{\text{fluc},lm}(-\boldsymbol{k},-\omega)\right\rangle = 2k_BT\eta. \tag{6.48}$$

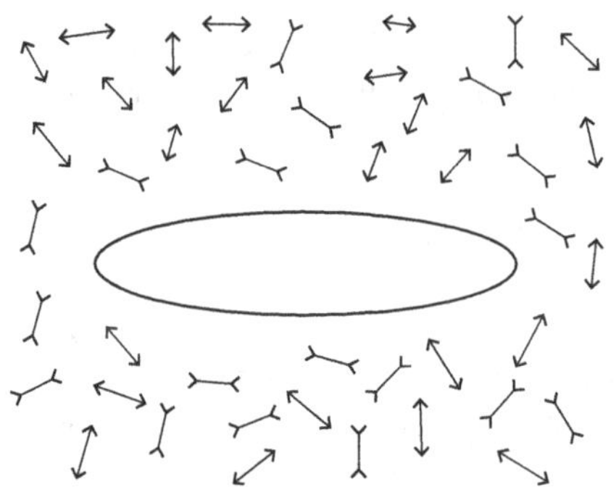

**Figure 6.3** A rigid particle in a thermally fluctuating fluid. The double-headed arrows represent thermally fluctuating stresslets.

The quantity $\left\langle \sigma_{\text{fluc},ij}(\boldsymbol{k},\omega)\sigma_{\text{fluc},lm}(-\boldsymbol{k},-\omega)\right\rangle$ is an isotropic fourth-order tensor and thus must have the general form (Segel 1987)

$$\left\langle \sigma_{\text{fluc},ij}(\boldsymbol{k},\omega)\sigma_{\text{fluc},lm}(-\boldsymbol{k},-\omega)\right\rangle = A\left(\delta_{il}\delta_{jm}+\delta_{im}\delta_{jl}\right)+B\left(\delta_{il}\delta_{jm}-\delta_{im}\delta_{jl}\right)+C\delta_{ij}\delta_{lm}. \tag{6.49}$$

The symmetry of $\boldsymbol{\sigma}_{\text{fluc}}$ requires that $B = 0$, because the term it multiplies changes sign upon exchange of $i$ and $j$ or $l$ and $m$. Inserting (6.49) into (6.48) yields that

$$A = 2k_BT\eta.$$

The term containing $C$ vanishes, so for incompressible flow it is indeterminate. Two arbitrary choices for $C$ are common in the literature. The condition that $\text{Tr}\,\boldsymbol{\sigma}_{\text{fluc}} = 0$ can be satisfied by setting $C = -4k_BT\eta/3$. (To see this, multiply (6.49) by $\delta_{ij}\delta_{lm}$, insert the previous values of $A$ and $B$, and recall that $\text{Tr}\,\boldsymbol{\sigma} = \sigma_{ij}\delta_{ij}$.) Alternatively one can simply set $C = 0$, which is what we will do here. In this case,

$$\left\langle \sigma_{\text{fluc},ij}(\boldsymbol{k},\omega)\sigma_{\text{fluc},lm}(-\boldsymbol{k},-\omega)\right\rangle = 2k_BT\eta\left(\delta_{il}\delta_{jm}+\delta_{im}\delta_{jl}\right), \tag{6.50}$$

and applying Wiener–Khinchin once again to revert to physical space gives

$$\left\langle \sigma_{\text{fluc},ij}(\boldsymbol{x},t)\sigma_{\text{fluc},lm}(\boldsymbol{x}',t')\right\rangle = 2k_BT\eta\delta(\boldsymbol{x}-\boldsymbol{x}')\delta(t-t')\left(\delta_{il}\delta_{jm}+\delta_{im}\delta_{jl}\right). \tag{6.51}$$

This expression is closely analogous to that for the fluctuating force on a particle. We again see the product of $\eta$ and $k_BT$: fluctuations balance dissipation.

## 6.4 Brownian Motion from Fluctuating Hydrodynamics

In Section 6.2, we noted that the fluctuating force $\boldsymbol{F}_{\text{fluc}}$ on a Brownian particle originates in the thermal fluctuations of the surrounding fluid. At that point, however, we had not yet characterized these fluctuations, and computed their net effect on the particle with a somewhat roundabout method. Now, having in hand the fluctuating Navier–Stokes equation and (6.51) for the fluctuating stresses, a direct computation of the time correlation function of $\boldsymbol{F}_{\text{fluc}}$ can be performed (Fox & Uhlenbeck 1970).

Central to this derivation is the Lorentz reciprocal relation, (1.55):

$$\int_V (\boldsymbol{v}' \cdot (\nabla \cdot \boldsymbol{\sigma}'') - \boldsymbol{v}'' \cdot (\nabla \cdot \boldsymbol{\sigma}'))\, dV = \int_S (\boldsymbol{v}' \cdot (\boldsymbol{n} \cdot \boldsymbol{\sigma}'') - \boldsymbol{v}'' \cdot (\boldsymbol{n} \cdot \boldsymbol{\sigma}'))\, dS.$$

Here, we will take the primed problem to be that of a moving rigid particle in a (nonfluctuating) unbounded Stokes flow: $\nabla \cdot \boldsymbol{\sigma}' = \mathbf{0}$, with $\boldsymbol{v}' = \boldsymbol{U}$ on the particle surface $S_p$ and $\boldsymbol{v}' \to \mathbf{0}$ and $\boldsymbol{\sigma}' \to \mathbf{0}$ as $r \to \infty$ (i.e., on the surface of a sphere $S_\infty$ whose radius tends to $\infty$). We focus on the case of a nonspherical, nonchiral particle, for which the drag force $\boldsymbol{F}_{\text{drag}}$ satisfies $\boldsymbol{F}_{\text{drag}} = -\boldsymbol{\zeta} \cdot \boldsymbol{U}$.

The double-primed problem is the one of interest. It will be for a particle held stationary, so $\boldsymbol{v}'' = \mathbf{0}$ on its surface, in a fluid that satisfies the *fluctuating* Stokes equation:

$$\nabla \cdot (\boldsymbol{\sigma}'' + \boldsymbol{\sigma}_{\text{fluc}}) = \mathbf{0}. \tag{6.52}$$

In other words, the particle is held still in an otherwise stationary fluid filled with fluctuating stresslets as illustrated in Figure 6.3. Fluid inertia is neglected. Recall from Section 5.3 that inertial effects on drag vanish faster with decreasing particle size than viscous effects – we assume that the same trend holds here. The as yet unknown fluctuating force exerted by the fluid on the particle is given by

$$\boldsymbol{F}_{\text{fluc}} = \int_{S_p} \hat{\boldsymbol{n}} \cdot (\boldsymbol{\sigma}'' + \boldsymbol{\sigma}_{\text{fluc}})\, dS, \tag{6.53}$$

$$= \int_{S_p} \hat{\boldsymbol{n}} \cdot \boldsymbol{\sigma}''\, dS, \tag{6.54}$$

where $\hat{\boldsymbol{n}} = -\boldsymbol{n}$ is the outward unit normal to the particle surface and the second equation was obtained from the first by noting that the integral of the fluctuating traction $\hat{\boldsymbol{n}} \cdot \boldsymbol{\sigma}_{\text{fluc}}$ over any nonzero area will be zero because of its spatially delta-correlated nature. (Sometimes the term $\hat{\boldsymbol{n}} \cdot \boldsymbol{\sigma}_{\text{fluc}}$ is neglected entirely in writing the force on a surface, but doing so risks local violation of momentum conservation (Hinch 1975).) Since $\langle \boldsymbol{\sigma}_{\text{fluc}} \rangle = \mathbf{0}$, $\langle \boldsymbol{F}_{\text{fluc}} \rangle = \mathbf{0}$ as well, and we seek to determine $\langle \boldsymbol{F}_{\text{fluc}}(t_1) \boldsymbol{F}_{\text{fluc}}(t_2) \rangle$.

We now insert the given properties of the primed and double-primed problem into the Lorentz reciprocal relation to yield

$$-\int_V \boldsymbol{v}' \cdot \nabla \cdot \boldsymbol{\sigma}_{\text{fluc}}\, dV = \int_{S_p} \boldsymbol{v}' \cdot (\boldsymbol{n} \cdot \boldsymbol{\sigma}'')\, dS \tag{6.55}$$

$$= -\boldsymbol{U} \cdot \boldsymbol{F}_{\text{fluc}}. \tag{6.56}$$

Evaluating (6.56) at times $t_1$ and $t_2$, multiplying the results, and ensemble-averaging yields that

$$U_i U_j \left\langle F_{\text{fluc},i}(t_1) F_{\text{fluc},j}(t_2) \right\rangle$$
$$= \int_V \int_V v_i'(\boldsymbol{x}) v_j'(\boldsymbol{y}) \frac{\partial}{\partial x_l} \frac{\partial}{\partial y_m} \left\langle \sigma_{\text{fluc},li}(\boldsymbol{x}, t_1) \sigma_{\text{fluc},mj}(\boldsymbol{y}, t_2) \right\rangle d\boldsymbol{x}\, d\boldsymbol{y},$$

where $\boldsymbol{x}$ and $\boldsymbol{y}$ are the dummy spatial variables associated with quantities evaluated at times $t_1$ and $t_2$, respectively. Insertion of (6.51) for the time correlation function of the fluctuating stress gives the right-hand side as

$$2k_BT\eta\delta(t_1-t_2)\int_V\int_V v_i'(\boldsymbol{x})v_j'(\boldsymbol{y})\left(\delta_{lm}\delta_{ij}+\delta_{lj}\delta_{im}\right)\frac{\partial}{\partial x_l}\frac{\partial}{\partial y_m}\delta(\boldsymbol{x}-\boldsymbol{y})\,d\boldsymbol{x}\,d\boldsymbol{y}.$$

Evaluating the integral over $\boldsymbol{y}$ using (A.59) leads to the result

$$U_iU_j\left\langle F_{\text{fluc},i}(t_1)F_{\text{fluc},j}(t_2)\right\rangle = 2k_BT\delta(t_1-t_2)\int_V\eta\left(\frac{\partial v_i'}{\partial x_j}+\frac{\partial v_j'}{\partial x_i}\right)\frac{\partial v_i'}{\partial x_j}\,dV. \tag{6.57}$$

This integral can be recognized as the rate of dissipation in the fluid in the primed problem (cf. Section 1.3.4 and (1.62)), which by the overall energy balance must equal the rate of work done by the particle on the fluid. This is given by $-\boldsymbol{U}\cdot\boldsymbol{F}_{\text{drag}}$, and since $\boldsymbol{F}_{\text{drag}}=-\boldsymbol{\zeta}\cdot\boldsymbol{U}$, the rate of work done on the fluid is $\boldsymbol{U}\cdot\boldsymbol{\zeta}\cdot\boldsymbol{U}$ and (6.57) can be rewritten

$$U_iU_j\left\langle F_{\text{fluc},i}(t_1)F_{\text{fluc},j}(t_2)\right\rangle = 2k_BT\delta(t_1-t_2)U_i\zeta_{ij}U_j. \tag{6.58}$$

From this equation, we can extract the desired result:

$$\langle \boldsymbol{F}_{\text{fluc}}(t_1)\,\boldsymbol{F}_{\text{fluc}}(t_2)\rangle = 2k_BT\boldsymbol{\zeta}\delta(t_1-t_2). \tag{6.59}$$

As it must be, this is the same expression that we found in Section 6.2, generalized to a particle of arbitrary shape.

## 6.5 Diffusion and Osmotic Stress

Now that we have some information about the fluctuating force exerted on a particle undergoing Brownian motion, let us consider now how its position will change with time. To do this, it is convenient to eliminate velocity and write a second-order equation for $\boldsymbol{R}$:

$$m\frac{d^2\boldsymbol{R}}{dt^2}+\zeta\frac{d\boldsymbol{R}}{dt}=\boldsymbol{F}_{\text{fluc}}. \tag{6.60}$$

With initial position $\boldsymbol{R}(0)=\boldsymbol{0}$ and velocity $\boldsymbol{U}(0)$, the solution to this equation is

$$\boldsymbol{R}(t)=\boldsymbol{c}(1-e^{-\lambda_v t})+\frac{1}{\zeta}\int_0^t(1-e^{\lambda_v(s-t)})\boldsymbol{F}_{\text{fluc}}(s)\,ds, \tag{6.61}$$

where $\boldsymbol{c}=\lambda_v^{-1}\boldsymbol{U}(0)$. We assume that the particle is in thermal equilibrium at $t=0$ so the velocity satisfies (6.3). By isotropy, $\langle\boldsymbol{R}(t)\rangle=\boldsymbol{0}$ so the primary characterization of the average particle motion is the mean squared displacement $\langle\boldsymbol{R}(t)\boldsymbol{R}(t)\rangle$.

At very short times, $\lambda_v t\ll 1$, the first term in (6.61) dominates over the second, and $\langle\boldsymbol{R}(t)\boldsymbol{R}(t)\rangle$ satisfies

$$\langle\boldsymbol{R}(t)\boldsymbol{R}(t)\rangle=\langle\boldsymbol{c}\boldsymbol{c}\rangle(\lambda_v t)^2=\langle\boldsymbol{U}\boldsymbol{U}\rangle t^2=\frac{k_BT}{m}t^2\boldsymbol{\delta}. \tag{6.62}$$

The quadratic dependence on time is characteristic of *ballistic* (projectile-like) motion with velocity $\boldsymbol{U}$.

For a colloidal particle, ballistic behavior persists only for a very short time. Of more interest is the regime $\lambda_v t \gg 1$, where (6.61) simplifies to

$$\boldsymbol{R}(t) = \boldsymbol{c} + \frac{1}{\zeta}\int_0^t \boldsymbol{F}_{\text{fluc}}(s)\, ds. \tag{6.63}$$

The mean squared displacement $\left\langle R_i(t)R_j(t)\right\rangle$ will satisfy:

$$\begin{aligned}\left\langle R_i(t)R_j(t)\right\rangle &= \left\langle c_i c_j\right\rangle \\ &+ \left\langle c_i \frac{1}{\zeta}\int_0^t F_{\text{fluc},j}(s)\, ds + c_j \frac{1}{\zeta}\int_0^t F_{\text{fluc},i}(s)\, ds\right\rangle \\ &+ \frac{1}{\zeta^2}\int_0^t\int_0^t \left\langle F_{\text{fluc},i}(s)F_{\text{fluc},j}(s')\right\rangle ds\, ds'.\end{aligned}$$

The second term averages to zero, so

$$\begin{aligned}\left\langle R_i(t)R_j(t)\right\rangle &= \left\langle c_i c_j\right\rangle + \frac{1}{\zeta^2}\int_0^t\int_0^t \left\langle F_{\text{fluc},i}(s)F_{\text{fluc},j}(s')\right\rangle ds\, ds' \\ &= \left\langle c_i c_j\right\rangle + \frac{1}{\zeta^2}\int_0^t\int_0^t 2\zeta kT\delta(s-s')\delta_{ij}\, ds\, ds' \\ &= \left\langle c_i c_j\right\rangle + \frac{2k_BT}{\zeta}\delta_{ij}\int_0^t ds \\ &= \left\langle c_i c_j\right\rangle + \frac{2k_BT}{\zeta}\delta_{ij}t.\end{aligned} \tag{6.64}$$

The second term, which increases linearly with time, dominates as $t \to \infty$. In fact, a linear relationship between mean squared displacement and time is a defining property of diffusive behavior, and the definition of the long-time diffusivity is

$$D_{ij} = \lim_{t\to\infty}\left\langle R_i(t)R_j(t)\right\rangle/2t. \tag{6.65}$$

For an isotropic system such as the one we consider here, $D_{ij} = D\delta_{ij}$, so (6.64) simply recovers the classical Stokes–Einstein formula for the diffusion coefficient of a spherical particle, $D = k_BT/\zeta$ (which we will derive from an argument closer to Einstein's in Section 6.9). Finally, if the particle is at thermal equilibrium at $t = 0$, then

$$\langle \boldsymbol{c}\cdot\boldsymbol{c}\rangle = \frac{\langle \boldsymbol{U}\cdot\boldsymbol{U}\rangle}{\lambda_v^2} = \frac{3k_BT}{m\lambda_v^2} = \frac{3k_BTm}{\zeta^2} = \frac{3D}{\lambda_v}.$$

This quantity is roughly the mean squared distance the particle will diffuse in one inertial-viscous relaxation time. This is vanishingly small for a micron-scale particle, in which case we can neglect it in (6.64) and write, for $t \gg \lambda_v^{-1}$,

$$\left\langle R_i(t)R_j(t)\right\rangle \to \frac{2k_BT}{\zeta}\delta_{ij}t. \tag{6.66}$$

This derivation would have been much simpler if we had neglected inertia by setting $m$ to zero and solving the inertialess problem

$$\zeta\frac{d\boldsymbol{R}}{dt} = \boldsymbol{F}_{\text{fluc}}. \tag{6.67}$$

For the simple case considered here, with a single particle with a position-independent friction coefficient $\zeta$, we can do this with no complications – (6.67) simply integrates to (6.63) with $\boldsymbol{c} = \boldsymbol{0}$ and (6.64) reduces to

$$\left\langle R_i(t)R_j(t)\right\rangle = \frac{2k_BT}{\zeta}\delta_{ij}t = 2Dt\delta_{ij} \tag{6.68}$$

at all times $t > 0$. For times much longer than $\lambda_v^{-1}$, the neglect of inertia is an excellent approximation. In general, however, taking $m$ to zero is a singular limit, as the highest derivative in the equation vanishes. Such situations need to be treated carefully, as discussed in Section 7.4.2.

Diffusion arises from the fluctuating velocity and associated kinetic energy imparted to a particle by the surrounding solvent. This fluctuating velocity contributes to momentum transport and thus drives a "kinetic" contribution to the stress in a suspension of Brownian particles. Recall from Section 1.2.2 that at the deterministic continuum level, there is a convective momentum flux $\rho\boldsymbol{vv}$ associated with the momentum that is transported by fluid motion. Additionally, in Section 1.2.2, we defined the stress vector $\boldsymbol{t} = \boldsymbol{n}\cdot\boldsymbol{\sigma}$ as the force per unit area exerted on a surface with unit normal $\boldsymbol{n}$ exerted by the material on the side to which $\boldsymbol{n}$ points. Because of this convention,[1] the stress tensor that is equivalent to the momentum flux becomes $-\rho\boldsymbol{vv}$. A Brownian particle has momentum $m\boldsymbol{U}$, so the analogous flux associated with the thermal momentum of Brownian particles is $nm\boldsymbol{UU}$, where $n$ the number density of particles and $nm$ is thus their mass density; the (ensemble-averaged) stress $\boldsymbol{\sigma}_{\text{kin}}$ is thus given by $-nm\langle\boldsymbol{UU}\rangle$. Taking the particle velocities to be at thermal equilibrium, we evaluate this expression using (6.3) to find that

$$\boldsymbol{\sigma}_{\text{kin}} = -nk_BT\boldsymbol{\delta} = -p_{\text{kin}}\boldsymbol{\delta}, \tag{6.69}$$

where $p_{\text{kin}} = nk_BT$. This is the classical thermodynamic expression for the osmotic stress in an ideal solution.

## 6.6 Diffusion and the Velocity Autocorrelation: The Taylor–Green–Kubo Formula for Diffusion

A very broadly applicable description of diffusion can be developed from the simple observation that

$$R_i(t) = R_i(0) + \int_0^t U_i(s)\,ds. \tag{6.70}$$

[1] We have followed the sign convention for stress that has been used historically in the continuum mechanics literature, where stresses are often thought to arise from forces exerted across interfaces and it is natural to think of a positive stress as resulting from pulling on a material. The opposite convention is used in the statistical mechanics and gas kinetic theory literature, where stresses arise from momentum fluxes across surfaces. In that convention, the stress associated with a momentum flux is $+\rho\boldsymbol{vv}$ and it is natural to think of a positive stress as resulting from imposing a pressure.

As earlier, we set $R_i(0) = 0$. If $\boldsymbol{U}$ is stationary and independent of $\boldsymbol{R}$, then using (6.70) to form the mean squared displacement yields

$$\begin{aligned}\left\langle R_i(t)R_j(t)\right\rangle &= \int_0^t \int_0^t \left\langle U_i(s)U_j(s')\right\rangle \, ds\, ds' \\ &= \int_0^t \int_0^t \phi_{\mathrm{v},ij}(s-s')\, ds\, ds' \\ &= 2\int_0^t \int_0^\tau \phi_{\mathrm{v},ij}(\tau')\, d\tau'\, d\tau \\ &= 2t\int_0^t \phi_{\mathrm{v},ij}(\tau)\, d\tau - 2\int_0^t \tau\phi_{\mathrm{v},ij}(\tau)\, d\tau. \end{aligned} \qquad (6.71)$$

Here the third line is obtained from the second by noting that $\boldsymbol{\phi}_\mathrm{v}$ is even in $s - s'$, and the fourth from the third by integration by parts. As long as $\boldsymbol{\phi}_\mathrm{v}(\tau)$ decays faster than $1/\tau$ at large $\tau$, the first term in (6.71) will grow linearly in time as $t \to \infty$ (the integral will converge to a constant), while the second will grow more slowly. In this case,

$$\lim_{t\to\infty} \frac{1}{t}\left\langle R_i(t)R_j(t)\right\rangle = 2\int_0^\infty \phi_{\mathrm{v},ij}(\tau)\, d\tau.$$

This result and the definition of the long-time diffusivity, (6.65), show very generally that

$$D_{ij} = \int_0^\infty \phi_{\mathrm{v},ij}(\tau)\, d\tau. \qquad (6.72)$$

This expression was originally derived by Taylor in the context of transport in turbulent flows (Taylor 1922). It is a special case of a class of relations known as Green–Kubo formulas (Kubo et al. 1991, Reichl 1998, McQuarrie 2000), which are expressions for a transport coefficient in terms of the integral of a correlation function. Green–Kubo relations also exist for many other transport properties, including viscosity, thermal conductivity, and linear viscoelastic properties. For an isotropic system,

$$D = \frac{1}{3}\int_0^\infty \mathrm{Tr}\,\boldsymbol{\phi}_\mathrm{v}(\tau)\, d\tau = \frac{1}{3}\int_0^\infty \langle \boldsymbol{U}(0)\cdot\boldsymbol{U}(\tau)\rangle\, d\tau. \qquad (6.73)$$

If $\boldsymbol{\phi}_\mathrm{v}$ does not decay sufficiently rapidly for the integrals in (6.71) to converge, then the mean squared displacement is not a linear function of time at long times, resulting in so-called *anomalous* diffusion. While many processes display such behavior, it is beyond our scope here. We see in Section 9.2 that (6.71) can be used to determine the rheological properties of a complex fluid by observing the Brownian motion of a particle suspended in it.

## 6.7 Time-Integration of the Inertialess Langevin Equation: "Basic Brownian Dynamics"

Up to this point, the Langevin equation has been used to determine ensemble-averaged properties of Brownian motion such as diffusivity and stress, while motions of individual

Brownian particles have not been directly addressed. In Chapter 7, we will systematically develop the background necessary for doing so. Nevertheless, at this point we have enough background to develop a simple approach for generating Brownian trajectories, a basic *Brownian dynamics* simulation method. Allowing now for fluid motion and external forces, the Langevin equation in terms of particle position is

$$m\frac{d^2\boldsymbol{R}}{dt^2} + \zeta\left(\frac{d\boldsymbol{R}}{dt} - \boldsymbol{v}(\boldsymbol{R})\right) = \boldsymbol{F}_{\text{ext}} + \boldsymbol{F}_{\text{fluc}}. \tag{6.74}$$

Neglecting inertia by dropping the first term, this equation becomes the inertialess Langevin equation

$$\frac{d\boldsymbol{R}}{dt} = \boldsymbol{v}(\boldsymbol{R}) + \frac{1}{\zeta}\left(\boldsymbol{F}_{\text{ext}} + \boldsymbol{F}_{\text{fluc}}\right). \tag{6.75}$$

For the moment, we note that $\lambda_{\text{v}}^{-1}$ is a very short time for small particles and that if our interest is in the motion of particle on time scales much longer than this, the inertialess Langevin equation (6.75) is an excellent approximation. Thus, in the present section, we develop a numerical scheme to find approximate solutions to this equation.

We begin with the case where the friction coefficient $\zeta$ is constant. Integrating between $t$ and $t + \Delta t$ gives

$$\boldsymbol{R}(t+\Delta t) - \boldsymbol{R}(t) = \int_t^{t+\Delta t}\left(\boldsymbol{v}(\boldsymbol{R}(s)) + \frac{1}{\zeta}\boldsymbol{F}_{\text{ext}}(\boldsymbol{R}(s), s)\right) ds + \frac{1}{\zeta}\int_t^{t+\Delta t}\boldsymbol{F}_{\text{fluc}}(s)\, ds. \tag{6.76}$$

As in a standard explicit Euler method, the first integral on the right-hand side can be approximated as follows:

$$\int_t^{t+\Delta t}\left(\boldsymbol{v}(\boldsymbol{R}(s)) + \frac{1}{\zeta}\boldsymbol{F}_{\text{ext}}(\boldsymbol{R}(s), s)\right) ds \approx \left(\boldsymbol{v}(\boldsymbol{R}(t)) + \frac{1}{\zeta}\boldsymbol{F}_{\text{ext}}(\boldsymbol{R}(t), t)\right)\Delta t. \tag{6.77}$$

The second integral is tricky: as yet we do not understand how $\boldsymbol{F}_{\text{fluc}}$ fluctuates in time (we will learn more in Chapter 7), so we cannot rely on our normal intuition about approximating integrals. Nevertheless, we do have (6.22) for the time correlation function of $\boldsymbol{F}_{\text{fluc}}$, which can be used to determine an expression for the covariance of this integral:

$$\begin{aligned}\left\langle\left[\int_t^{t+\Delta t}\boldsymbol{F}_{\text{fluc}}(s)\, ds\right]\left[\int_t^{t+\Delta t}\boldsymbol{F}_{\text{fluc}}(s')\, ds'\right]\right\rangle &= \int_t^{t+\Delta t}\int_t^{t+\Delta t}\left\langle\boldsymbol{F}_{\text{fluc}}(s)\boldsymbol{F}_{\text{fluc}}(s')\right\rangle ds\, ds' \\ &= \int_t^{t+\Delta t}\int_t^{t+\Delta t} 2k_BT\zeta\delta(s-s')\boldsymbol{\delta}\, ds\, ds' \\ &= 2k_BT\zeta\Delta t\boldsymbol{\delta}.\end{aligned} \tag{6.78}$$

Therefore, we must approximate the integral of the fluctuating force over the time interval $\Delta t$ with a term that has variance proportional to $\Delta t$. Defining a vector $\boldsymbol{\xi}$ whose components at each time step are independent random numbers that satisfy

$$\langle\xi_i\rangle = 0, \left\langle\xi_i\xi_j\right\rangle = \delta_{ij}, \tag{6.79}$$

we now let

$$\int_t^{t+\Delta t}\boldsymbol{F}_{\text{fluc}}(s)\, ds \approx \sqrt{2\zeta k_BT\Delta t}\boldsymbol{\xi}.$$

Note that this term is proportional to $\Delta t^{1/2}$, rather than the $\Delta t^1$ dependence of (6.77). Using this result, our time-integration scheme becomes

$$\boldsymbol{R}(t+\Delta t) = \boldsymbol{R}(t) + \left(\boldsymbol{v}(\boldsymbol{R}(t)) + \frac{1}{\zeta}\boldsymbol{F}_{\text{ext}}(\boldsymbol{R}(t),t)\right)\Delta t + \sqrt{\frac{2k_BT\Delta t}{\zeta}}\boldsymbol{\xi} \tag{6.80}$$

$$= \boldsymbol{R}(t) + \left(\boldsymbol{v}(\boldsymbol{R}(t)) + \frac{1}{\zeta}\boldsymbol{F}_{\text{ext}}(\boldsymbol{R}(t),t)\right)\Delta t + \sqrt{2D\Delta t}\boldsymbol{\xi}. \tag{6.81}$$

This is a stochastic version of the explicit Euler method. In the absence of external forces or flow, this simple approach automatically generates the proper diffusive result

$$\langle(\boldsymbol{R}(t) - \boldsymbol{R}(0)) \cdot (\boldsymbol{R}(t) - \boldsymbol{R}(0))\rangle = 2Dt\boldsymbol{\delta},$$

but does not necessarily match other more detailed features of Brownian motion, such as continuity of trajectories and a Gaussian probability distribution function, that will be addressed in Section 7.1. In particular, it will turn out that to capture these details the random numbers $\boldsymbol{\xi}$ should be drawn from a Gaussian distribution with zero mean and unit variance.

We now relax the assumption of constant friction coefficient. This situation can arise physically through gradients in viscosity, for example, and in the multiparticle case it occurs because the $N$-particle mobility depends on the positions of all the particles. Doing so presents a subtlety because $\zeta$ must be kept inside the integral over the fluctuating force in (6.76). Reconsidering this integral and Taylor-expanding $\zeta^{-1}$ around its value at time $t$ yields

$$\int_t^{t+\Delta t} \frac{1}{\zeta(\boldsymbol{R}(s))}\boldsymbol{F}_{\text{fluc}}\, ds \approx \zeta^{-1}|_{\boldsymbol{R}(t)} \int_t^{t+\Delta t} \boldsymbol{F}_{\text{fluc}}\, ds$$
$$+ \int_t^{t+\Delta t} \left.\frac{\partial \zeta^{-1}}{\partial \boldsymbol{R}}\right|_{\boldsymbol{R}(t)} \cdot (\boldsymbol{R}(s) - \boldsymbol{R}(t))\, \boldsymbol{F}_{\text{fluc}}(s)\, ds.$$

To evaluate the integral on the right-hand side (Lax 1966, Fixman 1978), we estimate $\boldsymbol{R}(s) - \boldsymbol{R}(t)$ in the integrand as follows:

$$\boldsymbol{R}(s) - \boldsymbol{R}(t) \approx \zeta^{-1}|_{\boldsymbol{R}(t)} \int_t^{t+s} \boldsymbol{F}_{\text{fluc}}(s')\, ds'.$$

(We could also include the term $\zeta^{-1}(\boldsymbol{R}(t))\boldsymbol{F}_{\text{ext}}\Delta t$ in this estimate, but it vanishes more rapidly as $\Delta t \to 0$ than the term we included.) Now we have

$$\int_t^{t+\Delta t} \left.\frac{\partial \zeta^{-1}}{\partial \boldsymbol{R}}\right|_{\boldsymbol{R}(t)} \cdot (\boldsymbol{R}(s) - \boldsymbol{R}(t))\, \boldsymbol{F}_{\text{fluc}}\, ds$$
$$\approx \zeta^{-1}|_{\boldsymbol{R}(t)} \left.\frac{\partial \zeta^{-1}}{\partial \boldsymbol{R}}\right|_{\boldsymbol{R}(t)} \cdot \int_t^{t+\Delta t} \int_t^{t+s} \boldsymbol{F}_{\text{fluc}}(s')\boldsymbol{F}_{\text{fluc}}(s)\, ds'\, ds.$$

We can further approximate the integral by replacing $\boldsymbol{F}_{\text{fluc}}(s')\boldsymbol{F}_{\text{fluc}}(s)$ with its ensemble average (6.22):

$$
\begin{aligned}
&\zeta^{-1}|_{\boldsymbol{R}(t)} \left.\frac{\partial \zeta^{-1}}{\partial \boldsymbol{R}}\right|_{\boldsymbol{R}(t)} \cdot \int_t^{t+\Delta t} \int_t^{t+s} \langle \boldsymbol{F}_{\text{fluc}}(s')\boldsymbol{F}_{\text{fluc}}(s)\rangle \, ds' \, ds \\
&\approx \zeta^{-1}|_{\boldsymbol{R}(t)} \left.\frac{\partial \zeta^{-1}}{\partial \boldsymbol{R}}\right|_{\boldsymbol{R}(t)} \cdot \int_t^{t+\Delta t} \int_t^{t+s} 2\zeta|_{\boldsymbol{R}(t)} k_B T \boldsymbol{\delta}\delta(s-s') \, ds' \, ds \\
&= 2k_B T \left.\frac{\partial \zeta^{-1}}{\partial \boldsymbol{R}}\right|_{\boldsymbol{R}(t)} \cdot \boldsymbol{\delta} \left( \int_t^{t+\Delta t} \int_t^{t+s} \delta(s-s') \, ds' \, ds \right) \\
&= 2k_B T \left.\frac{\partial \zeta^{-1}}{\partial \boldsymbol{R}}\right|_{\boldsymbol{R}(t)} \left( \frac{1}{2}\Delta t \right) \\
&= \frac{\partial D(\boldsymbol{R})}{\partial \boldsymbol{R}} \Delta t,
\end{aligned}
$$

where the factor of $1/2$ in the integration over the delta-function comes from noting that it is localized to the boundary $s = s'$ of the domain of integration and treating it as if only half of its area is within the domain. This is nonstandard (see Appendix A.4) but can be rationalized in the present case by first taking $\langle \boldsymbol{F}_{\text{fluc}}(s')\boldsymbol{F}_{\text{fluc}}(s)\rangle$ to be an even, very narrow function of $s - s'$ – that is, to have a small but nonzero correlation time – then to evaluate the integral and then to take the width to zero. Finally, our Euler step for the case of variable mobility becomes

$$
\boldsymbol{R}(t+\Delta t) = \boldsymbol{R}(t) + \left( \boldsymbol{v}(\boldsymbol{R}(t)) + \frac{1}{\zeta}\boldsymbol{F}_{\text{ext}}(\boldsymbol{R}(t), t) + \frac{\partial D(\boldsymbol{R}(t))}{\partial \boldsymbol{R}} \right) \Delta t + \sqrt{2D(\boldsymbol{R}(t))\Delta t}\,\boldsymbol{\xi}. \tag{6.82}
$$

This method is the one-particle version of the *Ermak–McCammon* algorithm (Ermak & McCammon 1978). The approximate analysis given here becomes exact as $\Delta t \to 0$.

At first sight, the term due to the spatial variation in $D$ might seem unphysical. At thermal equilibrium, the spatial distribution of the particle cannot depend on dynamic properties such as diffusivity, so in the absence of external forces it must be spatially uniform. But this gradient term would seem to drive particles up the gradient in diffusivity. However, this driving force is balanced by the random term, which leads to larger Brownian displacements in regions with larger diffusivity and thus will tend to drive particles preferentially out of those regions. The net result is that at equilibrium, the solution of this equation will yield a Boltzmann distribution as required by equilibrium statistical mechanics. Generalizations of this basic Brownian dynamics algorithm are discussed in Section 7.4.

## 6.8 Generalized Langevin Equations and Memory

At the end of Section 6.2, we observed that the delta-function nature of the autocorrelation function for the fluctuating force in (6.22) is not an assumption that can be made independent of the structure of the model for the drag on the particle. It is rather a consequence of the form of the model, following uniquely from the fact that we chose to model the friction on the bead with Stokes's law. If we had chosen a model other than Stokes's law (such as the generalized Stokes drag expression given by (5.6)–(5.9)), the fluctuating force would necessarily be different – otherwise, the proper velocity autocorrelation function (which is determined only by the nature of the friction model and the causality and stationarity conditions) would not be obtained. Here we briefly expand on this point.

A wide range of models can be put into the general form

$$\frac{du}{dt} = -\alpha u - \int_{-\infty}^{t} K(t-s)u(s)\, ds + \xi_{\text{fluc}}(t), \tag{6.83}$$

known as a generalized Langevin equation[2] (Kubo 1966, Zwanzig 2001, Snook 2007). The function $K(t-s)$ is known as a memory kernel. Generalizing the process that we used previously for the classical Langevin equation, and in particular by using Fourier and Laplace transforms, it can be shown that the fluctuating term must have the following time correlation function (Kubo 1966):

$$\langle \xi_{\text{fluc}}(t)\xi_{\text{fluc}}(t')\rangle = \left\langle u^2 \right\rangle_{\text{eq}} \left(2\alpha\delta(t-t') + K(|t-t'|)\right). \tag{6.84}$$

Here the noise at present and in the past is correlated, with the correlation function determined by the memory kernel. Such a process is said to be *non-Markovian*. By contrast, in a *Markov* process, the future evolution is determined only by the present state of the process – for example, a classical initial value problem governed by system of first-order differential equations satisfies this condition, as does a simple Langevin equation. Often a generalized Langevin equation can be associated with a system of conventional Langevin equations with auxiliary variables that capture the memory encapsulated in the memory function $K$. Exercise 6.7 illustrates one such case. Another is found in Section 9.2, where we describe the motion of a Brownian particle in a complex fluid.

## 6.9 Brownian Fluctuations as a Thermodynamic Driving Force: The Smoluchowski Equation

To conclude this chapter, we turn from a description of individual particle trajectories – Langevin equations – to a probabilistic one, starting from equilibrium statistical mechanics. We begin with a very simple situation, an infinitely dilute suspension of spherical particles subjected to an external potential field $\Phi_{\text{ext}}(\boldsymbol{R})$; the force exerted on each particle due to this field will be $\boldsymbol{F}_{\text{ext}} = -\frac{\partial}{\partial \boldsymbol{R}}\Phi_{\text{ext}}$. For concreteness, imagine a suspension of nonneutrally buoyant colloidal particles, whose sedimentation due to gravity will be resisted by Brownian motion. Because the system is infinitely dilute, the particles do not interact with one another and we can treat them as if they were alone in the fluid. At equilibrium, the probability density function for the particle velocity satisfies

[2] The standard form for a generalized Langevin equation does not include the term $-\lambda u$ on the right-hand side. Indeed, by giving the memory kernel $K(t-s)$ a piece proportional to $\delta(t-s)$, one can recover such a term. However, when working with Laplace or "one-sided Fourier" transforms, ambiguity can arise as noted in Appendix A.4 when the delta function has its spike at the boundary of a domain of integration (i.e., at $t-s=0$ in the present context). The formulation presented here avoids ambiguity as long as $\lambda$ and $K(t-s)$ are chosen so that $K(t-s)$ has no delta-function behavior at $t=s$.

the Maxwell distribution (6.1), while that for the positions of the particles satisfies a Boltzmann distribution

$$p(\boldsymbol{R}) = p_0 \exp\left(-\frac{\Phi_{\text{ext}}(\boldsymbol{R})}{k_B T}\right). \tag{6.85}$$

Here $p_0$ is determined by the normalization condition $\int_V p(\boldsymbol{R})\, d\boldsymbol{R} = 1$. Taking the logarithm and then the gradient of both sides yields that

$$\frac{\partial}{\partial \boldsymbol{R}} \ln p = -\frac{\partial}{\partial \boldsymbol{R}} \frac{\Phi_{\text{ext}}(\boldsymbol{R})}{k_B T} = \frac{1}{k_B T} \boldsymbol{F}_{\text{ext}}.$$

This equation can be viewed as a "velocity-space-averaged" force balance (i.e., where we have averaged over the Maxwell velocity distribution of the particle):

$$\boldsymbol{F}_{\text{ext}} + \breve{\boldsymbol{F}}_{\text{Brown}} = \boldsymbol{0}, \tag{6.86}$$

where

$$\breve{\boldsymbol{F}}_{\text{Brown}} = -k_B T \frac{\partial}{\partial \boldsymbol{R}} \ln p.$$

This is the "Brownian force" or "entropic force." In other words, the probability distribution that arises is as if each particle had a force $\breve{\boldsymbol{F}}_{\text{Brown}}$ acting on it. This force is not a literal microscopic driving force that acts on each particle but rather is an averaged statement of the fact that the probability distribution of particles will tend to increase its entropy due to Brownian motion; the breve mark ˘ serves to indicate this point. By contrast, the fluctuating force $\boldsymbol{F}_{\text{fluc}}$ of the previous sections *is* a literal force acting on each particle.

Now let the system deviate from equilibrium, evolving on time scales much longer than $\lambda_{\text{v}}^{-1}$. In this case, the velocity distribution remains Maxwellian and we need only consider the probability distribution $p(\boldsymbol{R}, t)$ of particle positions. The assumption of a separation of scales between the time scales for particle motion and the inertial-viscous relaxation time is further addressed in Section 7.4.2. Now we can generalize the force balance (6.86) to include a drag force on the particle, which we describe with Stokes's law based on the velocity-space averaged particle velocity $\breve{\boldsymbol{U}}$:

$$\breve{\boldsymbol{F}}_{\text{drag}} + \boldsymbol{F}_{\text{ext}} + \breve{\boldsymbol{F}}_{\text{Brown}} = \boldsymbol{0}, \tag{6.87}$$

where

$$\breve{\boldsymbol{F}}_{\text{drag}} = \zeta \left(\boldsymbol{v}(\boldsymbol{R}) - \breve{\boldsymbol{U}}\right). \tag{6.88}$$

Thus

$$\breve{\boldsymbol{U}} = \boldsymbol{v}(\boldsymbol{R}) + \frac{1}{\zeta}\left(\boldsymbol{F}_{\text{ext}} + \breve{\boldsymbol{F}}_{\text{Brown}}\right). \tag{6.89}$$

Again, the breve indicates that $\breve{\boldsymbol{U}}$ and $\breve{\boldsymbol{F}}$ are not the literal velocity or drag force for an individual particle.

This force balance cannot yet be solved because we do not yet know $p(\boldsymbol{R}, t)$. Whether the system is at equilibrium or not, we can write an evolution equation for $p(\boldsymbol{R}, t)$ as

$$\frac{\partial p}{\partial t} = -\frac{\partial}{\partial \boldsymbol{R}} \cdot \boldsymbol{J}_p, \tag{6.90}$$

where $\boldsymbol{J}_p$ is a flux of probability. This equation is simply a statement of conservation of probability. The probability flux is the probability density times the velocity-space averaged particle velocity:

$$\boldsymbol{J}_p = \check{\boldsymbol{U}}p = \boldsymbol{v}(\boldsymbol{R})p + \frac{1}{\zeta}\left(\boldsymbol{F}_{\text{ext}} + \check{\boldsymbol{F}}_{\text{Brown}}\right)p. \tag{6.91}$$

If $\boldsymbol{F}_{\text{ext}}$ and $\boldsymbol{v}_\infty$ are zero, this reduces to

$$\boldsymbol{J}_p = \frac{1}{\zeta}\check{\boldsymbol{F}}_{\text{Brown}}p = -\frac{k_BT}{\zeta}p\frac{\partial \ln p}{\partial \boldsymbol{R}} = -\frac{k_BT}{\zeta}\frac{\partial p}{\partial \boldsymbol{R}}. \tag{6.92}$$

For a dilute solute, each molecule of which obeys this equation, the number density $n$ at any point in solution is simply the probability density $p$ times the total number of molecules $N_{\text{mol}}$ in the system and the diffusive flux $\boldsymbol{J}_n$ is simply $\boldsymbol{J}_p N_{\text{mol}}$. Thus

$$\boldsymbol{J}_n = \boldsymbol{J}_p N_{\text{mol}} = -\frac{k_BT}{\zeta}\frac{\partial p N_{\text{mol}}}{\partial \boldsymbol{R}} = -\frac{k_BT}{\zeta}\frac{\partial n}{\partial \boldsymbol{R}}. \tag{6.93}$$

This is simply Fick's law for diffusion, where we see that the Stokes–Einstein diffusivity $D = k_BT/\zeta$ has naturally arisen. Finally, we note that this derivation of Fick's law is equivalent to the approach of Einstein (1998), who treated diffusion as a flux driven by the gradient in osmotic pressure $p_{\text{os}} = nk_BT$ in an ideal solution (recall (6.69)):

$$\boldsymbol{J}_n = -\frac{1}{\zeta}\frac{\partial p_{\text{os}}}{\partial \boldsymbol{R}} = -\frac{k_BT}{\zeta}\frac{\partial n}{\partial \boldsymbol{R}}. \tag{6.94}$$

To see the equivalence, observe that the Brownian force per unit volume $\check{\boldsymbol{F}}_{\text{Brown}}n$ is given by

$$\check{\boldsymbol{F}}_{\text{Brown}}n = -k_BT\frac{\partial p}{\partial \boldsymbol{R}}n = -k_BT\frac{\partial n}{\partial \boldsymbol{R}} = -\frac{\partial p_{\text{os}}}{\partial \boldsymbol{R}}. \tag{6.95}$$

Inserting (6.91) into the conservation equation for $p$ yields the so-called *Smoluchowski equation*:

$$\frac{\partial p}{\partial t} + \boldsymbol{v}\cdot\frac{\partial p}{\partial \boldsymbol{R}} = \frac{\partial}{\partial \boldsymbol{R}}\cdot\left(\frac{1}{\zeta}p\frac{\partial \Phi}{\partial \boldsymbol{R}}\right) + \frac{\partial}{\partial \boldsymbol{R}}\cdot D\frac{\partial p}{\partial \boldsymbol{R}}. \tag{6.96}$$

This is simply a convection-diffusion equation for the particle probability, with an additional term arising from the external forces exerted on the particle. As noted previously, for a solution that is sufficiently dilute that the particles do not interact, by replacing $p$ with the number density $n = p/N_{\text{mol}}$ this becomes the standard convection-diffusion equation for the evolution of particle concentration. In this simple case of a single particle with no internal degrees of freedom, the operator $\partial/\partial \boldsymbol{R}$ is simply the spatial gradient operator $\boldsymbol{\nabla}$, and we can write the Smoluchowski equation as

$$\frac{\partial p}{\partial t} + \boldsymbol{v}\cdot\boldsymbol{\nabla}p = \boldsymbol{\nabla}\cdot\left(\frac{1}{\zeta}p\boldsymbol{\nabla}\Phi\right) + \boldsymbol{\nabla}\cdot D\boldsymbol{\nabla}p. \tag{6.97}$$

For a flow with a characteristic velocity gradient $\dot{\gamma}_{\text{c}}$, the local velocity scale $\Delta v$ experienced by a particle in the flow is $a\dot{\gamma}_{\text{c}}$. In this flow, the relative importance of convective and diffusive transport of a particle can be estimated by forming the ratio between the time taken for a particle to diffuse its own size, $t_{\text{diff}} = a^2/D$, and to be convected its own size, $t_{\mathrm{conv}} = a/\Delta v = 1/\dot{\gamma}_{\mathrm{c}}$. This ratio is known as the *Péclet number*[3] Pé:

$$\mathrm{Pé} = \frac{t_{\mathrm{diff}}}{t_{\mathrm{conv}}} = \frac{\dot{\gamma}_{\mathrm{c}} a^2}{D} = \frac{\dot{\gamma}_{\mathrm{c}} a^2 \zeta}{k_B T}. \tag{6.98}$$

Since $\zeta \sim a$, $\mathrm{Pé} \sim a^3$, so for a given flow the importance of Brownian motion increases strongly as particle size decreases.

For an $N$-particle system with particle $\alpha$ at position $\boldsymbol{R}_\alpha$, the Brownian force on the $\alpha$th particle is

$$\breve{\boldsymbol{F}}_{\mathrm{Brown},\alpha} = -k_B T \frac{\partial \ln p(\boldsymbol{R}_1, \boldsymbol{R}_2, \ldots, \boldsymbol{R}_N)}{\partial \boldsymbol{R}_\alpha}$$

or more concisely,

$$\breve{\mathbf{F}}_{\mathrm{Brown}} = -k_B T \frac{\partial \ln p(\mathbf{R})}{\partial \mathbf{R}},$$

where $\breve{\mathbf{F}}_{\mathrm{Brown}} = \left[\breve{\boldsymbol{F}}_{\mathrm{Brown},1}, \breve{\boldsymbol{F}}_{\mathrm{Brown},2}, \ldots, \breve{\boldsymbol{F}}_{\mathrm{Brown},N}\right]^{\mathrm{T}}$. The evolution equation for $p(\mathbf{R})$ is now

$$\frac{\partial p}{\partial t} = -\frac{\partial}{\partial \mathbf{R}} \cdot \mathbf{J}_p.$$

The flux $\mathbf{J}_p$ can be written

$$\mathbf{J}_p = \left[\mathbf{V}_\infty + \mathbf{M} \cdot \left(\mathbf{F}_{\mathrm{ext}} + \breve{\mathbf{F}}_{\mathrm{Brown}}\right)\right] p, \tag{6.99}$$

where the term in the square brackets contains the velocity-space averaged velocities of each particle (cf. Section 2.5). Combining these two equations yields the $N$-particle Smoluchowski equation:

$$\frac{\partial p}{\partial t} + \mathbf{V}_\infty \cdot \frac{\partial p}{\partial \mathbf{R}} = -\frac{\partial}{\partial \mathbf{R}} \cdot (\mathbf{M} \cdot \mathbf{F}_{\mathrm{ext}} p) + \frac{\partial}{\partial \mathbf{R}} \cdot \left(k_B T \mathbf{M} \cdot \frac{\partial p}{\partial \mathbf{R}}\right). \tag{6.100}$$

The quantity $k_B T\mathbf{M}$ is often called the diffusion tensor and denoted $\mathbf{D}$. This is the fundamental equation for the dynamics of Brownian suspensions (Guazzelli & Morris 2012) and for coarse-grained models of dilute solution polymer dynamics (Bird et al. 1987). The latter case is further described in Chapter 8.

## Problems

**6.1** Compute $m/\zeta$ and $D$ for a 1 $\mu$m radius spherical polystyrene particle in water at room temperature. Estimate the distance this particle will travel in one inertial-viscous relaxation time.

**6.2** Derive (6.62) from (6.61) by Taylor-expanding the exponential terms.

[3] In flows with a characteristic velocity $V$ and overall length scale (e.g., tube diameter) $L$, a Péclet number based on transport on the scale of the whole system is also commonly used:

$$\mathrm{Pé}_L = \frac{L^2/D}{L/V} = \frac{VL}{D}.$$

**6.3** Recall that for a spheroidal particle, the friction coefficient is a tensor – $\boldsymbol{\zeta} = \zeta^{\parallel}\boldsymbol{uu} + \zeta^{\perp}(\boldsymbol{\delta} - \boldsymbol{uu})$ – and the Langevin equation for the linear velocity of the particle becomes

$$m\frac{d\boldsymbol{U}}{dt} = -\boldsymbol{\zeta} \cdot \boldsymbol{U} + \boldsymbol{F}_{\text{fluc}}(t).$$

Derive the expression

$$\langle \boldsymbol{F}_{\text{fluc}}(t)\boldsymbol{F}_{\text{fluc}}(t+\tau)\rangle = 2k_BT\delta(\tau)\boldsymbol{\zeta}$$

that maintains the equilibrium translational kinetic energy of the particle. Hint: Without loss of generality, one can set $\boldsymbol{u} = \boldsymbol{e}_x$.

**6.4** In the absence of flow or other torques, the Langevin equation for the rotational motion of a spheroid is

$$\boldsymbol{I} \cdot \frac{d\boldsymbol{\omega}}{dt} = -\boldsymbol{\zeta}_r \cdot \boldsymbol{\omega} + \boldsymbol{T}_{\text{fluc}},$$

where $\boldsymbol{I}$ is the moment of inertia tensor for the rod and $\boldsymbol{T}_{\text{fluc}}$ is the fluctuating torque exerted on the rod by the fluid. The moment of inertia tensor has the by now familiar structure $\boldsymbol{I} = I^{\parallel}\boldsymbol{uu} + I^{\perp}(\boldsymbol{\delta} - \boldsymbol{uu})$, and the three rotational degrees of freedom each get $\frac{1}{2}k_BT$ of energy. Derive the expression

$$\langle \boldsymbol{T}_{\text{fluc}}(t)\boldsymbol{T}_{\text{fluc}}(t+\tau)\rangle = 2k_BT\delta(\tau)\boldsymbol{\zeta}_r$$

that maintains the equilibrium rotational kinetic energy of the particle. The rotational dynamics of an anisotropic particle are further addressed in Sections 7.6 and 8.7.

**6.5** Derive (6.32) from (6.26).

**6.6** Given a random number generator that produces numbers $r$ chosen uniformly in the interval $[0, 1]$, determine how to generate random numbers $\xi$ that satisfy the conditions of (6.79).

**6.7** Consider the Langevin equation for a Brownian particle in a potential well. The only difference between this equation and that for the free particle is the presence of an additional (conservative) force due to the potential:

$$m\frac{dU}{dt} = -m\omega^2 X - \zeta U + F_{\text{fluc}}(t)$$
$$\frac{dX}{dt} = U,$$

where $F_{\text{fluc}}$ is unchanged from the form described earlier. Integrate the first equation to find $U$ as a function of the history of $X$ and $F_{\text{fluc}}$ (assume that $U(-\infty) = 0$), then insert into the equation for $X$ to get a non-Markovian evolution equation (i.e., a generalized Langevin equation) for it. What is the noise term that arises in this model?

**6.8** Consider a complex quantity $W$ that satisfies the equation

$$\frac{dW}{dt} = \gamma W + F_{\text{fluc}},$$

where $\gamma = -\alpha + i\omega$, and $\alpha \geq 0$. Here $\bar{\ }$ denotes complex conjugate. In the absence of fluctuations, $|W|$ will decay if $\alpha > 0$ but remain constant if $\alpha = 0$. The equations from Problem 6.7 can be cast into this form by transforming into the eigenvector basis for the system. Follow the process used in Sections 6.1 and 6.2 to find the velocity autocorrelation function $\left\langle W(0)\overline{W}(\tau)\right\rangle$ given that $\left\langle W(0)\overline{W}(0)\right\rangle = W_{\mathrm{eq}}^2$ (which is presumed to be known from equilibrium thermodynamics just as the mean squared velocity is). Also, given that the noise is delta-correlated, find its magnitude. I.e., given $\left\langle F_{\mathrm{fluc}}(t)\bar{F}_{\mathrm{fluc}}(t')\right\rangle = \sigma^2\delta(t-t')$, find $\sigma^2$. You will find that the noise magnitude only depends on $W_{\mathrm{eq}}^2$ and the real (dissipative) part of $\gamma$.

**6.9** Write a Brownian dynamics algorithm using (6.81) for a free particle with no external forces acting on it. Compute the mean squared displacement as a function of time. Set the code up so that it will perform simulations of an ensemble of independent particles (i.e., at each time step each particle sees different random numbers) – this will allow you to perform ensemble averages.

**6.10** Write a Brownian dynamics algorithm for a particle in a potential well, using (6.81) with $\boldsymbol{v} = \mathbf{0}$, $D = \zeta = 1$ and $\boldsymbol{F}_{\mathrm{ext}} = -\boldsymbol{R}$ ($\Phi_{\mathrm{ext}} = \frac{1}{2}\boldsymbol{R}\cdot\boldsymbol{R}$) and determine the time history and equilibrium value of $\langle \boldsymbol{R}\cdot\boldsymbol{R}\rangle$.

**6.11** Because of Brownian motion of solute molecules in a solution, the solute number density $n$ will fluctuate in space and time even if the solution is at equilibrium. For a solution of noninteracting solute particles with average number density $n_0$, the fluctuating part of the number density is given by $n' = n - n_0$, and the equilibrium spatial correlation function for the solute concentration fluctuations is given by

$$\langle n'(\boldsymbol{x})n'(\boldsymbol{x}')\rangle = n_0\delta(\boldsymbol{x}-\boldsymbol{x}') \tag{6.101}$$

(Landau & Lifschitz 1980). This equation is analogous to (6.26). Similarly, the evolution equation for the number density should contain a fluctuating flux term $\boldsymbol{j}_{\mathrm{fluc}}$ that is analogous to the fluctuating stress tensor in (6.27). That is, the usual diffusion equation that governs the solute motion becomes

$$\frac{dn'}{dt} = D\nabla^2 n' + \boldsymbol{\nabla}\cdot\boldsymbol{j}_{\mathrm{fluc}}. \tag{6.102}$$

Show that

$$\langle n(\boldsymbol{x},t)n(\boldsymbol{x}',t')\rangle = \frac{n_0}{\sqrt{4\pi D\,|t-t'|}}e^{-\frac{|\boldsymbol{x}-\boldsymbol{x}'|^2}{4D\,|t-t'|}} \tag{6.103}$$

and

$$\left\langle j_{\mathrm{fluc},i}(\boldsymbol{x},t)j_{\mathrm{fluc},j}(\boldsymbol{x}',t')\right\rangle = 2Dn_0\delta(\boldsymbol{x}-\boldsymbol{x}')\delta(t-t')\delta_{ij}. \tag{6.104}$$

Because the index of refraction of a solution depends on concentration, (6.103) and data from light scattering experiments can be used to determine $D$ for macromolecules in solution (Berne & Pecora 1976).

# 7 Stochastic Differential Equations

The Langevin equation (6.6) describes the dynamics of an individual particle in a fluid in response to the random fluctuating forces the fluid exerts on it. In Section 6.7, we saw an elementary method to approximate the time-integrated fluctuating force on the particle over a short time step and thus compute trajectories of particle motions in time. On the other hand, the Smoluchowski equation (6.96) describes the evolution of the probability distribution for the particle position as a function of time, rather than the motion of an individual particle – it does not directly generate predictions of individual particle displacements. In this chapter, we connect the Langevin and Smoluchowski descriptions of Brownian motion by introducing the basic concepts of stochastic differential equations. This background will enable us to better understand the relationship between inertial and inertialess descriptions of Brownian motion, develop numerical simulation methods for Brownian motion, and describe processes such as rotational diffusion.

## 7.1 The Diffusion Equation and the Wiener Process

To begin, we illustrate how the Smoluchowski equation can be used to generate trajectories of individual Brownian particles. Consider an individual particle in the absence of any external forces, boundaries, or imposed flow. In one dimension, the Smoluchowski equation (6.96) is simply a transient diffusion equation in an unbounded domain:

$$\frac{\partial p}{\partial t} = D\frac{\partial^2 p}{\partial x^2}.$$

For a particle that is at position $x = x_0$ at time $t = t_0$, the initial condition for the probability density is $p(x, t_0) = \delta(x - x_0)$. The solution to this problem is a Gaussian or normal distribution with mean $\mu = x_0$ and variance $\sigma^2 = 2D(t - t_0)$:

$$p(x,t) = \frac{1}{\sqrt{4D\pi(t-t_0)}} e^{-\frac{(x-x_0)^2}{4D(t-t_0)}}. \tag{7.1}$$

(See Appendix A.5 for background information on random variables in general and Gaussian random variables in particular.) Thus the mean squared displacement of the particle will be

$$\left\langle (x-x_0)^2 \right\rangle = \int_{-\infty}^{\infty} (x-x_0)^2 p(x,t)\, dx = 2D(t-t_0). \tag{7.2}$$

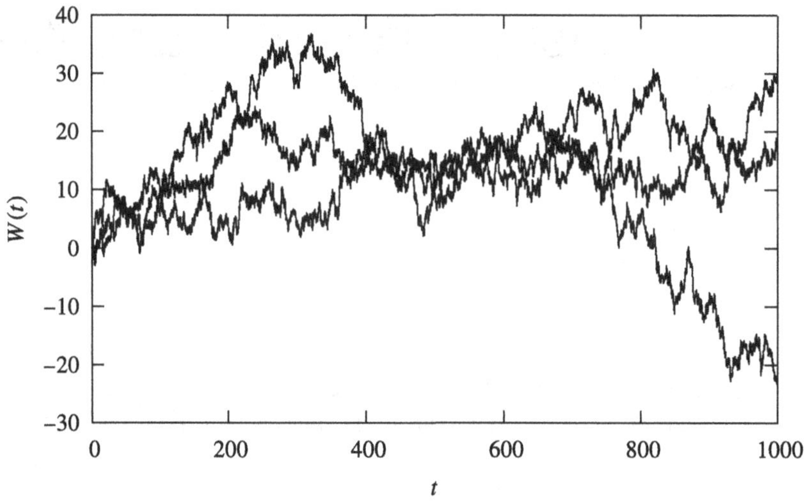


**Figure 7.1** Three trajectories of the Wiener process $W(t)$ with initial condition $W(0) = 0$.

We now use (7.1) with $D = 1/2$ to generate possible trajectories for the Brownian particle; specifically, we will determine particle positions at a set of increasing times $t_1, t_2, t_3, \ldots, t_n$, given that the particle is at position $W_0 = W(t_0)$ at time $t = t_0$. To determine a specific possible position for the particle at time $t_1 = t_0 + \Delta t_1$, we choose a random displacement $\Delta W(\Delta t_1)$ from the Gaussian probability distribution

$$p(\Delta W, \Delta t) = \frac{1}{\sqrt{2\pi\Delta t}} e^{-\frac{(\Delta W)^2}{2\Delta t}} \tag{7.3}$$

and set $W(t_1) = W(t_0) + \Delta W(\Delta t_1)$. Alternatively, we can write that

$$\Delta W(\Delta t) \sim \mathcal{N}(0, \Delta t). \tag{7.4}$$

To generate the next step of the trajectory, let $\Delta t_2 = t_2 - t_1$ and again choose a value $\Delta W(\Delta t_2)$ according to (7.3) to yield $W(t_2) = W(t_1) + \Delta W(\Delta t_2)$. Continuing this process generates a trajectory that satisfies

$$W(t_n) = W(t_0) + \sum_{i=1}^{n} \Delta W(\Delta t_i), \tag{7.5}$$

where $\Delta W(\Delta t_i) \sim \mathcal{N}(0, \Delta t_i)$. A different trajectory can be generated by repeating the process but using different samples from (7.3) at each time step. Figure 7.1 shows three realizations of $W(t)$.

As constructed, each step of this process corresponds to a motion governed by the diffusion equation, so we expect the result $W(t_n)$ to do so as well. To verify this, we recall (A.104), which states that a linear combination of Gaussian random variables is also Gaussian. Applying this result to (7.5) indicates that

$$W(t_n) - W(t_0) \sim \mathcal{N}\left(0, \sum_{i=1}^{n} \Delta t_i\right) \sim \mathcal{N}(0, t_n - t_0). \tag{7.6}$$

Therefore, for any choice of increasing time instants $t_0, t_1, t_2, t_3, \ldots, t_n$, (7.5) generates a trajectory of particle positions whose probability density at each time $t_i$ satisfies the diffusion equation with $D = 1/2$. Such a trajectory is called a *Wiener process* (Wiener 1976) or *standard Brownian motion* or *Gaussian random walk*, and the quantities $\Delta W$ are called *Wiener increments*. Letting $\Delta t_i \to 0$, (7.5) becomes

$$W(t) = W(t_0) + \int_{t_0}^{t} dW(t), \tag{7.7}$$

where $dW(t)$ is the differential of the Wiener process, also called a *white noise* process. This process satisfies

$$dW(t) \sim \mathcal{N}(0, dt). \tag{7.8}$$

Stochastic integrals will be further discussed in Section 7.2.

As an aside, consider a general random walk process

$$Z(t_n) = Z(t_0) + \sum_{i=1}^{n} \Delta Z(\Delta t),$$

where the $\Delta Z(\Delta t)$ are chosen from a *non-Gaussian* distribution with zero mean and variance $B^2 \Delta t$. For example, at each time step $\Delta Z$ could be $\pm B$ with the sign chosen by a coin toss. The central limit theorem result (A.99) in Appendix A.5.3 shows that after a large number of steps, $Z(t_n)$ will nevertheless approach a normal distribution. This observation provides a rationale separate from the diffusion equation for constructing a model of Brownian motion from Gaussian-distributed displacements – even if on very short time scales the displacements are not Gaussian, on longer time scales they are.

The Wiener process and its increments have a number of important properties that follow from their defining features, (7.3)–(7.6) and the properties of sums of Gaussian random variables described in Appendix A.5.3:

1. Wiener increments have zero mean:

$$\langle \Delta W(\Delta t) \rangle = \int_{-\infty}^{\infty} \Delta W p(\Delta W, \Delta t) \, d\Delta W = 0. \tag{7.9}$$

2. Wiener increments corresponding to distinct time intervals are independent:

$$\left\langle \Delta W(\Delta t_i) \Delta W(\Delta t_j) \right\rangle = \Delta t_i \delta_{ij}. \tag{7.10}$$

3. Wiener increments that overlap in time satisfy, for $t, t' > t_0$,

$$\langle (W(t) - W(t_0)) \left( W(t') - W(t_0) \right) \rangle = \min(t, t') - t_0. \tag{7.11}$$

4. The Wiener process is self-similar, satisfying the following scaling relation: For all $c > 0$, the process

$$W_c(t) = c^{-1/2} W(ct) \tag{7.12}$$

is also a Wiener process. That is, to within a constant, the Wiener process looks the same at all scales.

5. Integer powers of the Wiener increment satisfy

$$\langle \Delta W(\Delta t_i)^n \rangle = \begin{cases} 0 & \text{for } n \text{ odd,} \\ \Delta t_i^{n/2}(n-1)!! & \text{for } n \text{ even,} \end{cases} \tag{7.13}$$

where $p!! = (1 \cdot 3 \cdot 5 \cdot \ldots \cdot p)$.

6. Consider a stochastic process

$$X(t_n) = X(t_0) + \sum_i \Delta X(\Delta t_i) \tag{7.14}$$

with increments $\Delta X = B\Delta W(\Delta t_i)$, where $B$ is constant. This process satisfies

$$X(t_n) - X(t_0) \sim \mathcal{N}(0, B^2 \Delta t). \tag{7.15}$$

That is, it is Gaussian with zero mean and variance $B^2\Delta t$, and corresponds to diffusion with diffusivity $D = B^2/2$.

7. Consider a stochastic process (7.14) with increments

$$\Delta X(\Delta t_i) = B_\alpha \Delta W_\alpha(\Delta t_i) + B_\beta \Delta W_\beta(\Delta t_i), \tag{7.16}$$

where $\Delta W_\alpha$ and $\Delta W_\beta$ are increments of independent Wiener processes and $B_\alpha$ and $B_\beta$ are constant. This process can be written in terms of increments of a third Wiener process $W_\gamma$:

$$\Delta X = \left(B_\alpha^2 + B_\beta^2\right)^{1/2} \Delta W_\gamma. \tag{7.17}$$

In other words, the sum of two diffusion processes with diffusivities $D_\alpha = B_\alpha^2/2$ and $D_\beta = B_\beta^2/2$ is equivalent to a different diffusion process whose diffusivity $D_\gamma = D_\alpha + D_\beta$.

8. The Wiener process $W(t)$ is continuous in the sense that $\Pr(|\Delta W| > \epsilon) \to 0$ as $\Delta t \to 0$ for all $\epsilon > 0$. This can be shown with a simple calculation:

$$\begin{aligned} \lim_{\Delta t \to 0} \Pr(|\Delta W| > \epsilon) &= \lim_{\Delta t \to 0} 2 \int_\epsilon^\infty \frac{1}{\sqrt{2\pi\Delta t}} e^{-\frac{(\Delta W)^2}{2\Delta t}} \, d\Delta W \\ &= \lim_{\Delta t \to 0} \operatorname{erfc}\left(\frac{\epsilon}{\sqrt{2\Delta t}}\right) \\ &= 0. \end{aligned}$$

9. On the other hand, $W(t)$ is *not* differentiable. Specifically,

$$\left\langle \frac{|\Delta W|}{\Delta t} \right\rangle = \int_{-\infty}^{\infty} |\Delta W| \frac{1}{\sqrt{2\pi\Delta t}} e^{-\frac{(\Delta W)^2}{2\Delta t}} \, d\Delta W = \sqrt{\frac{2}{\pi \Delta t}},$$

which diverges as $\Delta t \to 0$. Intuitively we expect this result and the previous one by observing that the size of a Wiener increment $|\Delta W| \propto \sqrt{\Delta t}$ as $\Delta t \to 0$.

We now know how to generate particle trajectories whose behavior is consistent with the diffusion equation, by constructing steps in the trajectory that satisfy

$$\Delta x = \sqrt{2D}\Delta W. \tag{7.18}$$

Thus we might expect to be able to recover the inertialess Langevin equation for the evolution of particle position by dividing by $\Delta t$ and taking the limit $\Delta t \to 0$:

$$\frac{dx}{dt} = \sqrt{2D}\frac{dW}{dt},$$

where we now identity the random displacement with $\sqrt{2D}\frac{dW}{dt}$. But we just found that $\frac{dW}{dt}$ does not exist. Therefore, the more mathematically appropriate way to view the Langevin equation is as the *stochastic differential equation* (SDE)

$$dx = \sqrt{2D}dW, \tag{7.19}$$

where $dW$ is the differential that appears in (7.7).

Generalizing slightly, let

$$dx = A(x,t)\, dt + B(x,t)\, dW. \tag{7.20}$$

The first and second terms on the right-hand side are referred to as *drift* and *diffusion* terms, respectively, and we consider the general case where these terms can depend on the instantaneous value of $x$. If $B = 0$, then we can divide by $dt$ to recover a classical (deterministic) differential equation. Formally, we can integrate (7.20) to yield

$$x(t) = x(t_0) + \int_{t_0}^{t} A(x(t'),t')\, dt' + \int_{t_0}^{t} B(x(t'),t')\, dW(t'). \tag{7.21}$$

The first integral is classical. If $\frac{dW}{dt}$ existed, the second would be as well, and we would write that

$$\int_{t_0}^{t} B(x(t'),t')\, dW(t') = \int_{t_0}^{t} B(x(t'),t')\frac{dW}{dt}\, dt'.$$

However, since $\frac{dW}{dt}$ does not exist, this integral is nontrivial, and to understand it we need to learn a little bit about the calculus of stochastic processes.

## 7.2 Elementary Stochastic Calculus

We wish to understand how to evaluate integrals, like the second integral in (7.21), that have the form:

$$S = \int_{t_0}^{t} G(x(t'),t')dW(t').$$

As with conventional integrals, we divide the interval $[t_0, t]$ into $n$ subintervals defined by $t_0 \le t_1 \le t_2 ... \le t_{n-1} \le t$, and choose intermediate time points $\tau_i$ such that $t_{i-1} \le \tau_i \le t_i$. Now the integral $S$ is approximated by the sum

$$S_n = \sum_{i=1}^{n} G(x(\tau_i),\tau_i)(W(t_i) - W(t_{i-1})) = \sum_{i=1}^{n} G(x(\tau_i),\tau_i)\Delta W(\Delta t_i). \tag{7.22}$$

Now it remains to choose $\tau_i$. In normal calculus, $S_n$ converges to the same value independent of this choice, but in stochastic calculus this is not the case. We choose $\tau_i = t_{i-1}$, which defines the *Itô stochastic integral*. Specifically,

$$\int_{t_0}^{t} G(x(t'),t')dW(t') = \underset{n\to\infty}{\text{ms-lim}}\left(\sum_{i=1}^{n} G(x(t_{i-1}),t_{i-1})\Delta W(\Delta t_i)\right), \tag{7.23}$$

where ms-lim denotes mean square convergence: I.e., $\text{ms-lim}_{n\to\infty} S_n = S$ if

$$\lim_{n\to\infty}\left\langle (S_n - S)^2\right\rangle = 0.$$

Other stochastic integrals can be defined, the most common of which is the *Stratonovich* integral. One definition (Schuss 2010, Lasota & Mackey 1994) of this integral is

$$\int_{t_0}^{t} G(x(t'),t')d_S W(t') = \underset{n\to\infty}{\text{ms-lim}}\left(\sum_{i=1}^{n} G\left(x\left(\frac{t_{i-1}+t_i}{2}\right),t_{i-1}\right)\Delta W(\Delta t_i)\right). \tag{7.24}$$

Note that, in $G(x(t),t)$, the stochastic process $x(t)$ is evaluated at the midpoint of the interval: $\tau_i = (t_i + t_{i-1})/2$. Other equivalent forms of this integral can also be defined (Schuss 2010, Gardiner 1985). The rules of calculus with Itô and Stratonovich stochastic integrals are different, although given a stochastic differential equation based on Stratonovich integrals, one can determine the corresponding Itô SDE and vice versa (Schuss 2010). Gardiner (1985), Lasota & Mackey (1994), Öttinger (1996), and Schuss (2010) provide further discussion of Itô and Stratonovich integrals and stochastic calculus more generally.

We exclusively consider Itô calculus. Thus the integral in (7.21) is evaluated according to (7.23), and we call (7.20) an Itô stochastic differential equation. One practical reason for this choice is that it is the one most straightforwardly applied in numerical solutions of stochastic differential equations. The Itô stochastic integral corresponds to a stochastic "rectangle rule," with the function value chosen at the left side of the subinterval. The *Euler–Maruyama* scheme generalizes the explicit Euler method to the stochastic case, using this rectangle rule approximation. To advance (7.20) by one time step one would approximate (7.21) by

$$x(t+\Delta t) = x(t) + A(x(t),t)\Delta t + B(x(t),t)\Delta W(\Delta t). \tag{7.25}$$

This is the standard method for finding trajectories of (Itô) SDEs and is often referred to as *Brownian dynamics simulation*; it is not highly accurate, but higher-order schemes for SDEs are very complex to implement (Kloeden & Platen 1992).

A further reason for working with the Itô integral is that its expected value is zero:

$$\left\langle \int_{t_0}^{t} G(x(t'),t')\, dW(t')\right\rangle = 0, \tag{7.26}$$

or in differential form,

$$\langle G(x(t),t)\, dW(t)\rangle = 0. \tag{7.27}$$

In the context of (7.20), this property implies that the stochastic term does not change the expected value of $x$ – any "drift" or "bias" in the evolution of $x$ comes from the deterministic term $A\,dt$. To derive (7.26), consider the discrete sum (7.22) and use the facts that for the Itô integral, $G(\tau_i)$ and $\Delta W(\Delta t_i)$ are independent and that $\langle \Delta W(\Delta t_i)\rangle = 0$:

$$\begin{aligned}\langle S_n\rangle &= \sum_{i=1}^{n} \langle G(x(\tau_i), t_{i-1})\Delta W(\Delta t_i)\rangle \\ &= \sum_{i=1}^{n} \langle G(x(t_{i-1}), t_{i-1})\Delta W(\Delta t_i)\rangle \\ &= \sum_{i=1}^{n} \langle G(x(t_{i-1}), t_{i-1})\rangle\langle \Delta W(\Delta t_i)\rangle \\ &= 0.\end{aligned}$$

This calculation makes clear that the choice of $\tau_i$ matters: if $\tau_i > t_{i-1}$, then $G$ and $\Delta W = W(t_i) - W(t_{i-1})$ are generally not independent in the interval $t_{i-1} < t < \tau_i$ and thus $\langle S_n\rangle$ is not necessarily zero. To see a simple example of this point, let $G(x(t), t) = W(t)$ and recall (7.11). Hereafter, we simply denote $G(x(t), t)$ as $G(t)$, because for Itô integrals we do not need to distinguish the separate dependencies of $G$ on $x(t)$ and $t$.

For the special case where $G(t)$ is a deterministic square-integrable function $f(t)$, the Itô stochastic integral satisfies the following relation:

$$\int_0^t f(t')dW(t') \sim \mathcal{N}\left(0, \int_0^t f^2(t')dt'\right). \tag{7.28}$$

This expression follows from the definition of the Itô integral and (A.104): just as the sum of independent Gaussian random variables is Gaussian, so is the integral.

Now consider integrals of the form

$$\int G(t')[dW(t')]^{2+N} = \underset{n\to\infty}{\text{ms-lim}}\left(\sum_{i=1}^{n} G(t_{i-1})[\Delta W(\Delta t_i)]^{2+N}\right),$$

for integer $N$. In the limit,

$$[\Delta W]^2 \to dW^2 = dt, \qquad [\Delta W]^{2+N} \to dt\, dW^N = 0.$$

Equivalently,

$$\int_{t_0}^{t} G(t')[dW(t')]^{2+N} = \begin{cases} \int_{t_0}^{t} G(t')dt' & N = 0 \\ 0 & N > 0 \end{cases}.$$

If $dW_\alpha$ and $dW_\beta$ are different white-noise processes, e.g., corresponding to different components of a vector of such processes, then

$$dW_\alpha dW_\beta = \delta_{\alpha\beta}dt. \tag{7.29}$$

In summary, unlike in regular calculus, when working with differentials of $W$ in Itô calculus, one must keep terms up to $dW^2$. To understand why, simply recall that $\langle|\Delta W|\rangle \propto \sqrt{\Delta t}$.

An important consequence of the result $dW^2 = dt$ is the *Itô stochastic chain rule*. Let $F$ be a function of $t$ and $W(t)$. Expanding in differentials $dt$ and $dW$ yields that

$$dF(t, W(t)) = \left(\frac{\partial F}{\partial t} + \frac{1}{2}\frac{\partial^2 F}{\partial W^2}\right) dt + \frac{\partial F}{\partial W}\, dW(t). \tag{7.30}$$

The second term in the parentheses arises from the result $dW^2 = dt$; it is absent from the conventional chain rule.

As an example, let $F(t, W(t)) = W(t)^2 - t$, with $W(0) = 0$. This function is the difference of $W(t)^2$ from its expected value $t$. Here $\partial F/\partial t = -1, \partial F/\partial W = 2W, \partial^2 F/\partial W^2 = 2$, so

$$dF = 2W(t)dW(t). \tag{7.31}$$

Therefore,

$$F(t) = \int_0^t 2W(t')\, dW(t'). \tag{7.32}$$

But by definition, $F(t, W(t)) = W(t)^2 - t$, so this exercise reveals that

$$\int_0^t 2W(t')\, dW(t') = W(t)^2 - t, \tag{7.33}$$

which differs from the classical integral by the term $-t$.

Another important result of stochastic calculus comes from considering a function $f(x(t))$, where $x(t)$ evolves according to the Itô stochastic differential equation (7.20). The differential of $f$ can be written

$$\begin{aligned} df(x(t)) &= f(x(t + dt)) - f(x(t)) \\ &= f'(x(t))dx(t) + \frac{1}{2}f''(x(t))(dx(t))^2 \\ &= f'(x(t))(A\, dt + B\, dW) + \frac{1}{2}f''(x(t))(A\, dt + B\, dW)^2, \end{aligned}$$

where we have kept quadratic terms in $dt$ and $dW$. Now, noting that $dt^2 = 0$ and $dW^2 = dt$, we have *Itô's formula*:

$$df(x(t)) = (Af' + \frac{1}{2}B^2 f'')\, dt + Bf'\, dW. \tag{7.34}$$

As an application of Itô's formula, let $f(F(t)) = F(t)^2$, where $F(t)$ satisfies (7.31), which is an SDE with $A = 0, B = 2W(t)$. Now $f' = 2F, f'' = 2$ and Itô's formula evaluates to

$$df = 4W(t)^2 dt + 4W(t)F(t)dW(t).$$

The variance of $F(t)$ is $\langle f(t)\rangle$, which we can now evaluate from this expression and (7.27):

$$\begin{aligned} \langle df\rangle &= \left\langle 4W(t)^2 dt\right\rangle \\ &= 4t\, dt \end{aligned}$$

and therefore

$$\left\langle F(t)^2 \right\rangle = 2t^2.$$

## 7.3 Time Evolution of the Probability Density for a Stochastic Differential Equation

At the beginning of this chapter, we used the diffusion equation – an evolution equation for the probability distribution function of a diffusing particle – as a starting point for generating trajectories of a stochastic process, the Wiener process, which in turn led us to the concept of the Itô stochastic differential equation. Now we will use stochastic calculus to come full circle to the evolution equation for probability that corresponds to a given stochastic differential equation.

Applying Itô's formula to an arbitrary function $f(x(t))$, where $x$ satisfies (7.20), and taking the ensemble average yields

$$\begin{aligned} \langle df(x(t)) \rangle &= \left\langle (Af' + \frac{1}{2}B^2 f'')\, dt + Bf'\, dW \right\rangle \\ &= \left\langle Af' + \frac{1}{2}B^2 f'')\, dt \right\rangle. \end{aligned} \tag{7.35}$$

If $x$ can take on any real value, this can be rewritten as

$$\frac{d}{dt}\int_{-\infty}^{\infty} f(x)p(x,t)\, dx = \int_{-\infty}^{\infty} (Af' + \frac{1}{2}B^2 f'')p(x,t)\, dx, \tag{7.36}$$

where $p(x,t)$ is the probability density function for $x(t)$. This function must vanish as $|x| \to \infty$ because it is nonnegative and satisfies

$$\int_{-\infty}^{\infty} p(x,t)\, dx = 1. \tag{7.37}$$

Rearranging, integrating by parts, and applying the condition that $p(x)$ vanishes at infinity yields

$$\begin{aligned} \int_{-\infty}^{\infty} f(x)\frac{\partial p(x,t)}{\partial t}\, dx = \int_{-\infty}^{\infty} f(x)\Bigg(\frac{\partial}{\partial x}&(-Ap(x,t)) \\ &+ \frac{1}{2}\frac{\partial}{\partial x^2}\left(B^2 p(x,t)\right)\Bigg)\, dx. \end{aligned} \tag{7.38}$$

Finally, since $f$ is arbitrary, this result can only hold in general if

$$\frac{\partial p(x,t)}{\partial t} = \frac{\partial}{\partial x}(-A(x,t)p(x,t)) + \frac{\partial^2}{\partial x^2}\left(\frac{1}{2}B^2(x,t)p(x,t)\right). \tag{7.39}$$

This is the evolution equation for $p(x,t)$; for a trajectory starting at $x = x_0$, the initial condition is $p(x,0) = \delta(x - x_0)$. The same result holds if $x$ lives in a periodic domain such that $p(x+L,t) = p(x,t)$ for some value $L$. In the mathematics literature, (7.39) is called the *forward Kolmogorov equation* or the *Fokker–Planck equation* (FPE). In the physics literature, the latter term is often used to specify the particular physical context of the dynamics of a Brownian particle, including inertia (see Section 7.4.1), and the equation for the dynamics when inertia is neglected is called a Smoluchowski equation. Indeed, if $x$ is a component of the position of a Brownian particle, then (7.39) can be put in the same form as the Smoluchowski equation we derived from completely different, nonstochastic, considerations in Section 6.9.

The Fokker–Planck equation can be put into conservation form

$$\frac{\partial p(x,t)}{\partial t} = -\frac{\partial}{\partial x}\left\{A(x,t)p(x,t) - \frac{\partial}{\partial x}\left(\frac{1}{2}B^2(x,t)p(x,t)\right)\right\}, \tag{7.40}$$

where the term inside the curly brackets is the flux of probability density. One can show using Itô's formula (Problem 7.2) that for a particle at position $x'$ at time $t'$

$$\left\langle \frac{d(x-x')^2}{dt} \right\rangle_{t=t'} = 2D(x',t'), \tag{7.41}$$

where $D(x,t) = B^2(x,t)/2$ is called the *short-time* diffusivity, because it characterizes the intantaneous rate of change of the mean squared displacement. Similarly, the instantaneous drift velocity of the trajectory is (as in the deterministic case)

$$\left\langle \frac{d(x-x')}{dt} \right\rangle_{t=t'} = A(x',t'). \tag{7.42}$$

Finally, it can be shown that any continuous Markov process that satisfies (7.41) and (7.42) is governed by (7.39) (Gardiner 1985).

We also can generalize the preceding analysis to an $n$-vector random process $\boldsymbol{x}(t)$, with components $x_i$, $i = 1, 2, \ldots, n$. The SDE for this process is

$$dx_i = A_i(\boldsymbol{x},t)dt + B_{ij}(\boldsymbol{x},t)\, dW_j, \tag{7.43}$$

where $dW_1, dW_2, \ldots$ are all independent white noise processes: $dW_i dW_j = dt\delta_{ij}$. The corresponding *multidimensional Itô formula* is

$$df(\boldsymbol{x}) = \left(A_i\frac{\partial f}{\partial x_i} + \frac{1}{2}B_{ik}B_{jk}\frac{\partial}{\partial x_i}\frac{\partial}{\partial x_j}f\right)\, dt + B_{ij}\frac{\partial f}{\partial x_i}dW_j. \tag{7.44}$$

Repeating the derivation of the single-variable Fokker–Planck equation, (7.35)–(7.39), but now using (7.44) and the expected value definition

$$\langle f(\boldsymbol{x},t)\rangle = \int f(\boldsymbol{x},t)p(\boldsymbol{x})\, dx_1\, dx_2 \ldots dx_n \tag{7.45}$$

yields the multivariable Fokker–Planck equation for the probability density $p(\boldsymbol{x},t)$:

$$\frac{\partial}{\partial t}p(\boldsymbol{x},t) = -\frac{\partial}{\partial x_i}(A_i(\boldsymbol{x},t)p(\boldsymbol{x},t)) + \frac{\partial^2}{\partial x_i \partial x_j}(D_{ij}(\boldsymbol{x},t)p(\boldsymbol{x},t)). \tag{7.46}$$

Here $D_{ij} = \frac{1}{2}B_{ik}B_{jk}$ are the elements of the diffusion coefficient *matrix*, which is symmetric positive definite by construction. In vector/matrix notation, the equations are written

$$d\boldsymbol{x} = \boldsymbol{A}(\boldsymbol{x},t)dt + \mathbf{B}(\boldsymbol{x},t)\cdot d\boldsymbol{W}, \tag{7.47}$$

$$\frac{\partial}{\partial t}p(\boldsymbol{x},t) = -\frac{\partial}{\partial \boldsymbol{x}} \cdot (\boldsymbol{A}(\boldsymbol{x},t)p(\boldsymbol{x},t)) + \frac{\partial}{\partial \boldsymbol{x}}\frac{\partial}{\partial \boldsymbol{x}} : (\mathbf{D}(\boldsymbol{x},t)p(\boldsymbol{x},t)), \tag{7.48}$$

with

$$\mathbf{D}(\boldsymbol{x},t) = \frac{1}{2}\mathbf{B} \cdot \mathbf{B}^{\mathrm{T}}. \tag{7.49}$$

As in the single-variable case, this result holds when the domain for each component of $\boldsymbol{x}$ is unbounded or periodic. In Sections 7.5, 7.6 and, A.5, we address Fokker–Planck equations in cylindrical and spherical coordinates.

For numerical integration of multidimensional SDEs, the Euler–Maruyama scheme extends straightforwardly:

$$\boldsymbol{x}(t+\Delta t) = \boldsymbol{x}(t) + \boldsymbol{A}(\boldsymbol{x}(t),t)\Delta t + \mathbf{B}(\boldsymbol{x}(t),t) \cdot \Delta \boldsymbol{W}(\Delta t), \tag{7.50}$$

where $\Delta \boldsymbol{W}(\Delta t)$ is a vector of independent Wiener increments.

## 7.4 The Langevin Equation Revisited

### 7.4.1 Inertial Dynamics

With the preceding framework for stochastic processes in hand, we now return to the Langevin equation (6.6) for the motion of a Brownian particle, restricting attention for the moment to one component of the motion:

$$\frac{dU}{dt} = -\frac{\zeta(R)}{m}(U - v(R)) + \frac{1}{m}F_{\text{ext}} + \frac{1}{m}F_{\text{fluc}}, \tag{7.51}$$

$$\frac{dR}{dt} = U, \tag{7.52}$$

with

$$\langle F_{\text{fluc}}(s)F_{\text{fluc}}(s')\rangle = 2\zeta(R)k_BT\delta(s-s'). \tag{7.53}$$

In Chapter 6, we did not specify anything beyond the mean and variance of $F_{\text{fluc}}$, using the result $\frac{1}{2}m\langle \boldsymbol{U}\boldsymbol{U}\rangle = \frac{1}{2}k_BT\boldsymbol{\delta}$. As a result, although we showed that the Brownian particle would have the correct kinetic energy at equilibrium, we did not force it to satisfy the Maxwell–Boltzmann distribution

$$p_{\text{eq}}(U,R) = p_0 \exp\left(\frac{-\frac{1}{2}mU^2 - \Phi_{\text{ext}}(R)}{k_BT}\right). \tag{7.54}$$

The proper formulation of (7.51)–(7.52) will have to satisfy this requirement. For the moment, we convert the Langevin equations into a pair of stochastic differential equations by multiplying by $dt$ and replacing $F_{\text{fluc}}/m\ dt$ with $B\ dW$: i.e., we are assuming that treating the fluctuating force as a Wiener differential will lead to the correct distribution. Leaving $B$ arbitrary as yet, we can write these equations in the form of (7.47) with $\boldsymbol{x} = [U, R]^{\mathrm{T}}$ and

$$\boldsymbol{A} = \begin{bmatrix} -\frac{\zeta(R)}{m}(U - v(R)) + \frac{1}{m}F_{\text{ext}} \\ U \end{bmatrix}, \quad \mathbf{B} = \begin{bmatrix} B & 0 \\ 0 & 0 \end{bmatrix}.$$

The forward Kolmogorov equation, (7.46) (which is usually called a Fokker–Planck equation in the present context), thus becomes

$$\frac{\partial p(U,R,t)}{\partial t} = -\frac{\partial}{\partial U}\left(\left(-\frac{\zeta(R)}{m}(U - v(R)) + \frac{1}{m}F_{\text{ext}}\right)p(U,R,t)\right) - \frac{\partial}{\partial R}(Up(U,R,t)) + \frac{1}{2}\frac{\partial^2}{\partial U^2}\left(B^2 P(U,R,t)\right). \tag{7.55}$$

At equilibrium, $v = 0$ and the probability distribution will have Maxwell–Boltzmann form as long as

$$B = \sqrt{\frac{2\zeta(R)k_B T}{m^2}}. \tag{7.56}$$

This result is the fluctuation-dissipation theorem in another form and can also be seen by recasting (7.53) in terms of Wiener increments:

$$\left\langle m^2 B^2 dW(dt_i) dW(dt_j)\right\rangle = 2\zeta k_B T dt \delta_{ij}. \tag{7.57}$$

Therefore, we see that the Wiener differential provides the correct way to treat the fluctuating force in Langevin equations. Finally, in SDE form the equations of motion for a Brownian particle are

$$dU = \left(-\frac{\zeta(R)}{m}(U - v(R)) + \frac{1}{m}F_{\text{ext}}\right)dt + \sqrt{\frac{2\zeta(R)k_B T}{m^2}}dW, \tag{7.58}$$

$$dR = U\,dt. \tag{7.59}$$

The Euler–Maruyama scheme is straightforwardly applied to (7.58):

$$U(t + \Delta t) = U(t) + \left(-\frac{\zeta(R(t))}{m}(U(t) - v(R(t))) + \frac{1}{m}F_{\text{ext}}(R(t),t)\right)\Delta t + \sqrt{\frac{2\zeta(R(t)k_B T}{m^2}}\Delta W(\Delta t), \tag{7.60}$$

$$R(t + \Delta t) = R(t) + U(t)\Delta t. \tag{7.61}$$

For accuracy on the time scale of the inertial-viscous relaxation, the time step must satisfy $\lambda_v \Delta t \ll 1$, where $\lambda_v = \zeta/m$, and even for numerical stability the condition $\lambda_v \Delta t < 2$ is required. These are fairly restrictive conditions, given that in many problems the dynamics on the time scale $\lambda_v^{-1}$ are not of interest. In this situation, the inertialess approach described in Sections 6.7 and 7.4.2 is highly advantageous because it does not require resolution of this time scale. Sometimes the stochastic simulation of the Langevin equation including inertia is called *Langevin dynamics* simulation, with the term *Brownian dynamics* indicating simulation of the inertialess formulation.

For a system of $N$ identical point particles in three dimensions, the preceding results generalize fairly naturally. Recalling the notation introduced in Section 2.5 and at the same level of description, the equation of motion becomes

$$d\mathbf{U} = \frac{1}{m}\left(\mathbf{M}(\mathbf{R})\cdot(\mathbf{U} - \mathbf{V}_\infty) + \mathbf{F}_{\text{ext}}\right)dt + \frac{1}{m}\mathbf{C}(\mathbf{R})\cdot\Delta\mathbf{W},$$

$$d\mathbf{R} = \mathbf{U}\,dt, \tag{7.62}$$

where

$$\mathbf{C} \cdot \mathbf{C}^{\mathrm{T}} = 2k_B T \mathbf{M}^{-1} \tag{7.63}$$

is the SDE version of (6.24). The primary subtlety in stochastic simulations of this system is that in general (7.63) must be solved for $\mathbf{C}$ at each time step, because $\mathbf{M}$ depends on $\mathbf{R}$. This can be done, for example, by eigenvalue decomposition: $\mathbf{M}^{-1} = \mathbf{Q}\boldsymbol{\Lambda}\mathbf{Q}^{\mathrm{T}}$, where $\boldsymbol{\Lambda} = \mathrm{diag}(\lambda_1, \lambda_2, \ldots, \lambda_{3N})$. Then we can take

$$\mathbf{C} = \sqrt{2k_B T}\mathbf{Q}\boldsymbol{\Lambda}^{1/2}\mathbf{Q}^{\mathrm{T}}, \tag{7.64}$$

where $\boldsymbol{\Lambda}^{1/2} = \mathrm{diag}(\lambda_1^{1/2}, \lambda_2^{1/2}, \ldots, \lambda_{3N}^{1/2})$. Another slightly faster approach uses Cholesky decomposition (the symmetric version of LU decomposition for factorizing a matrix): $\mathbf{M}^{-1} = \mathbf{L} \cdot \mathbf{L}^{\mathrm{T}}$, and thus

$$\mathbf{C} = \sqrt{2k_B T}\mathbf{L}. \tag{7.65}$$

For large systems, this step of the simulation can be the most time consuming because, in general, both the eigenvalue and Cholesky decompositions require $O(N^3)$ operations. There are, however, highly accurate approximate methods for more rapidly evaluating these terms (Fixman 1986, Jendrejack, Graham & de Pablo 2000, Ando et al. 2012, Fiore et al. 2017). In a nutshell, because the operation of $\mathbf{M}$ on a vector is equivalent to the solution of a Stokes problem, if we have a rapid Stokes solver for a particular flow geometry, then we can do rapid Brownian dynamics simulations for systems of point particles in that geometry (Hernández-Ortiz et al. 2007).

### 7.4.2 From Inertial to Inertialess: The High Friction Limit

For processes that are slow compared to the inertial-viscous relaxation time, we have argued that the velocity distribution for a Brownian system will stay very close to Maxwellian and thus it is only necessary to keep track of the particle positions. We begin this section with derivations of these results, focusing for clarity on a single particle in one dimension and taking $\zeta$ to be independent of position (Schuss 2010).

The inertial problem, (7.58)–(7.59), can be be integrated to yield

$$\begin{aligned} U(t) &= U(0)e^{-\lambda_v t} \\ &+ \int_0^t e^{-\lambda_v(t-s)}\left(\left(\lambda_v v(R(s)) + \frac{1}{m}F_{\mathrm{ext}}(R(s))\right)ds + \sqrt{\frac{2k_B T}{m}\lambda_v}dW(s)\right), \end{aligned} \tag{7.66}$$

$$\begin{aligned} R(t) &= R(0) + \int_0^t U(s')ds' \\ &= R(0) + \frac{U(0)}{\lambda_v}\left(1 - e^{-\lambda_v t}\right) \\ &+ \frac{1}{\lambda_v}\int_0^t \left(1 - e^{-\lambda_v(t-s)}\right)\left(\left(\lambda_v v(R(s)) + \frac{1}{m}F_{\mathrm{ext}}(R(s))\right) ds + \sqrt{\frac{2k_B T}{m}\lambda_v}dW(s)\right). \end{aligned} \tag{7.67}$$

In Section 6.7, the limit $\lambda_v^{-1} = m/\zeta \to 0$ was addressed by naively setting the particle mass $m$ to zero. Alternatively, one could take $\zeta$ to infinity, so this limit is sometimes called the high-friction limit. But what is most appropriate is to take this limit by considering a time scale *ratio* between $\lambda_v^{-1}$, which characterizes the relaxation of the particle velocity, and the much slower time scales that characterize the dynamics of the paricle position.

Consider a characteristic time scale $t_c$ for the particle, which is much larger than $\lambda_v^{-1}$ but much smaller than the slow processes of particle motion due to diffusion, flow, or an external force. Defining a small parameter $\epsilon = (\lambda_v t_c)^{-1}$, we will rescale time by introducing a dimensionless slow time scale $\tau = \epsilon t/t_c$ (and dimensionless dummy variables $\sigma = \epsilon s/t_c$ and $\sigma' = \epsilon s'/t_c$); now $\tau$ changes by an amount of $O(\epsilon)$ when $t$ changes by $t_c$ and an amount of $O(\epsilon^2)$ when $t$ changes by $\lambda_v^{-1}$. To write the fluctuating term in the new time variable, we recall the scaling property of the Wiener process, (7.12), and write that

$$dW(s) = dW\left(\frac{\sigma t_c}{\epsilon}\right) = \left(\frac{t_c}{\epsilon}\right)^{1/2} dW_\epsilon(\sigma), \tag{7.68}$$

where $W_\epsilon(\sigma)$ is also a Wiener process. With these scalings, the Stokes drag and fluctuating forces balance on the dimensionless time scale $\epsilon^2$ and everything else happens much more slowly.

First we address the evolution of the velocity. Assume that $v$ varies weakly enough with position that we can treat it as a constant on the time scale $\lambda_v^{-1}$ (dimensionless time scale $\epsilon^2$). In this case, we can move $v$ outside the integral in (7.66) and write an equation for $V = U - v$:

$$V(t) = V(0)e^{-\lambda_v t} + \int_0^t e^{-\lambda_v(t-s)}\left(\frac{1}{m}F_{\text{ext}}\,ds + \sqrt{\frac{2k_BT}{m}\lambda_v}\,dW(s)\right). \tag{7.69}$$

Using the scaled variables, this becomes

$$V(\tau) = V(0)e^{-\tau/\epsilon^2} + \int_0^\tau e^{-(\tau-\sigma)/\epsilon^2}\left(\frac{t_c}{m\epsilon}F_{\text{ext}}\,d\sigma + \sqrt{\frac{2k_BT}{m\epsilon^2}}\,dW_\epsilon(\sigma)\right). \tag{7.70}$$

Now we nondimensionalize the velocity $V$ with the thermal velocity, defining $\Upsilon = V/\sqrt{k_BT/m}$ and rewriting (7.70) as

$$\Upsilon(\tau) = \Upsilon(0)e^{-\tau/\epsilon^2} + \int_0^\tau e^{-(\tau-\sigma)/\epsilon^2}\left(\frac{t_c}{\epsilon\sqrt{k_BTm}}F_{\text{ext}}\,d\sigma + \sqrt{\frac{2}{\epsilon^2}}\,dW_\epsilon(\sigma)\right). \tag{7.71}$$

Consider the integral containing $F_{\text{ext}}$. This can be treated as follows:

$$\begin{aligned}
\left|\int_0^\tau e^{-(\tau-\sigma)/\epsilon^2}\frac{t_c}{\epsilon\sqrt{k_BTm}}F_{\text{ext}}\,d\sigma\right| &\le \frac{t_c}{\epsilon\sqrt{k_BTm}}\max_{0\le\sigma\le\tau}|F_{\text{ext}}(\sigma)|\int_0^\tau e^{-(\tau-\sigma)/\epsilon^2}\,d\sigma \\
&\le \frac{t_c}{\sqrt{k_BTm}}\max_{0\le\sigma\le\tau}|F_{\text{ext}}(\sigma)|\,\epsilon\left(1-e^{-\tau/\epsilon^2}\right) \\
&\le \left(\frac{\max_{0\le\sigma\le\tau}|F_{\text{ext}}(\sigma)|}{\zeta}\right)\left(\frac{k_BT}{m}\right)^{-1/2}.
\end{aligned}$$

The second line reveals that this term is $O(\epsilon)$ as long its prefactor is not too large. The third line shows what this means physically: this term is small if the Stokes settling velocity due to $F_{\text{ext}}$ is much less than the thermal velocity.

The stochastic integral in (7.71) has a deterministic integrand (i.e., one that does not depend on $W_\epsilon(s)$), so from (7.28) we have that

$$\int_0^\tau e^{-(\tau-\sigma)/\epsilon^2}\sqrt{\frac{2}{\epsilon^2}}\,dW_\epsilon(\sigma) \sim \mathcal{N}\left(0, \int_0^\tau \left(e^{-(\tau-\sigma)/\epsilon^2}\sqrt{\frac{2}{\epsilon^2}}\right)^2 d\sigma\right)$$
$$\sim \mathcal{N}(0,1).$$

As $\epsilon \to 0$, the integral in (7.71) will be dominated by this term because the term containing $F_{\text{ext}}$ is $O(\epsilon)$. Furthermore, in this limit the term $\Upsilon(0)e^{-\tau/\epsilon^2} \to 0$ for any finite $\tau$. Thus (7.70) becomes

$$V(\tau) \sim \mathcal{N}\left(0, \frac{k_B T}{m}\right) \tag{7.72}$$

or

$$U(\tau) \sim \mathcal{N}\left(v, \frac{k_B T}{m}\right). \tag{7.73}$$

This is the Maxwellian velocity distribution centered on the local fluid velocity $v$.

Now we turn to the evolution of the particle position. As before, we take $v$ to vary with position weakly enough that we can treat it as constant on the time scale $\lambda_{\text{v}}^{-1}$. We also define a new position variable $X$ such that

$$\frac{dX}{dt} = V = U - v = \frac{dR}{dt} - v \tag{7.74}$$

and rewrite the evolution equation for $R$ in terms of $X$. In the rescaled variables, and replacing $\lambda_{\text{v}}$ by $(\epsilon t_{\text{c}})^{-1}$, (7.67) becomes

$$X(\tau) = X(0) + \epsilon t_{\text{c}} U(0)\left(1 - e^{-\tau/\epsilon^2}\right)$$
$$+ \int_0^\tau \left(1 - e^{-(\tau-\sigma)/\epsilon^2}\right)\left(\frac{t_{\text{c}}^2}{m} F_{\text{ext}}(R(\sigma))\,d\sigma + t_{\text{c}}\sqrt{\frac{2k_B T}{m}}\,dW_\epsilon(\sigma)\right).$$

As $\epsilon \to 0$, $1 - e^{-(\tau-\sigma)/\epsilon^2} \to 1$ for all $\sigma < \tau$, so this equation becomes

$$X(\tau) = X(0) + \int_0^\tau \left(\frac{t_{\text{c}}^2}{m} F_{\text{ext}}(R(\sigma))\,d\sigma + t_{\text{c}}\sqrt{\frac{2k_B T}{m}}\,dW_\epsilon(\sigma)\right).$$

In differential form, this is

$$dX(\tau) = \frac{t_{\text{c}}^2}{m} F_{\text{ext}}(R(\tau))\,d\tau + t_{\text{c}}\sqrt{\frac{2k_B T}{m}}\,dW_\epsilon(\tau).$$

Reverting back to the original variables and applying the Wiener process scaling relation, (7.68), in reverse yields

$$dR(t) = \left(v(R(t)) + \frac{1}{\zeta} F_{\text{ext}}(R(t))\right) dt + \sqrt{\frac{2k_B T}{\zeta}}\,dW(t).$$

This is the SDE for a Brownian particle with constant $\zeta$ in the inertialess (high-friction) limit. See Ermak & McCammon (1978) for a similar derivation that does not assume fixed $\zeta$. Recall from Section 6.7 that if $\zeta$ depends on $R$, then an additional term appears:

$$dR(t) = \left(v(R(t)) + \frac{1}{\zeta(R(t))}F_{\text{ext}}(R(t)) + \frac{dD(R(t))}{dR}\right)dt + \sqrt{2D(R(t))}\,dW(t), \tag{7.75}$$

where here we have applied the Stokes–Einstein relation $D = k_BT/\zeta$. This is equivalent to (6.82). The corresponding Smoluchowski equation,

$$\frac{\partial p}{\partial t} = -\frac{\partial}{\partial R}\left(\left(v + \frac{1}{\zeta(R)}F_{\text{ext}}\right)p\right) + \frac{\partial}{\partial R}D(R)\frac{\partial p}{\partial R}, \tag{7.76}$$

can also be derived using asymptotic methods from the Fokker–Planck equation (7.55) (Bocquet 2004, Schuss 2010). This equation differs in form from (7.48) in that the diffusivity is only under one derivative; in standard form for stochastic processes, it becomes

$$\frac{\partial p}{\partial t} = -\frac{\partial}{\partial R}\left(\left(v + \frac{1}{\zeta(R)}F_{\text{ext}} + \frac{dD(R)}{dR}\right)p\right) + \frac{\partial^2}{\partial R}D(R)p. \tag{7.77}$$

For the $N$-particle case, the inertialess (high-friction limit) SDE is

$$d\mathbf{R} = \left(\mathbf{V}_\infty + \mathbf{M}\cdot\mathbf{F}_{\text{ext}} + k_BT\frac{\partial}{\partial\mathbf{R}}\cdot\mathbf{M}\right)dt + \mathbf{B}\cdot d\mathbf{W}, \tag{7.78}$$

with

$$\mathbf{B}\cdot\mathbf{B}^{\text{T}} = 2k_BT\mathbf{M}. \tag{7.79}$$

The corresponding Smoluchowski equation is (6.100). Stochastic simulations with the inertialess Brownian formulation can be performed directly with the discretized form of (7.78):

$$\mathbf{R}(t+\Delta t) = \mathbf{R}(t) + \left(\mathbf{V}_\infty(t) + \mathbf{M}\cdot\mathbf{F}_{\text{ext}}(t) + k_BT\frac{\partial}{\partial\mathbf{R}}\cdot\mathbf{M}\right)\Delta t + \mathbf{B}(\mathbf{R})\cdot\Delta\mathbf{W}, \tag{7.80}$$

with

$$\mathbf{B}\cdot\mathbf{B}^{\text{T}} = 2k_BT\mathbf{M}. \tag{7.81}$$

This is sometimes called the Ermak–McCammon algorithm (Ermak & McCammon 1978). It is quite straightforward but has the potential disadvantage that it requires the computation of $\frac{\partial}{\partial\mathbf{R}}\cdot\mathbf{M}$ at each time step. In some important cases, this quantity vanishes, for example when $\mathbf{M}$ describes interactions between point particles or regularized point particles in an unbounded domain. This result follows from the divergence-free nature of the Green's function. For a confined system, however, $\frac{\partial}{\partial\mathbf{R}}\cdot\mathbf{M}$ does not vanish (Jendrejack et al. 2003) and its computation can be cumbersome.

To address this issue, Fixman (1978) gave a method that avoids computing this term. We present here a variant of this approach that is simpler to understand, less expensive to implement, and equivalent to the Fixman approach as $\Delta t \to 0$. We begin in one dimension, (7.75) with position-dependent $D$ (and thus $B$, since $B = \sqrt{2D}$). An Euler step is simply:

$$R(t + \Delta t) = R(t) + \left(v + \frac{1}{\zeta}F_{\text{ext}} + \frac{dD}{dR}\right)\Delta t + B\,\Delta W,$$

where the right-hand side is evaluated at time $t$. We can approximate the derivative with a forward difference:

$$\Delta R \approx (v + \frac{1}{\zeta}F_{\text{ext}} + \frac{D(R + \Delta R) - D(R)}{\Delta R})\,\Delta t + B\,\Delta W.$$

This is now of course an implicit method, because $D(R + \Delta R)$ must be evaluated. To do so, an intermediate step can be defined:

$$\Delta R^* = R^* - R(t) = B(R(t))\Delta W.$$

This is just a Brownian dynamics step without the drift term. Using this equation, the derivative term is then estimated as:

$$\frac{D(R(t) + \Delta R) - D(R(t))}{\Delta R(t)}\Delta t \approx \frac{D(R^*) - D(R(t))}{\Delta R^*}\Delta W^2 = \frac{D(R^*) - D(R(t))}{B(R(t))\Delta W}\Delta W^2,$$

where we have also used the approximation $\Delta t \approx \Delta W^2$. The final expression for a time step is then

$$R(t + \Delta t) = R(t) + \left(v + \frac{1}{\zeta}F_{\text{ext}}\right)\Delta t + \left(B(R(t)) + \frac{D(R^*) - D(R(t))}{B(R(t))}\right)\Delta W. \quad (7.82)$$

To confirm the validity of this approach, we work backward from the discrete version of the last term to the continuous:

$$\begin{aligned}
\frac{D(R^*) - D(R(t))}{B(R(t))}\Delta W &= \frac{D(R(t)) + \frac{dD}{dR}\left(\Delta R^* + O(\Delta R^{*2})\right) - D(R(t))}{B(R(t))}\Delta W \\
&= \frac{\frac{dD}{dR}\,(B(R(t))\Delta W + O(\Delta t))}{B(R(t))}\Delta W \\
&= \frac{dD}{dR}\Delta W^2 + O(\Delta W \Delta t) \\
&= \frac{dD}{dR}dW^2 = \frac{dD}{dR}dt \quad \text{as } \Delta t \to 0.
\end{aligned}$$

The multiparticle version of (7.82), with $\mathbf{D} = k_B T\mathbf{M} = \frac{1}{2}\mathbf{B}\cdot\mathbf{B}^{\mathrm{T}}$ is given by

$$\begin{aligned}
\mathbf{R}(t + \Delta t) &= \mathbf{R}(t) + (\mathbf{V}_\infty(t) + \mathbf{M}\cdot\mathbf{F}_{\text{ext}}(t))\,\Delta t \\
&\quad + \left(\mathbf{B}(\mathbf{R}(t)) + (\mathbf{D}(\mathbf{R}^*) - \mathbf{D}(\mathbf{R}(t)))\cdot\mathbf{B}^{-1^{\mathrm{T}}}(\mathbf{R}(t))\right)\cdot\Delta\mathbf{W}, \quad (7.83)
\end{aligned}$$

where $\mathbf{D}(\mathbf{R}^*)$ is evaluated at

$$\mathbf{R}^* = \mathbf{R}(t) + \mathbf{B}(\mathbf{R}(t))\cdot\Delta\mathbf{W}. \quad (7.84)$$

The only difference between this algorithm and that of Fixman (1978) is that he chooses

$$\mathbf{R}^* = \mathbf{R}(t) + (\mathbf{V}_\infty(t) + \mathbf{M}\cdot\mathbf{F}_{\text{ext}}(t))\,\Delta t + \mathbf{B}(\mathbf{R}(t))\cdot\Delta\mathbf{W}. \quad (7.85)$$

As $\Delta t \to 0$ both algorithms reduce to (7.78). At each time step, evaluation of $\mathbf{B}$ can be accomplished as described earlier for $\mathbf{C}$ in inertial Brownian dynamics.

An even simpler variant of Fixman's approach is the so-called *random finite difference* method (Balboa Usabiaga, Delmotte & Donev 2017). Here the following approximation is used:

$$k_BT\frac{\partial}{\partial \mathbf{R}} \cdot \mathbf{M}\,\Delta t \approx \frac{k_BT\,\Delta t}{\delta}\left(\mathbf{M}\left(\mathbf{R}(t) + \frac{\delta}{2}\mathbf{X}\right) - \mathbf{M}\left(\mathbf{R}(t) - \frac{\delta}{2}\mathbf{X}\right)\right) \cdot \mathbf{X},$$

where $\delta \ll 1$ and $\mathbf{X}$ is a $3N$-vector of independent Gaussian random variables with unit variance: $\langle X_i X_j \rangle = \delta_{ij}$. At each time step, a new (independent) realization $\mathbf{X}$ is used. By Taylor-expanding in $\delta$, one can see that the expected value of the right-hand side equals the left-hand side as $\delta \to 0$.

These methodologies and generalizations of them are widely used to study diffusion and rheology of suspensions, dilute polymer solutions, and other complex fluids. Of particular note is the incorporation of Brownian motion into the Stokesian dynamics methodology (Banchio & Brady 2003).

## 7.5 Coordinate Transformations and Constraints

To the extent that our formulations have relied on specific coordinate systems, we have generally focused on Cartesian descriptions. In this section, we illustrate through a number of examples the process of changing coordinate systems in the description of stochastic processes and the closely related topic of the description of processes with geometric constraints. In the latter case, the Cartesian description is not necessarily the most natural for formulating or understanding the problem but may be the most straightforward for implementation of a stochastic simulation algorithm.

### 7.5.1 Diffusion on a Plane in Cartesian and Polar Coordinate Systems

Neglecting inertia, a Brownian particle diffusing in the $x - y$ plane with constant diffusivity $D$ moves according to

$$dx = B\,dW_x, \tag{7.86}$$

$$dy = B\,dW_y, \tag{7.87}$$

where $W_x$ and $W_y$ are independent Wiener processes and $B = \sqrt{2D}$. A sample trajectory of this random walk is shown in Figure 7.2. If the particle starts at the origin, how do its radial and angular position evolve with time?

Considering first the radial position, we apply the multidimensional Itô formula, (7.44):

$$dr(x,y) = \frac{1}{2}B^2\left(\frac{\partial^2 r}{\partial x^2} + \frac{\partial^2 r}{\partial y^2}\right)dt + B\left(\frac{\partial r}{\partial x}dW_x + \frac{\partial r}{\partial y}dW_y\right).$$

(This is equivalent to computing $dr(x, y)$ up to quadratic terms in $dx$ and $dy$, then

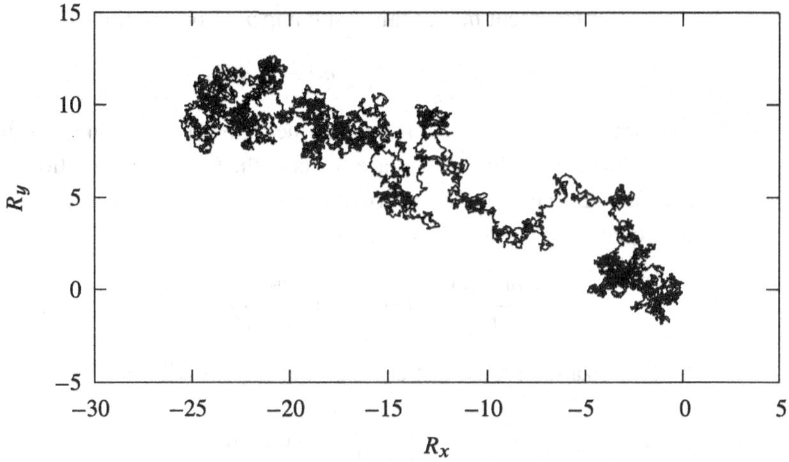


**Figure 7.2** A two-dimensional Brownian random talk of 100 time units with $B = 1$ and initial condition $R_x(0) = R_y(0) = 0$.

applying (7.86)–(7.87).) Here all the partials can be evaluated from the formulas $r = \sqrt{x^2 + y^2}$ and $\theta = \tan^{-1}\left(\frac{y}{x}\right)$, yielding

$$dr = \cos\theta B\, dW_x + \sin\theta B\, dW_y + \frac{1}{2r}B^2\, dt.$$

Now, using Eqs. (7.16)–(7.17) we note that $\cos\theta\, dW_x + \sin\theta\, dW_y$ is a diffusion process with variance $dt$. We will denote this process as $dW_r$, so

$$dr = \frac{B^2}{2r}\, dt + B\, dW_r. \tag{7.88}$$

Thus the evolution of $r$ is independent of $\theta$ (as it must be, by isotropy). To find the evolution equation for $r^2$, we can use Itô's formula, (7.34):

$$dr^2 = 2B^2\, dt + 2Br\, dW_r.$$

Ensemble averaging yields

$$\left\langle dr^2 \right\rangle = 2B^2\, dt.$$

Since $B^2 = 2D$, we find that $\left\langle r^2 \right\rangle = 4Dt$.

Similarly, applying (7.44) to angular position we have

$$\begin{aligned} d\theta(x, y) &= \frac{1}{2}B^2\left(\frac{\partial^2\theta}{\partial x^2} + \frac{\partial^2\theta}{\partial y^2}\right) dt + B\left(\frac{\partial\theta}{\partial x}dW_x + \frac{\partial\theta}{\partial y}dW_y\right) \\ &= \frac{B}{r}(-\sin\theta\, dW_x + \cos\theta\, dW_y). \end{aligned}$$

Using Eqs. (7.16)–(7.17) again we can replace $-\sin\theta\, dW_x + \cos\theta\, dW_y$ with $dW_\theta$:

$$d\theta = \frac{B}{r}\, dW_\theta. \tag{7.89}$$

A slight rearrangement makes this result appear more intuitive:

$$r\,d\theta = B\,dW_\theta. \tag{7.90}$$

In this form, we see that the differential linear displacement in the $\theta$-direction, $r\,d\theta$, has the same form as the linear displacements in the Cartesian directions.

The Fokker–Planck equation corresponding to Eqs. (7.86)–(7.87) is the familiar diffusion equation in Cartesian coordinates:

$$\frac{\partial p(x,y)}{\partial t} = D\left(\frac{\partial^2}{\partial x^2} + \frac{\partial^2}{\partial y^2}\right)p(x,y) = D\nabla^2 p(x,y) \tag{7.91}$$

with normalization (conservation of probability) condition

$$\int_{-\infty}^{\infty}\int_{-\infty}^{\infty} p(x,y)\,dx\,dy = 1. \tag{7.92}$$

This result can be found by direct application of (7.46). In polar coordinates, and using the conventional definition of area element $r\,d\theta\,dr$, these equations become

$$\frac{\partial p(r,\theta)}{\partial t} = D\left(\frac{1}{r}\frac{\partial}{\partial r}\left(r\frac{\partial}{\partial r}\right) + \frac{1}{r^2}\frac{\partial^2}{\partial \theta^2}\right)p(r,\theta) = D\nabla^2 p(r,\theta) \tag{7.93}$$

and

$$\int_0^{2\pi}\int_0^{\infty} p(r,\theta) r\,dr\,d\theta = 1. \tag{7.94}$$

To arrive at the polar coordinate FPE (7.93) starting from the polar coordinate SDEs (7.88) and (7.89) requires a slight modification of the development shown in Section 7.3 (Graham & Rawlings 2013). Specifically, in taking the ensemble average of Itô's formula, we must use the appropriate definition of expected value for a function on the plane in polar coordinates:

$$\langle f(r,\theta,t)\rangle = \int_0^{2\pi}\int_0^{\infty} f(r,\theta,t)p(r,\theta)J(r,\theta)\,dr\,d\theta, \tag{7.95}$$

where $J(r,\theta) = r$ is the Jacobian determinant for the area element in polar coordinates. Because it does not incorporate $J(\boldsymbol{x})$ into the definition of ensemble average, one can view the derivation of (7.46) as giving the evolution equation for $p(\boldsymbol{x},t)J(\boldsymbol{x})$ rather than $p(\boldsymbol{x},t)$. See Section A.5 and Problem 7.7 for further discussion of this point.

### 7.5.2 Stochastic Processes with Constraints: Diffusion on a Circle or Sphere

Constraints arise naturally in modeling motion of rigid objects or assemblages thereof. For example, consider the rotational Brownian fluctuations of an axisymmetric particle whose axis is defined by a unit vector $\boldsymbol{u}$, a situation that will be further described in Sections 7.6 and 8.7. Here we will consider motion restricted to a plane, where the tip of $\boldsymbol{u}$ lies on the unit circle. One way to treat this constraint is to work in polar coordinates – the angular variable $\theta$ fully defines the rod orientation. Alternatively, the evolution of the tip of the vector $\boldsymbol{u}$ is constrained to lie on the unit circle. The evolution equation for $\theta$ is (7.89) with $r$ set to unity and $B$ replaced by $\beta$:

$$d\theta = \beta\, dW_\theta. \tag{7.96}$$

Given (7.96), what is the Cartesian evolution equation for $\boldsymbol{u}$?

Applying Itô's formula to $\boldsymbol{u}(\theta)$ and (7.96) yields

$$d\boldsymbol{u}(\theta(t)) = \frac{1}{2}\beta^2 \frac{d^2\boldsymbol{u}}{d\theta^2}\, dt + \beta \frac{d\boldsymbol{u}}{d\theta}\, dW_\theta.$$

Since $\boldsymbol{u} = \cos\theta \boldsymbol{e}_x + \sin\theta \boldsymbol{e}_y$, the derivatives are straightforward to evaluate:

$$\frac{d\boldsymbol{u}}{d\theta} = \boldsymbol{t}, \qquad \frac{d^2\boldsymbol{u}}{d\theta^2} = -\boldsymbol{u},$$

where $\boldsymbol{t}$ is the unit vector tangent to the unit circle and is thus equal to the polar coordinate basis vector $\boldsymbol{e}_\theta$ at angular position $\theta$. Therefore,

$$d\boldsymbol{u}(\theta(t)) = -\frac{1}{2}\beta^2 \boldsymbol{u}\, dt + \beta \boldsymbol{t}\, dW_\theta.$$

To complete the transformation to Cartesian coordinates, we must rewrite $\boldsymbol{t} dW_\theta$ in terms of a two-dimensional Wiener process on the plane, $d\boldsymbol{W} = dW_x \boldsymbol{e}_x + dW_y \boldsymbol{e}_y$. This is accomplished by noting that $\boldsymbol{t} dW_\theta$ is simply the component of $d\boldsymbol{W}$ in the $\boldsymbol{e}_\theta$ direction, i.e., orthogonal to $\boldsymbol{u}$. Therefore, using the orthogonal projection operator $\boldsymbol{\delta} - \boldsymbol{u}\boldsymbol{u}$, we write $\boldsymbol{t}\, dW_\theta = (\boldsymbol{\delta} - \boldsymbol{u}\boldsymbol{u}) \cdot d\boldsymbol{W}$, giving the final result

$$d\boldsymbol{u} = -\frac{1}{2}\beta^2 \boldsymbol{u}\, dt + \beta(\boldsymbol{\delta} - \boldsymbol{u}\boldsymbol{u}) \cdot d\boldsymbol{W}. \tag{7.97}$$

In the polar coordinate expression for the orientation, there was only diffusion, while in the Cartesian coordinate version there is also drift (i.e., a term multiplied by $dt$). This seemingly strange result occurs because of the curvature of the constraint surface. We can understand this geometrically. Let $u_x = 0, u_y = 1$ at $t = 0$. Near this point, we can Taylor-expand the shape of the circle as $y = 1 - \frac{1}{2}x^2 + O(x^4)$. The solution has to remain on this curve; this fact is the origin of the drift. If $u_x$ changes by an amount $\beta dW_\theta$, then $u_y$ must change by an amount $-\frac{1}{2}(\beta dW_\theta)^2 = -\frac{1}{2}\beta^2 dt$ as illustrated in Figure 7.3. For Brownian motion on a sphere rather than a circle, the factor of $\frac{1}{2}$ in this geometric drift term becomes 1. In general, we can write

$$d\boldsymbol{u} = -\frac{d_s}{2}\beta^2 \boldsymbol{u}\, dt + \beta(\boldsymbol{\delta} - \boldsymbol{u}\boldsymbol{u}) \cdot d\boldsymbol{W}, \tag{7.98}$$

where $d_s = 1$ for diffusion on a circle (a 1-sphere) an $d_s = 2$ for diffusion on a (2-) sphere (Öttinger 1996).

If an angular drift term is included, then (7.96) changes to

$$d\theta = \alpha\, dt + \beta\, dW. \tag{7.99}$$

Taking $\alpha$ to be the $\theta$-component of a vector $\boldsymbol{\alpha}$ (which depends on $\boldsymbol{u}$ if $\alpha$ depends on $\theta$), then the equation for $\boldsymbol{u}$ becomes

$$d\boldsymbol{u} = \left( (\boldsymbol{\delta} - \boldsymbol{u}\boldsymbol{u}) \cdot \boldsymbol{\alpha} - \frac{1}{2}\beta^2 \boldsymbol{u} \right) dt + \beta(\boldsymbol{\delta} - \boldsymbol{u}\boldsymbol{u}) \cdot d\boldsymbol{W}. \tag{7.100}$$

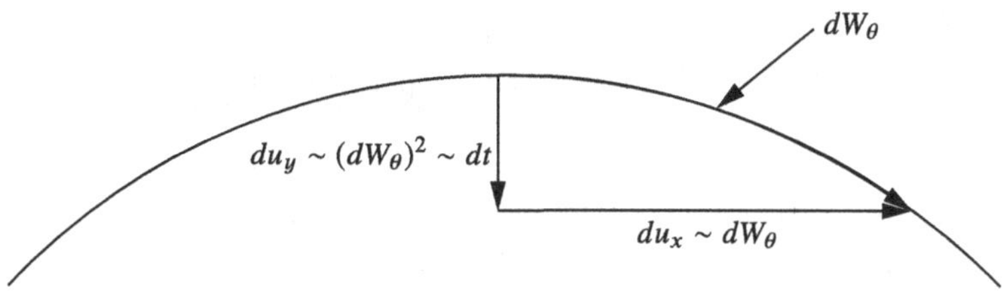


**Figure 7.3** Origin of the drift terms in the Cartesian representation of Brownian motion on a circle.

This has the form of (7.47) with

$$\boldsymbol{A} = (\boldsymbol{\delta} - \boldsymbol{uu}) \cdot \boldsymbol{\alpha} - \frac{1}{2}\beta^2 \boldsymbol{u},$$
$$\mathbf{B} = \beta\,(\boldsymbol{\delta} - \boldsymbol{uu})\,.$$

Naively, we can imagine performing a stochastic simulation using a simple Euler step:

$$\boldsymbol{u}(t + \Delta t) = \boldsymbol{u}(t) + \left((\boldsymbol{\delta} - \boldsymbol{u}(t)\boldsymbol{u}(t)) \cdot \boldsymbol{\alpha}(\boldsymbol{u}(t)) - \frac{1}{2}\beta^2 \boldsymbol{u}(t)\right) \Delta t + \beta(\boldsymbol{\delta} - \boldsymbol{u}(t)\boldsymbol{u}(t)) \cdot \Delta \boldsymbol{W}. \tag{7.101}$$

This approach, however, does not guarantee that $\boldsymbol{u}(t + \Delta t)$ remains precisely a unit vector. On the other hand, simply taking an Euler step that ignores the presence of the constraint, then normalizing this intermediate result, yields an algorithm that gives the proper dynamics:

$$\boldsymbol{u}^* = \boldsymbol{u}(t) + \boldsymbol{\alpha}(\boldsymbol{u}(t))\,\Delta t + \beta \Delta \boldsymbol{W},$$
$$\boldsymbol{u}(t + \Delta t) = \frac{\boldsymbol{u}^*}{||\boldsymbol{u}^*||}. \tag{7.102}$$

For more complex systems such as linked rigid rods, generalizations of this approach can be used, but the second step becomes more complicated, involving determination of Lagrange multipliers (Doi & Edwards 1986, Öttinger 1996).

The Fokker–Planck equation corresponding to (7.99) is

$$\frac{\partial p(\theta, t)}{\partial t} = -\frac{\partial}{\partial \theta}\left(\alpha p(\theta, t)\right) + D_{\mathrm{r}} \frac{\partial^2 p(\theta, t)}{\partial \theta^2}, \tag{7.103}$$

where $D_{\mathrm{r}} = \frac{1}{2}\beta^2$ is the *rotational diffusivity*. To write this in Cartesian coordinates, we note that

$$\frac{\partial^2}{\partial \theta^2} = \boldsymbol{e}_\theta \frac{\partial}{\partial \theta} \cdot \boldsymbol{e}_\theta \frac{\partial}{\partial \theta}$$

and that $\boldsymbol{e}_\theta \frac{1}{r}\frac{\partial}{\partial \theta}$ is just the component of $\boldsymbol{\nabla}$ in the $\theta$-direction:

$$\boldsymbol{e}_\theta \frac{1}{r}\frac{\partial}{\partial \theta} = \frac{1}{r}\boldsymbol{e}_\theta\,(\boldsymbol{e}_\theta \cdot \boldsymbol{\nabla}) = \frac{1}{r}\boldsymbol{tt} \cdot \boldsymbol{\nabla} = \frac{1}{r}(\boldsymbol{\delta} - \boldsymbol{uu}) \cdot \boldsymbol{\nabla}.$$

The surface gradient operator is this quantity evaluated at $r = 1$ (Bird et al. 1987):

$$\frac{\partial}{\partial \boldsymbol{u}} = (\boldsymbol{\delta} - \boldsymbol{u}\boldsymbol{u}) \cdot \boldsymbol{\nabla}.$$

Now we can rewrite the Fokker–Planck equation in Cartesian coordinates as

$$\begin{aligned}\frac{\partial p(\boldsymbol{u},t)}{\partial t} &= -(\boldsymbol{\delta} - \boldsymbol{u}\boldsymbol{u}) \cdot \boldsymbol{\nabla} \cdot ((\boldsymbol{\delta} - \boldsymbol{u}\boldsymbol{u}) \cdot \boldsymbol{\alpha} p(\boldsymbol{u},t)) \\ &\quad + D_{\mathrm{r}}(\boldsymbol{\delta} - \boldsymbol{u}\boldsymbol{u}) \cdot \boldsymbol{\nabla} \cdot (\boldsymbol{\delta} - \boldsymbol{u}\boldsymbol{u}) \cdot \boldsymbol{\nabla} p(\boldsymbol{u},t) && (7.104) \\ &= -\frac{\partial}{\partial \boldsymbol{u}} \cdot ((\boldsymbol{\delta} - \boldsymbol{u}\boldsymbol{u}) \cdot \boldsymbol{\alpha} p(\boldsymbol{u},t)) + D_{\mathrm{r}} \frac{\partial}{\partial \boldsymbol{u}} \cdot \frac{\partial}{\partial \boldsymbol{u}} p(\boldsymbol{u},t). && (7.105)\end{aligned}$$

As must be the case, this result can also be obtained using (7.100) and the relationship between SDE and FPE given in (7.47) and (7.48). The same result holds for diffusion on the unit sphere if the three-dimensional gradient operator is used.

## 7.6 Rotational Diffusion and the Wormlike Random Walk

Having in hand the preceding results, we elaborate now on the important problem of rotational diffusion. We begin in two dimensions, where the probability distribution is governed by (7.96) or (7.97). In polar coordinates, the corresponding Fokker–Planck equation is (7.103) with $A_\theta = 0$:

$$\frac{\partial p(\theta,t)}{\partial t} = D_{\mathrm{r}} \frac{\partial^2 p(\theta,t)}{\partial \theta^2} \tag{7.106}$$

and $p(\theta,t)$ must satisfy the normalization condition

$$\int_0^{2\pi} p(\theta,t)\, d\theta = 1.$$

We will now address the rotational evolution of a particle that has orientation $\theta = 0$ (i.e., is pointed in the $x$-direction) at $t = 0$. The initial condition for the probability density is

$$p(\theta,0) = \delta(\theta).$$

To solve for the time-dependent density, we write it as a Fourier cosine series:

$$p(\theta,t) = p_{\mathrm{eq}} + \sum_{n=1}^{\infty} p_n(t) \cos n\theta,$$

where $p_{\mathrm{eq}} = 1/2\pi$ is the uniform equilibrium density and we can use a cosine series because by symmetry $p(\theta,t)$ will be even in $\theta$. The coefficients $p_n(t)$ satisfy

$$\frac{dp_n}{dt} = -n^2 D_{\mathrm{r}} p_n$$

$$p_n(0) = \frac{1}{\pi}$$

so the solution is

$$p(\theta,t) = \frac{1}{2\pi} + \sum_{n=1}^{\infty} \frac{1}{\pi} e^{-n^2 D_r t} \cos n\theta. \tag{7.107}$$

The relaxation of the orientational distribution from completely oriented to perfectly uniform is dominated by the first, slowest decaying term in the series, i.e., with a rate given by $D_r$.

Another characterization of this relaxation process is given by the time correlation function of the orientation. Recalling that $\boldsymbol{u}(t) = \cos\theta(t)\boldsymbol{e}_x + \sin\theta(t)\boldsymbol{e}_y$, we have that

$$\begin{aligned}\langle \boldsymbol{u}(t)\cdot\boldsymbol{u}(0)\rangle &= \langle \cos\theta\rangle \\ &= \int_0^{2\pi} \cos\theta p(\theta,t)\, d\theta \\ &= \int_0^{2\pi} \cos\theta \left(\frac{1}{2\pi} + \sum_{n=1}^{\infty}\frac{1}{\pi}e^{-n^2 D_r t}\cos n\theta\right) d\theta \\ &= e^{-D_r t}. \end{aligned} \tag{7.108}$$

This result can also be obtained by applying Itô's formula, (7.34) with $x = \theta$ and $f = \cos\theta$ to give

$$d\cos\theta = -\frac{1}{2}\beta^2\cos\theta\, dt - \beta\sin\theta\, dW_\theta.$$

Taking the ensemble average of this expression yields

$$d\langle\cos\theta\rangle = -\frac{1}{2}\beta^2\langle\cos\theta\rangle\, dt,$$

which integrates to (7.108). Finally, if we assume that we have chosen this particular particle from a stationary distribution, then we can use the arguments of Section 6.2 to conclude that

$$\langle \boldsymbol{u}(t+\tau)\boldsymbol{u}(t)\rangle = \frac{1}{2}e^{-D_r|\tau|}\boldsymbol{\delta}. \tag{7.109}$$

Now instead of a passive particle undergoing rotational diffusion, consider an active particle that moves with speed $U$ in the direction of its axis $\boldsymbol{u}$ while this axis undergoes rotational diffusion. This is a good model of a swimming microorganism whose axis wobbles randomly as it swims (Berg 1993), and we will see it again with a different physical interpretation in Chapter 8. For simplicity, we continue to focus on the two-dimensional case. The equations of motion for this particle are

$$d\boldsymbol{R} = U\boldsymbol{u}\, dt, \tag{7.110}$$

$$d\theta = \beta\, dW_\theta. \tag{7.111}$$

In this model, we have neglected translational diffusion; we show later when doing this is a good approximation. Figure 7.4 illustrates a trajectory of such a swimmer. Because the evolution of $\theta$ is independent of that of $\boldsymbol{R}$, the results for the orientation distribution

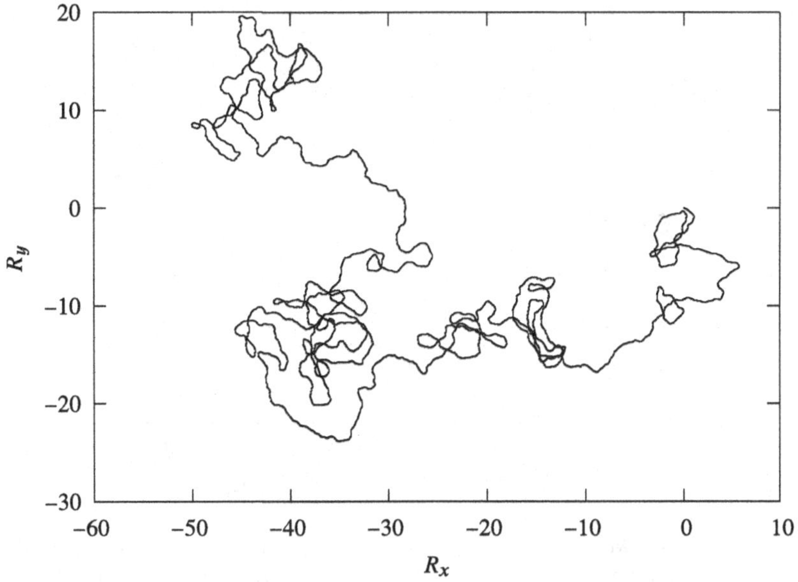


**Figure 7.4** A two-dimensional wormlike random talk of 500 time units with $U = \beta = 1$ and initial condition $R_x(0) = R_y(0) = \theta(0) = 0$.

remain unchanged from the previous example; in particular, (7.109) still holds. Since the velocity of the particle is simply $\boldsymbol{U} = U\boldsymbol{u}$, the velocity autocorrelation function is

$$\boldsymbol{\phi}_v(\tau) = \frac{U^2}{2} e^{-D_r|\tau|} \boldsymbol{\delta}. \tag{7.112}$$

Recalling the results of Section 6.6 and in particular (6.71), which relates the mean squared displacement to the velocity autocorrelation function, we can use (7.112) to find that

$$\left\langle R_i(t) R_j(t) \right\rangle = \delta_{ij} \frac{U^2}{D_r} \left(1 - e^{-D_r t}\right) t + \delta_{ij} \frac{U^2}{D_r^2} \left((1 + D_r t)\, e^{-D_r t} - 1\right). \tag{7.113}$$

As $t \to \infty$, the second term becomes a constant while the first increases linearly with $t$ so that

$$\lim_{t \to \infty} \frac{1}{t} \left\langle R_i(t) R_j(t) \right\rangle = \frac{U^2}{D_r} \delta_{ij}.$$

Thus the long-time behavior is diffusive with an effective translational diffusivity

$$D_{\text{eff}} = \frac{U^2}{2D_r}. \tag{7.114}$$

Even though there is no translational diffusivity included in the model, the long-time behavior is diffusive anyway. Indeed, as long as the bare translational diffusivity is much smaller than $D_{\text{eff}}$, neglecting it is a good approximation except at very short times. It is interesting to note that $D_{\text{eff}}$ varies inversely with $D_r$, diverging as $D_r \to 0$. This result

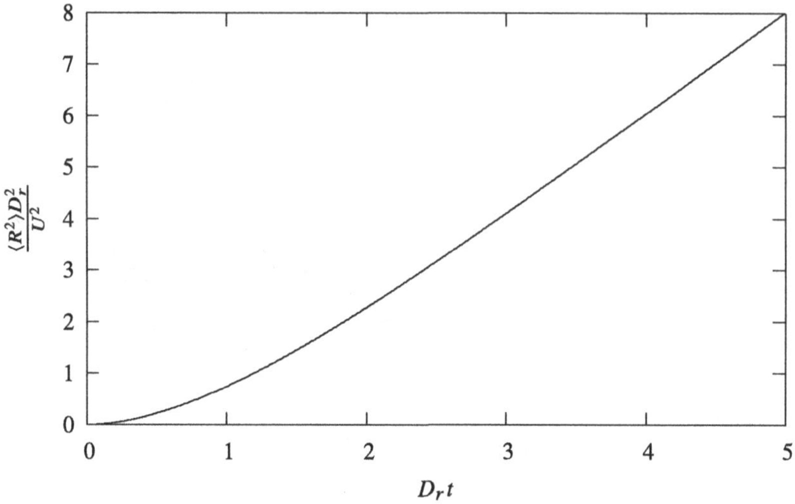


**Figure 7.5** Mean squared displacement for a two-dimensional wormlike random walk; this plot is the trace of (7.113). Observe the quadratic behavior for $D_r t \ll 1$ and linear behavior for $D_r t \gg 1$.

arises because as $D_r$ decreases the particle will continue swimming in the same direction for longer and longer before reorienting and changing direction. In the limit $D_r = 0$, the swimmer never changes direction so its behavior is ballistic, not diffusive, in which case

$$\left\langle R_i(t)R_j(t)\right\rangle = \frac{1}{2}U^2t^2\delta_{ij} \tag{7.115}$$

or

$$\left\langle R^2\right\rangle = U^2t^2, \tag{7.116}$$

where $R^2 = \boldsymbol{R}\cdot\boldsymbol{R}$. The mean squared displacement reduces to that for an object moving at constant speed $U$ without changing direction.

In the Wiener process described at the beginning of the chapter, the mean squared displacement scales linearly with time over all time intervals and the process itself is not differentiable. These properties are related, both following from the fact that $dW^2 = dt$. By contrast, the displacement variable $\boldsymbol{R}$ in the wormlike random walk *is* differentiable: from (7.110), we see that

$$\frac{d\boldsymbol{R}}{dt} = U\boldsymbol{u}.$$

Since $\boldsymbol{u}$ remains correlated with itself on time scales smaller than $D_r$, we expect that for short times the particle will continue to move more or less in the same direction, a property that can be observed in Figure 7.4. This property is also reflected in the short-time behavior of the mean squared displacement; in this limit, (7.113) becomes

$$\left\langle R_i(t)R_j(t)\right\rangle = \frac{1}{2}\left(U^2t^2 + O(t^3)\right)\delta_{ij}. \tag{7.117}$$

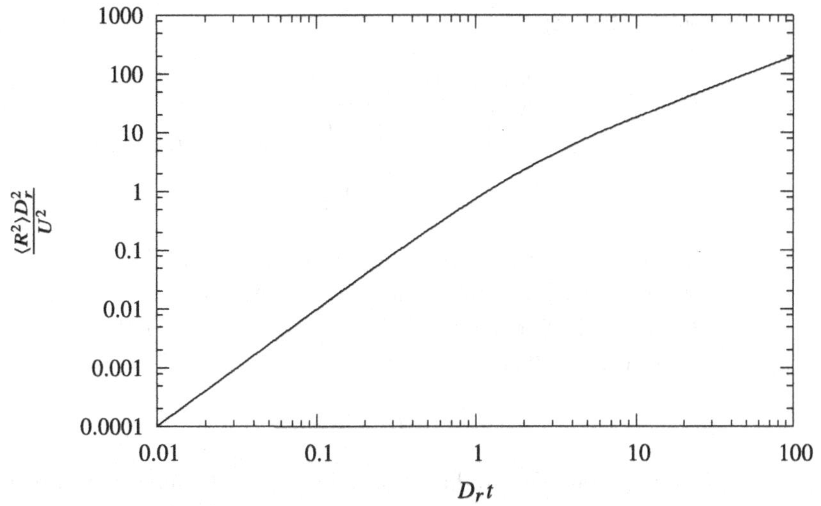


**Figure 7.6** log-lot plot of mean squared displacement for a two-dimensional wormlike random walk; this plot is the trace of (7.113). The slope is two for $D_r t \ll 1$ and one for $D_r t \gg 1$.

Thus at short times, the behavior of the wormlike walk is ballistic even when $D_r \neq 0$. The behavior crosses over from ballistic to diffusive transport when $D_r t \sim 1$. Figure 7.5 shows this on a linear scale, while Figure 7.6 shows it on a log-lot plot, where the slopes of two and one at short and long times indicate ballistic and diffusive behaviors, respectively.

In three dimensions, where the tip of the orientation vector $\boldsymbol{u}$ diffuses on the surface of a sphere, very similar results arise. In spherical coordinates, with polar angle $\theta$ and azimuthal angle $\phi$, the probability distribution $p(\theta, \phi, t)$ satisfies

$$\frac{\partial p(\theta,\phi,t)}{\partial t} = D_r \left( \frac{1}{\sin\theta} \frac{\partial}{\partial \theta} \left( \sin\theta \frac{\partial p(\theta,\phi,t)}{\partial \theta} \right) + \frac{1}{\sin^2\theta} \frac{\partial^2 p(\theta,\phi,t)}{\partial \phi^2} \right) \tag{7.118}$$

(the right-hand side is the angular part of the Laplacian), subject to the normalization condition

$$\int_0^\pi \int_0^{2\pi} p(\theta,\phi,t) \sin\theta \, d\phi \, d\theta = 1. \tag{7.119}$$

With the initial condition

$$p(\theta,\phi,0) = \frac{1}{2\pi}\delta(\theta)$$

corresponding to alignment with the polar axis, this problem has an axisymmetric Legendre series solution

$$p(\theta,t) = \sum_{n=0}^{\infty} \frac{1}{2\pi} \frac{2n+1}{2} e^{-D_r n(n+1)t} P_n(\cos\theta), \tag{7.120}$$

where $P_n(x)$ is the $n$th Legendre polynomial. The orientational correlation function can be found as before:

$$\begin{aligned}\langle \boldsymbol{u}(t)\cdot\boldsymbol{u}(0)\rangle &= \langle\cos\theta\rangle \\ &= \int_0^{\pi}\int_0^{2\pi}\cos\theta p(\theta,\phi,t)\sin\theta\, d\phi\, d\theta \\ &= e^{-2D_r t}. \end{aligned} \tag{7.121}$$

Evaluation of this integral is simplified by observing that $P_1(\cos\theta) = \cos\theta$ and applying the orthogonality properties of the Legendre polynomials (see, e.g., Deen 2012). This result can also be found with Itô's formula (Problem 7.6). By stationarity and isotropy, these results imply that

$$\langle \boldsymbol{u}(t+\tau)\boldsymbol{u}(t)\rangle = \frac{1}{3}e^{-2D_r|\tau|}\boldsymbol{\delta}. \tag{7.122}$$

This differs from the two-dimensional result only by the factor of two in the exponent and the prefactor of $1/3$ rather than $1/2$. Accordingly, all of the two-dimensional results that involve $\langle \boldsymbol{u}(t+\tau)\boldsymbol{u}(t)\rangle$ hold in three dimensions with very little change. Specifically, the velocity autocorrelation function is

$$\boldsymbol{\phi}_v(\tau) = \frac{U^2}{3}e^{-2D_r|\tau|}\boldsymbol{\delta}, \tag{7.123}$$

and the mean squared displacement is

$$\left\langle R_i(t)R_j(t)\right\rangle = \delta_{ij}\frac{U^2}{3D_r}\left(1-e^{-2D_r t}\right)t + \delta_{ij}\frac{U^2}{6D_r^2}\left((1+2D_r t)\,e^{-2D_r t}-1\right). \tag{7.124}$$

The short-time behavior of the mean squared displacement is still ballistic with $\left\langle R^2(t)\right\rangle = U^2t^2 + O(t^3)$ and the long-time diffusivity becomes

$$D_{\text{eff}} = \frac{U^2}{6D_r}. \tag{7.125}$$

## 7.7 Coupled Rotational and Translational Diffusion

For an anisotropic particle such as a rod or disc with friction coefficient $\boldsymbol{\zeta} = \zeta^{\parallel}\boldsymbol{uu} + \zeta^{\perp}(\boldsymbol{\delta} - \boldsymbol{uu})$, the anisotropy of the drag couples the translational dynamics to the orientation of the particle and thus to the rotational dynamics that we just described. We can write the inertialess evolution equation for the center of mass $\boldsymbol{R}$ as

$$d\boldsymbol{R} = \mathbf{B}\cdot d\boldsymbol{W}. \tag{7.126}$$

By separately considering the motions along $\boldsymbol{u}$ and orthogonal to it, the analysis of Section 6.2 can be used to show (Problem 6.3) that

$$\mathbf{B}\cdot\mathbf{B}^{\mathrm{T}} = 2k_BT\boldsymbol{\zeta}^{-1}. \tag{7.127}$$

Because of the special form of $\boldsymbol{\zeta}$, its inverse (the tensor-valued mobility of the rod) is easily found to be

$$\zeta^{-1} = \frac{1}{\zeta^{\|}}\boldsymbol{uu} + \frac{1}{\zeta^{\perp}}(\boldsymbol{\delta} - \boldsymbol{uu}). \tag{7.128}$$

Additionally, $\boldsymbol{uu}$ and $\boldsymbol{\delta} - \boldsymbol{uu}$ are orthogonal projection operators and every such operator $\mathbf{P}$ satisfies $\mathbf{P} \cdot \mathbf{P}^{\mathrm{T}} = \mathbf{P}$, so we can take $\mathbf{B}$ to be

$$\mathbf{B} = \sqrt{2D^{\|}}\boldsymbol{uu} + \sqrt{2D^{\perp}}(\boldsymbol{\delta} - \boldsymbol{uu}), \tag{7.129}$$

where $D^{\|} = k_B T/\zeta^{\|}$ and $D^{\perp} = k_B T/\zeta^{\perp}$ are the short-time diffusivities of the object in the directions parallel and perpendicular to $\boldsymbol{u}$, respectively. Using Itô's formula (7.44), the evolution of the mean squared displacement of the particle is found to be

$$d\langle \boldsymbol{RR}\rangle = \left(2D^{\|}\langle \boldsymbol{uu}\rangle + 2D^{\perp}(\boldsymbol{\delta} - \langle \boldsymbol{uu}\rangle)\right) dt. \tag{7.130}$$

Obviously we cannot determine the mean squared displacement without information about the orientations.

In three dimensions, the orientation vector $\boldsymbol{u}$ is governed by (7.98) with $d_s = 2$. Application of Itô's formula to this equation yields that

$$d\langle \boldsymbol{uu}\rangle = (2D_{\mathrm{r}}(\boldsymbol{\delta} - 3\langle \boldsymbol{uu}\rangle))\, dt, \tag{7.131}$$

where as before we let $D_{\mathrm{r}} = \beta^2/2$. This equation is easily solved with initial condition $\langle \boldsymbol{uu}\rangle(0)$:

$$\langle \boldsymbol{uu}\rangle(t) = \frac{1}{3}\boldsymbol{\delta} + \left(\langle \boldsymbol{uu}\rangle(0) - \frac{1}{3}\boldsymbol{\delta}\right) e^{-6D_{\mathrm{r}}t}. \tag{7.132}$$

As $t \to \infty$, this approaches the equilibrium result

$$\langle \boldsymbol{uu}\rangle = \frac{1}{3}\boldsymbol{\delta}. \tag{7.133}$$

Now we can solve (7.130) for an arbitrary initial orientation distribution, but rather than present the general result, we focus on two important cases, the short- and long-time limits. First, consider the short-time diffusion of a particle whose orientation is known at $t = 0$ to be $\boldsymbol{u}_0$. Here (7.130) simply becomes

$$\frac{d\langle \boldsymbol{RR}\rangle}{dt} = 2\mathbf{D}, \tag{7.134}$$

where

$$\mathbf{D} = \frac{1}{2}\mathbf{B}(0) \cdot \mathbf{B}^{\mathrm{T}}(0) = k_B T\boldsymbol{\zeta}^{-1}(0) = D^{\|}\boldsymbol{u}_0\boldsymbol{u}_0 + D^{\perp}(\boldsymbol{\delta} - \boldsymbol{u}_0\boldsymbol{u}_0). \tag{7.135}$$

This quantity is the diffusivity tensor for the particle, evaluated at its initial orientation $\boldsymbol{u}_0$. The diffusion at short times reflects the anisotropy of the mobility. On the other hand, at long times, when the orientational distribution has equilibrated, we can evaluate (7.130) using the equilibrium result (7.133) to yield that

$$\frac{d\langle \boldsymbol{RR}\rangle}{dt} = 2\left(\frac{D^{\|}}{3} + \frac{2D^{\perp}}{3}\right)\boldsymbol{\delta}. \tag{7.136}$$

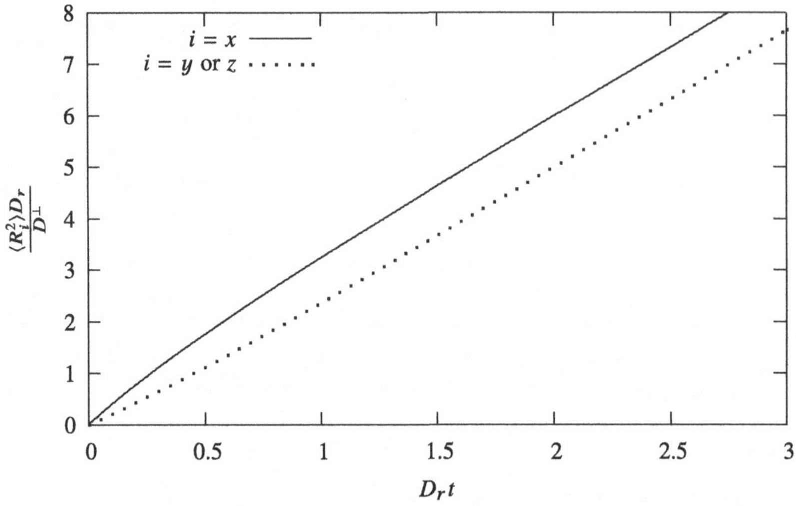


**Figure 7.7** Mean squared displacements in the $x$-, $y$-, and $z$-directions for a particle with $D^{\|}/D^{\perp} = 2$ initially aligned with the $x$-axis. The off-diagonal components of $\langle \boldsymbol{RR} \rangle$ are zero.

That is, the long-time diffusion of the particle is isotropic, with diffusivity given by

$$D_{\text{eff}} = \left( \frac{D^{\|}}{3} + \frac{2D^{\perp}}{3} \right). \tag{7.137}$$

The isotropy arises from the fact that even though the particle itself is anisotropic, it has no preferred orientation, so given enough time it will sample all possible orientations with equal probability. The perpendicular diffusivity $D^{\perp}$ is more strongly weighted than the parallel diffusivity simply because there are two directions perpendicular to the particle axis over which it can diffuse. In the limiting case $D^{\|} = D^{\perp} = D$, we recover the proper result $D_{\text{eff}} = D$. The crossover from anisotropic to isotropic diffusion is determined by the rotational diffusivity. Figure 7.7 shows the diagonal components of $\langle \boldsymbol{RR} \rangle$ vs. $t$ for a particle with initial orientation $\boldsymbol{u}_0 = \boldsymbol{e}_x$ in the case $D^{\|}/D^{\perp} = 2$. The initial displacements in the directions parallel and perpendicular to $\boldsymbol{u}_0$ are somewhat faster and slower, respectively, than at long times. Perhaps surprisingly, the *scalar* mean squared displacement $\langle \boldsymbol{R} \cdot \boldsymbol{R} \rangle$ is given by $6D_{\text{eff}}t$ for *all* times.

## Problems

**7.1** Consider two freely diffusing particles in three dimensions. Their equations of motion are

$$d\boldsymbol{X}_1 = B_1 \, d\boldsymbol{W}_1,$$
$$d\boldsymbol{X}_2 = B_2 \, d\boldsymbol{W}_2.$$

What is the SDE for the distance between the particles $\boldsymbol{X}_2 - \boldsymbol{X}_1$?

**7.2** (a) Use Itô's formula to derive (7.41) and (7.42).

(b) Derive the multidimensional form of Itô's formula, (7.44), for an SDE in the form

$$dx_i = A_i(x,t)dt + B_{ij}(x,t)dW_j.$$

Recall (7.29).

(c) Use this formula to derive the multidimensional versions of (7.41) and (7.42):

$$\left\langle \frac{d(x_i - x_i')(x_j - x_j')}{dt} \right\rangle_{t=t'} = 2D_{ij}(x',t'),$$

$$\left\langle \frac{d(x_i - x_i')}{dt} \right\rangle_{t=t'} = A_i(x',t').$$

**7.3** Derive the Maxwell–Boltzmann expression for $p_{\text{eq}}(U,R)$ from (7.55) and (7.56). Hint: Seek a solution of the form $p_{\text{eq}}(U,R) = \Upsilon(U)\rho(R)$.

**7.4** Show that (7.83)–(7.84) reduce to (7.78) as $\Delta t \to 0$.

**7.5** Diffusion in three dimensions is represented in Cartesian coordinates by

$$dx = B\,dW_x,$$
$$dy = B\,dW_y,$$
$$dz = B\,dW_z.$$

Write the SDE in spherical coordinates, recalling that $r = \sqrt{x^2+y^2+z^2}$, $\theta = \cos^{-1}(z/r)$, $\phi = \tan^{-1}(y/x)$.

**7.6** If $r$ is fixed at unity, the previous exercise yields the equations for diffusion on the surface of the unit sphere. Use the solution to the previous exercise and Itô's formula to derive (7.121).

**7.7** Derive (7.93) from the polar coordinate SDEs (7.88) and (7.89) using the ensemble average definition (7.95). Compare it to the equation that arises from direct application of (7.46); in particular, show that the solution to the latter, denoted $p_P(r,\theta,t)$ is related to the solution to the former, $p(r,\theta,t)$, as follows:

$$p(r,\theta,t) = crp_P(r,\theta,t),$$

where $c$ is a normalization constant.

**7.8** Show that, for the algorithm given by (7.102), $\boldsymbol{u}(t+\Delta t) - \boldsymbol{u}(t)$ reduces to the right-hand side of (7.100) as $\Delta t \to dt$.

**7.9** Add translational diffusion to the wormlike random walk model by adding a term $B\,dW_R$ to (7.110). Derive an expression for the mean squared displacement and show that for small but nonzero $B$, there is a short-time diffusive regime, then a ballistic regime, and then finally a diffusive regime, again at long times.

**7.10** Derive an expression for the full time dependence of the mean squared displacement of a particle with anisotropic mobility and given initial orientation $\boldsymbol{u}_0$. Write a Brownian dynamics code to simulate the motion of such a particle and use it to find the mean squared displacement.

# 8 Coarse-Grained Models of Polymers in Dilute Solution

The dynamics of polymer molecules in solution exemplify phenomena that are common to a wide range of complex fluid systems. Many aspects of these dynamics can be understood by applying the concepts and tools that have been developed in the previous chapters. In this chapter, we introduce some basic facts about polymer structure and use these to develop a coarse-grained model in the form of a system of point particles that interact by various forces and by the fluid motion that the particles generate as they move. This model will be analyzed to illustrate some key aspects of the dynamical behavior (diffusion, stress, response to flow) of polymers in solution. Two important highly simplified cases, the bead–spring dumbbell and rigid rod models, will be described in some detail. We focus exclusively on solutions of linear polymer molecules in the dilute limit where there are no interactions between different polymer molecules: each molecule behaves independently of all the others.

## 8.1 Models of Equilibrium Polymer Structure

A typical polymer molecule in solution is much larger than the surrounding solvent molecules. The slowest dynamics of the molecule will then be controlled by scales much larger than the solvent scale, and it is these slow dynamics that are the most important for many applications. At a qualitative level, this point can be understood from an argument based on diffusion. Recall that the time scale for diffusion, $t_{\mathrm{diff}}$, scales as $R^2/D$, where $R$ is the length scale of interest and $D$ is a diffusivity. Thus diffusion drives small scales toward equilibrium much more rapidly than large scales. Since the small scales are not dynamically active, many important processes in polymer dynamics can be captured using coarse-grained models that capture the large-scale motions while dramatically simplifying the polymer structure at small scales. In this chapter, we touch on basic elements of these models and illustrate how they shed light on important dynamical phenomena in polymer solutions. The following sources provide a great deal of additional information on polymer structure, physical chemistry and dynamics: de Gennes (1979), Doi & Edwards (1986), Bird et al. (1987), Öttinger (1996), Strobl (1996), Larson (1999), Rubinstein & Colby (2003).

### 8.1.1 Wormlike Chain

At the first level of coarse-graining we consider, the structure of the molecular backbone of the polymer is treated as an inextensible thin rod with a total length or *contour length* $L$ and flexural rigidity $K_{\text{bend}}$. This a particularly realistic description of many biopolymers such as double-stranded DNA (Phillips, Kondev & Theriot 2009). Throughout this chapter, we will take the mass of the polymer to be uniformly distributed along the chain. At every position along the backbone of the molecule, we can define a unit tangent vector $\boldsymbol{u}(s)$, where $s$ is the distance along the molecule's backbone as measured from one end. If we place that end of the molecule at the origin, then the three-dimensional position $\boldsymbol{R}(s)$ of a point a distance $s$ along the backbone is determined by integrating the equation

$$\frac{d\boldsymbol{R}}{ds} = \boldsymbol{u} \tag{8.1}$$

with initial condition $\boldsymbol{R}(0) = \boldsymbol{0}$. To evaluate this requires us to specify $\boldsymbol{u}(s)$, which we will do by assuming that at equilibrium $\boldsymbol{u}$ changes orientation continuously and randomly with $s$ due to thermal fluctuations, and that the change in $\boldsymbol{u}$ at position $s$ is independent of that at $s + ds$. That is, $\boldsymbol{u}$ executes an uncorrelated random walk in orientation, a Wiener process on the surface of the unit sphere. Recalling the results of Section 7.5.2, we can write that

$$d\boldsymbol{u} = -B_u^2 \boldsymbol{u}\, ds + B_u(\boldsymbol{\delta} - \boldsymbol{u}\boldsymbol{u}) \cdot d\boldsymbol{W}. \tag{8.2}$$

This is an evolution equation for $\boldsymbol{u}$ not in the time variable $t$ but in the arclength variable $s$, so here $d\boldsymbol{W}$ is a differential Wiener increment in arclength rather than time. The quantity $B_u$ is a property of the bending stiffness of the molecule that we will determine later in this section. For the moment, we note that $B_u$ has units of (length)$^{-1/2}$ and define a quantity $L_{\text{p}} = B_u^{-2}$ called the *persistence length*.

This model is called the *wormlike chain* (WLC). It is precisely the three-dimensional version of the wormlike random walk model analyzed in Section 7.6 but with time replaced by arclength, $V$ replaced by unity and rotational diffusivity $D_{\text{r}}$ replaced by $(2L_{\text{p}})^{-1}$. Thus the equilibrium orientational correlation function of the WLC is

$$\langle \boldsymbol{u}(s)\boldsymbol{u}(s+s')\rangle = \frac{1}{3} e^{-|s'|/L_{\text{p}}} \boldsymbol{\delta}. \tag{8.3}$$

Additionally, when $L \gg L_{\text{p}}$, the mean squared distance between the two ends of the molecule, commonly denoted as $R_0^2$, is

$$R_0^2 = \left\langle R^2(s)\right\rangle\Big|_{s=L} = 2L_{\text{p}}L. \tag{8.4}$$

These expressions follow directly from (7.122) and (7.125), respectively. Note the linear, that is, diffusive, scaling of $R_0$ with $L$. Effectively, a flexible polymer molecule occupies a volume that scales with $R_0^3$ so a polymer solution with number density of chains $n$ can be considered dilute if $nR_0^3 \ll 1$. The crossover from dilute to *semidilute* solution occurs when $nR_0^3 \approx 1$ and a solution is considered *concentrated* if $nR_0^3 \gg 1$.

To address the determination of $B_u$, or equivalently, of $L_p$, we observe that at equilibrium the only energy available for bending the polymer is thermal energy. The bending energy of a differential element $ds$ of the rod with curvature $\kappa$ is

$$dE_{\text{bend}} = \frac{1}{2}K_{\text{bend}}\kappa^2 ds = \frac{1}{2}\frac{K_{\text{bend}}}{r_c^2(s)}ds, \tag{8.5}$$

where $r_c = \kappa^{-1}$ is the radius of curvature of the element. Consider a polymer whose length is $L_p$. Now the problem has only one length scale, $L_p$, which we can use to estimate the radius of curvature of the rod and thus its total bending energy:

$$E_{\text{bend}} \approx \frac{1}{2}\frac{K_{\text{bend}}}{L_p^2}L_p \approx \frac{1}{2}\frac{K_{\text{bend}}}{L_p}. \tag{8.6}$$

This bending degree of freedom has $k_BT/2$ of thermal energy at equilibrium, so

$$\frac{1}{2}\frac{K_{\text{bend}}}{L_p} \approx \frac{1}{2}k_BT. \tag{8.7}$$

Solving for $L_p$, we find that

$$L_p \approx \frac{K_{\text{bend}}}{k_BT}. \tag{8.8}$$

Remarkably, an exact statistical mechanical calculation (Phillips et al. 2009) yields the same result:

$$L_p = \frac{K_{\text{bend}}}{k_BT}. \tag{8.9}$$

The persistence length is the polymer length at which bending energy balances thermal energy. Thus, at equilibrium the local radius of curvature of a wormlike chain is highly unlikely to be much larger than $L_p^{-1}$.

Although the wormlike chain is a good physical model for large biomolecules such as double-stranded DNA, whose molecular structure looks like a rod, it is also useful for synthetic polymers such as polyethylene oxide, $(CH_2 - CH_2 - O-)_n$, whose flexibility arises from rotations of the single bonds forming the backbone of the chain rather than from bending elasticity. In this case as well, the orientational correlation along the backbone decays exponentially (Doi & Edwards 1986) and the correlation length can be identified with $L_p$. Depending on the molecule, $L_p$ can vary by orders of magnitude. For polyethylene oxide in water, an important system for turbulent drag reduction (Graham 2014), $L_p \approx 0.9$ nm (Kawaguchi et al. 1997). For single- and double-stranded DNA, $L_p \approx 1$ and 50 nm, respectively (Smith, Cui & Bustamante 1996, Shaqfeh 2005). For a bacterial flagellum, $L_p \approx 7 \cdot 10^5$ nm (Darnton et al. 2007).

Based on the contour length $L$ and persistence length $L_p$, we can characterize a linear polymer as lying in one of three distinct regimes: *rigid*, when $L \ll L_p$; *semiflexible*, when $L \sim L_p$ and *flexible*, when $L \gg L_p$. Up to this point, we have neglected the possibility of the chain intersecting itself; this effect can be important when $L \gg L_p$, and we further address it in Section 8.2.

**Figure 8.1** An ideal chain. The dots connected by thick lines represent the coarse-graining of the ideal chain into 10 subchains connecting 11 beads.

### 8.1.2 Freely Jointed Chain and Gaussian Chain

From the wormlike chain model, we see that orientations of different segments of the polymer backbone become uncorrelated from one another on the length scale $L_p$. The *freely jointed chain* or *Kramers chain* model is a more highly idealized model that explicitly incorporates this lack of correlation. Here a polymer backbone of contour length $L$ is coarse-grained into $N_K$ rigid rods each with length $L_K$, linked at freely rotating joints (called "sites" or "beads") that are at positions $\boldsymbol{R}_1, \boldsymbol{R}_2, \ldots, \boldsymbol{R}_{N_{K+1}}$. The rods are also known as *Kuhn segments*, and the quantity $L_K = L/N_K$ is called the *Kuhn length*. As before, we continue to neglect the possible effects of self-intersections of the chain. Therefore, at equilibrium the orientation of each rod is completely independent of every other, so each conformation of the chain is a discrete random walk with $N_K$ steps of length $L_K$. Thus the mean squared end-to-end distance $R_0^2$ is given by the simple formula

$$R_0^2 = N_K L_K^2 = L_K L. \tag{8.10}$$

Comparing this result to (8.4), we see that for a flexible polymer

$$L_K = 2L_p. \tag{8.11}$$

Consider a polymer comprised of $N_m \gg 1$ monomers for which the sum of squares of the backbone bond lengths is $L_b^2$. The *characteristic ratio* $C_\infty$ for this polymer is given by

$$C_\infty = \frac{R_0^2}{N_m L_b^2}. \tag{8.12}$$

Since $L \sim N_m L_b$, we can use (8.10) to write that

$$C_\infty \sim \frac{L_K}{L_b}. \tag{8.13}$$

Thus $C_\infty$ is a measure of the apparent stiffness of the polymer backbone. For synthetic polymers whose backbone consists of single covalent bonds, $C_\infty \sim 4 - 10$ (Rubinstein

& Colby 2003). For more complex biopolymers, it can be much larger. Double-stranded DNA, for example, has $L_K \approx 100$ nm, but a monomer spacing of only $\approx 0.3$ nm, for a characteristic ratio of $\approx 300$.

Now consider the extreme limit where $L_K/L \to 0$ or $N_K \to \infty$. This leads to a random walk of infinitely many independent steps, each of infinitesimal length relative to $L$: a Brownian motion process in arclength rather than time. This limiting model, illustrated in Figure 8.1, is called the *ideal* or *Gaussian* chain. In three dimensions, the mean squared displacement of Brownian motion with diffusivity $D$ is $6Dt$; comparing this result with (8.10), we see that to generate the correct Brownian motion in arclength, we simply replace $t$ with $L$ and $D$ with $L_K/6$ to find that

$$d\boldsymbol{R}(s) = B\,d\boldsymbol{W}(s), \tag{8.14}$$

where $B = \sqrt{2D} = \sqrt{L_K/3}$. Equivalently, we can write the equation for the probability density $p(\boldsymbol{R}, s)$:

$$\frac{\partial p}{\partial s} = \frac{L_K}{6}\frac{\partial^2 p}{\partial \boldsymbol{R}^2}. \tag{8.15}$$

This is a diffusion equation for the evolution of $\boldsymbol{R}$ with $s$, and its solution for a polymer with one end at the origin, i.e., $p(\boldsymbol{R}, 0) = \delta(\boldsymbol{R})$, is

$$p(\boldsymbol{R}, s) = \left(\frac{3}{2\pi L_K s}\right)^{3/2} \exp\left(-\frac{3R^2}{2L_K s}\right). \tag{8.16}$$

As long as properties that only depend on chain structure at scales much larger than $L_K$ are considered, this is a very useful model.

Now consider a section or "subchain" of an ideal chain extending from $s$ to $s + \Delta s$, and let $\Delta\boldsymbol{R}(\Delta s) = \boldsymbol{R}(s + \Delta s) - \boldsymbol{R}(s)$. For this subchain, ideal chain statistics are still valid as long as its contour length $\Delta s$ remains much greater than $L_K$. Therefore,

$$p(\Delta\boldsymbol{R}, \Delta s) = \left(\frac{3}{2\pi L_K \Delta s}\right)^{3/2} \exp\left(-\frac{3\Delta R^2}{2L_K \Delta s}\right) \tag{8.17}$$

and

$$\Delta R^2(\Delta s) = L_K \Delta s. \tag{8.18}$$

This observation allows us to coarse-grain the chain into $N_s$ subchains each of contour length $\Delta s = L/N_s$ as illustrated in Figure 8.1. We will consider there to be a "bead" at each end of each subchain, for a total of $N_b = N_s + 1$ beads located at arclength values $s_\alpha = (\alpha - 1)\Delta s$ and positions $\boldsymbol{R}_{b,\alpha}$, where $\alpha = 1, 2, \ldots, N_b$. That is, each bead is connected to its neighbors by a subchain of contour length $L_s = L/(N_s)$ consisting of $N_{K,s} = L_s/L_K$ Kuhn segments. Letting $\boldsymbol{R}_{\alpha\beta} = \boldsymbol{R}_\beta - \boldsymbol{R}_\alpha$ and $s_{\alpha\beta} = s_\beta - s_\alpha$, the preceding results imply that

$$p(\boldsymbol{R}_{\alpha\beta}, s_{\alpha\beta}) = \left(\frac{3}{2\pi L_K s_{\alpha\beta}}\right)^{3/2} \exp\left(-\frac{3R_{\alpha\beta}^2}{2L_K s_{\alpha\beta}}\right). \tag{8.19}$$

Furthermore, defining the connector vector $\boldsymbol{Q}_\alpha = \boldsymbol{R}_{\alpha+1} - \boldsymbol{R}_\alpha$ between neighboring beads and noting that $s_{\alpha+1} - s_\alpha = N_{K,s}L_K$, we can write the connector vector probability distribution

$$p(\boldsymbol{Q}_\alpha) = \left(\frac{3}{2\pi L_K^2 N_{K,s}}\right)^{3/2} \exp\left(-\frac{3Q_\alpha^2}{2L_K^2 N_{K,s}}\right). \tag{8.20}$$

Finally, observe that the Gaussian distribution for an arbitrary subchain can be recast in the form of a Boltzmann distribution

$$p(\Delta\boldsymbol{R}, \Delta s) = p_0 \exp\left(-\frac{\Phi_{\text{sub}}(\Delta R, \Delta s)}{k_B T}\right) \tag{8.21}$$

where

$$p_0 = \left(\frac{H}{2\pi k_B T}\right)^{3/2} \tag{8.22}$$

and

$$\Phi_{\text{sub}}(\Delta\boldsymbol{R}, \Delta s) = \frac{3k_B T}{2L_K \Delta s}\Delta R^2. \tag{8.23}$$

The latter is the potential associated with a Hooke's law (or "Hookean") spring with a rest length of zero and spring constant

$$H_{\Delta s} = \frac{3k_B T}{L_K \Delta s}. \tag{8.24}$$

Therefore, at equilibrium a subchain behaves as if its two ends were connected by a simple linear spring rather than a random trajectory of polymer segments, with the spring force balanced by the Brownian forces exerted on each end of the subchain to yield the equilibrium distribution as discussed in Section 6.9. The effective spring force is entirely of entropic origin because in this model there are no energetic interactions, so the spring in this model is often said to be an *entropic* spring.

## 8.2 Bead–Spring Chain

Based on the preceding observations, we can consider the coarse-grained ideal chain with $N_b$ beads as a *bead–spring chain*, where each bead is connected to its neighbors by a spring with spring constant

$$H = \frac{3k_B T}{L_K L_s} = \frac{3k_B T}{L_K^2 N_{K,s}}. \tag{8.25}$$

Now (8.20) becomes

$$p(\boldsymbol{Q}_\alpha) = \left(\frac{H}{2\pi k_B T}\right)^{3/2} \exp\left(-\frac{\frac{1}{2}HQ_\alpha^2}{k_B T}\right). \tag{8.26}$$

Because the segments comprising the springs are independent, the conformational distribution function for the entire chain, $p(\boldsymbol{Q}_1, \boldsymbol{Q}_2, \ldots, \boldsymbol{Q}_{N_s})$, can be written as

$$p(\boldsymbol{Q}_1, \boldsymbol{Q}_2, \ldots, \boldsymbol{Q}_{N_s}) = p(\boldsymbol{Q}_1)p(\boldsymbol{Q}_2)\ldots p(\boldsymbol{Q}_{N_s}). \tag{8.27}$$

Therefore, in the Gaussian chain limit

$$p(\boldsymbol{Q}_1, \boldsymbol{Q}_2, \ldots, \boldsymbol{Q}_{N_s}) = \left(\frac{H}{2\pi k_B T}\right)^{3N_s/2} \exp\left(-\frac{\Phi_{\text{spring}}}{k_B T}\right), \tag{8.28}$$

where

$$\Phi_{\text{spring}}(\boldsymbol{Q}_1, \boldsymbol{Q}_2, \ldots, \boldsymbol{Q}_{N_s}) = \frac{1}{2} H \sum_{\alpha=1}^{N_s} Q_\alpha^2. \tag{8.29}$$

This potential is simply the sum of the energies associated with the stretching of each spring.

The bead–spring chain model is the most widely used description of flexible polymer molecules. At equilibrium in situations where self-interactions of the chain are negligible, it provides an excellent description of the structure of the chain. In nonequilibrium situations such as flow, a number of extensions must be made. These will be presented here; they culminate in an $N_b$-particle Smoluchowski equation in the form of (6.100). First we examine the model for the spring forces. For Hookean springs, as introduced in Section 8.1.2, the force on bead $\alpha$ by the springs attaching it to beads $\alpha \pm 1$ is given by

$$\boldsymbol{F}_{\text{spring},\alpha} = -\frac{\partial \Phi_{\text{spring}}}{\partial \boldsymbol{R}_\alpha} = H(\boldsymbol{Q}_\alpha - \boldsymbol{Q}_{\alpha-1}). \tag{8.30}$$

This can be rewritten in terms of *connector forces*: $\boldsymbol{F}_{c,\alpha}$ is the force exerted on bead $\alpha$ by bead $\alpha + 1$, so for Hookean springs

$$\boldsymbol{F}_{\text{c},\alpha} = H\boldsymbol{Q}_\alpha \tag{8.31}$$

and thus

$$\boldsymbol{F}_{\text{spring},\alpha} = \boldsymbol{F}_{\text{c},\alpha} - \boldsymbol{F}_{\text{c},\alpha-1}. \tag{8.32}$$

(Since there is no bead zero and no bead $N_b + 1$, $\boldsymbol{F}_{\text{c},0} = \boldsymbol{F}_{\text{c},N_b} = \boldsymbol{0}$.)

The Hookean model is fine near equilibrium, but does not account for the fact that the polymer subchain connecting each bead has finite length, so the spring force must diverge as $Q_\alpha = \sqrt{\boldsymbol{Q}_\alpha \cdot \boldsymbol{Q}_\alpha} \to L_s$. For the freely jointed chain, there is an exact expression for the spring force, the *inverse Langevin function*, but a simpler approximate form that is widely used is the "Warner" or Finitely Extensible Nonlinearly Elastic (FENE) spring force (Bird et al. 1987):

$$\boldsymbol{F}_{\text{c},\alpha} = \frac{H\boldsymbol{Q}_\alpha}{1 - Q_\alpha^2/L_s^2}. \tag{8.33}$$

For the wormlike chain model, Marko & Siggia (1994, 1995) proposed a simple spring model that, for $L_s \gg L_p$, captures the correct low- and high-stretch asymptotes of the WLC model (as determined analytically) and accurately approximates data in between:

$$\boldsymbol{F}_{\text{c},\alpha} = \frac{k_B T}{4L_p}\left(\left(1 - \frac{Q_\alpha}{L_s}\right)^{-2} + \frac{4Q_\alpha}{L_s} - 1\right)\frac{\boldsymbol{Q}_\alpha}{Q_\alpha}. \tag{8.34}$$

For small $Q_\alpha$, this reduces as it should to a Hookean spring with $H$ given by (8.25). In contrast to the Warner spring force, which goes as $(1 - Q_\alpha/L_s)^{-1}$ as $Q_\alpha \to L_s$, the

Marko–Siggia force goes as $(1 - Q_\alpha/L_s)^{-2}$, though this difference does not seem to lead to substantial differences in the dynamics of model molecules in flow. For example, the scaling arguments we develop in Section 9.3 to understand shear and extensional viscosities at high deformation rate are insensitive to the specific form of the spring law.

A polymer chain whose contour length is not much larger than $L_p$ is unlikely to turn back on itself. As chain length increases, however, the chain is increasingly likely to do so, allowing segments that are distant from one another in arclength to closely approach one another in position. The extent to which these segments are likely to interact is strongly dependent on the liquid in which the polymer is dissolved: that is, on the *solvent quality* (Doi & Edwards 1986, Rubinstein & Colby 2003). In an *ideal* or $\theta$ solvent, the polymer chain has no enthalpic preference for the solvent over itself or vice versa, self-interactions are negligible, and the preceding models provide a good description of the equilibrium chain structure. In a *poor* solvent, the polymer collapses into a compact globule and is often better treated as a single colloidal particle than as a chain. In a *good* solvent, the chain enthalpically prefers the solvent to itself, leading to a repulsion between chain segments called *excluded volume interaction*. In this case, at equilibrium the polymer backbone undergoes a *self-avoiding random walk*. Rather than scaling as $L^1$, the mean squared end-to-end distance $R_0^2$ scales as $L^{2\nu}$, where $\nu \approx 0.588$ and the value $\nu = 3/5$ is commonly used. For bead–spring chain models, a useful approximation to this repulsion is a potential that is derived from the free energy of two Gaussian chains forced to overlap (Jendrejack, de Pablo & Graham 2002):

$$\Phi_{\text{ev}}(\boldsymbol{R}_1, \boldsymbol{R}_2, \ldots, \boldsymbol{R}_{N_b}) = \sum_{\alpha=1}^{N_b} \sum_{\beta=1,\beta\neq\alpha}^{N_b} \frac{1}{2} \upsilon k_B T N_{\text{K,s}}^2 \left(\frac{3}{4\pi R_{g,s}^2}\right)^{3/2} \exp\left(-\frac{3R_{\alpha\beta}^2}{4R_{g,s}^2}\right), \quad (8.35)$$

where $\upsilon$ is an empirically determined excluded volume parameter and $R_{g,s}^2 = N_{\text{K,s}} L_{\text{K}}^2/6$ is the equilibrium radius of gyration of the subchain in the Gaussian chain approximation. Each bead then experiences a force

$$\boldsymbol{F}_{\text{ev},\alpha} = -\frac{\partial \Phi_{\text{ev}}}{\partial \boldsymbol{R}_\alpha}. \quad (8.36)$$

Because the spring and excluded volume forces act pairwise (there is no three-body interaction) and depend only on the relative positions of the beads, we can write in general that

$$\boldsymbol{F}_{\text{spring},\alpha} + \boldsymbol{F}_{\text{ev},\alpha} = \sum_{\beta\neq\alpha} \boldsymbol{F}_{\alpha\beta}(\boldsymbol{R}_{\alpha\beta}), \quad (8.37)$$

where $\boldsymbol{F}_{\alpha\beta}$ is the sum of the spring and excluded volume forces exerted by bead $\beta$ on bead $\alpha$. By Newton's third law, $\boldsymbol{F}_{\alpha\beta} = -\boldsymbol{F}_{\beta\alpha}$. Finally, strictly speaking a chain should be prevented from crossing itself even under $\theta$ conditions where there is no thermodynamic preference of a chain for the solvent. This strict topological constraint is rarely enforced in dilute solution bead–spring polymer simulations, although approaches for doing so have been proposed (Kumar & Larson 2001).

To incorporate the viscous drag exerted by the fluid on the polymer chain, each bead in the classical bead–spring chain model is treated as a sphere subject to Stokes's law with a friction coefficient $\zeta = 6\pi\eta a$, where $a$ is proportional to the equilibrium subchain size $R_{0,s} = L_K N_{K,s}{}^{1/2}$. As such, the chain is simply a system of $N_b$ point particles that interact hydrodynamically via Stokeslets or regularized Stokeslets as described in Section 2.5.

As discussed briefly in Section 3.6, hydrodynamic interactions promote collective motions – this point will be seen in a dramatic way in our treatment later of the diffusivity of a polymer chain. Confinement can significantly reduce hydrodynamic interactions because of the screening effects described in Section 4.3; additionally, in a confined system hydrodynamic interactions between and chain and a wall can lead to migration of the molecule during flow as described in Sections 4.4 and 9.3.1 (Graham 2011).

To complete the description of the bead–spring chain model, it remains to introduce the Brownian or fluctuating forces, as well as any external forces (buoyant, gravitational, confinement, etc.) that may be present in the system of interest. The assumption of rapid equilibration of the bead velocities to a Maxwellian is valid, and the Brownian forces are just as described in Chapter 6, so the $N$-particle Smoluchowski equation that we presented there is directly applicable in the present case. (We present the corresponding Langevin equation in the next section.) All that needs to be done is replace $\mathbf{F}_{\text{ext}}$ in (6.99) and (6.100) with $\mathbf{F}_{\text{spring}} + \mathbf{F}_{\text{ev}} + \mathbf{F}_{\text{ext}}$, to yield

$$\mathbf{J}_p = \left[\mathbf{V}_\infty + \mathbf{M} \cdot \left(\mathbf{F}_{\text{spring}} + \mathbf{F}_{\text{ev}} + \mathbf{F}_{\text{ext}} + \check{\mathbf{F}}_{\text{Brown}}\right)\right] p \tag{8.38}$$

and

$$\frac{\partial p}{\partial t} + \mathbf{V}_\infty \cdot \frac{\partial p}{\partial \mathbf{R}} = -\frac{\partial}{\partial \mathbf{R}} \cdot \left(\mathbf{M} \cdot \left(\mathbf{F}_{\text{spring}} + \mathbf{F}_{\text{ev}} + \mathbf{F}_{\text{ext}}\right) p\right) + \frac{\partial}{\partial \mathbf{R}} \left(k_B T \mathbf{M} \cdot \frac{\partial p}{\partial \mathbf{R}}\right). \tag{8.39}$$

In some situations, it is more convenient to keep track of the center of mass

$$\boldsymbol{R}_c = \frac{1}{N_b} \sum_{\alpha=1}^{N_b} \boldsymbol{R}_\alpha \tag{8.40}$$

and the connector vectors, writing the probability distribution for the molecule as $p(\boldsymbol{R}_c, \mathbf{Q}, t)$, where $\mathbf{Q}$ is a $3N_s$-dimensional vector containing the connector vectors $\boldsymbol{Q}_\alpha$. The transformation between the coordinate systems is as follows:

$$\boldsymbol{Q}_\alpha = \sum_{\beta=1}^{N_b} \bar{B}_{\alpha\beta} \boldsymbol{R}_\beta, \tag{8.41}$$

$$\boldsymbol{R}_\beta = \boldsymbol{R}_c + \sum_{\alpha=1}^{N_b-1} B_{\beta\alpha} \boldsymbol{Q}_\alpha. \tag{8.42}$$

Here

$$\bar{B}_{\alpha\beta} = \delta_{\alpha+1,\beta} - \delta_{\alpha\beta} \tag{8.43}$$

and

$$B_{\beta\alpha} = \begin{cases} \frac{\alpha}{N_b} & \alpha < \beta \\ -\left(1 - \frac{\alpha}{N_b}\right) & \alpha \geq \beta. \end{cases} \tag{8.44}$$

Strictly speaking, (8.39) governs a single molecule in solution. In dilute solution, however, each molecule behaves independently and the probability of finding a molecule at a particular position is proportional to the number density at that position. Consider a dilute polymer solution containing a total of $N_{\mathrm{mol}}$ molecules in a volume $V$. Because the solution is dilute, each of the molecules will obey the same probability distribution $p(\boldsymbol{R}_{\mathrm{c}}, \mathbf{Q}, t)$; this satisfies the normalization condition

$$\int\int p(\boldsymbol{R}_{\mathrm{c}}, \mathbf{Q}, t)\, d\boldsymbol{R}_{\mathrm{c}}\, d\mathbf{Q} = 1. \tag{8.45}$$

As usual,

$$\langle f \rangle = \int\int f(\boldsymbol{R}_{\mathrm{c}}, \mathbf{Q}) p(\boldsymbol{R}_{\mathrm{c}}, \mathbf{Q}, t)\, d\boldsymbol{R}_{\mathrm{c}}\, d\mathbf{Q}. \tag{8.46}$$

At each position in the fluid, the number density $n$ of polymer centers of mass will be given by

$$n(\boldsymbol{R}_{\mathrm{c}}, t) = N_{\mathrm{mol}} \int p(\boldsymbol{R}_{\mathrm{c}}, \mathbf{Q}, t)\, d\mathbf{Q}, \tag{8.47}$$

where the integral simply accounts for all molecules at position $\boldsymbol{R}_{\mathrm{c}}$ regardless of conformation $\mathbf{Q}$. Now we define the probability density of conformation, $\psi(\boldsymbol{R}_{\mathrm{c}}, \mathbf{Q}, t)$, for polymers residing at position $\boldsymbol{R}_{\mathrm{c}}$:

$$\psi(\boldsymbol{R}_{\mathrm{c}}, \mathbf{Q}, t) = \frac{N_{\mathrm{mol}}}{n(\boldsymbol{R}_{\mathrm{c}}, t)} p(\boldsymbol{R}_{\mathrm{c}}, \mathbf{Q}, t), \tag{8.48}$$

where

$$\int \psi(\boldsymbol{R}_{\mathrm{c}}, \mathbf{Q}, t)\, d\mathbf{Q} = 1 \tag{8.49}$$

at all $\boldsymbol{R}_{\mathrm{c}}$. In many situations, we will be interested in a quantity $f(\boldsymbol{R}_{\mathrm{c}}, \mathbf{Q}, t)$ (stress, for example) averaged over conformation at a given position $\boldsymbol{R}_{\mathrm{c}}$. We denote such a quantity as

$$\langle f(\boldsymbol{R}_{\mathrm{c}}, t) \rangle_Q = \int f(\boldsymbol{R}_{\mathrm{c}}, \mathbf{Q}, t)\, d\mathbf{Q}. \tag{8.50}$$

If $n$ is constant, then

$$\psi(\boldsymbol{R}_{\mathrm{c}}, \mathbf{Q}, t) = V p(\boldsymbol{R}_{\mathrm{c}}, \mathbf{Q}, t). \tag{8.51}$$

Finally, if both $n$ and $\psi$ are independent of $\boldsymbol{R}_{\mathrm{c}}$, then

$$\langle f \rangle = \langle f \rangle_Q. \tag{8.52}$$

## 8.3 Diffusivity of a Polymer Chain in Solution

The diffusivity of a polymer in dilute solution is a fundamental measure of the dynamics of the molecule and a sensitive indicator of both the structure of the polymer chain and the solvent in which it resides. Using the bead–spring chain model, we derive here an important classical formula for the short-time diffusivity. (It turns out that this quantity is almost identical to the long-time diffusivity but is much easier to derive (Liu & Dünweg 2003).) At the end of this section, we will use this result to address some broader issues in the equilibrium dynamics of polymer chains in solution.

At equilibrium in the absence of external forces, the (inertialess) SDE for a bead–spring chain can be written as

$$d\mathbf{R} = \left(\mathbf{M}\cdot\left(\mathbf{F}_{\text{spring}} + \mathbf{F}_{\text{ev}}\right) + kT\frac{\partial}{\partial\mathbf{R}}\cdot\mathbf{M}\right)dt + \mathbf{B}(\mathbf{R}(t))\cdot d\mathbf{W}, \tag{8.53}$$

$$\mathbf{B}\cdot\mathbf{B}^{\mathrm{T}} = 2k_BT\mathbf{M}. \tag{8.54}$$

The term $\frac{\partial}{\partial\mathbf{R}}\cdot\mathbf{M}$ vanishes for a system of point or regularized point particles in an unbounded domain, because $\boldsymbol{\nabla}\cdot\mathbf{G} = \mathbf{0}$. This SDE is equivalent to the Smoluchowski equation given by (8.39). Given (8.40), we can write the evolution equation for the change in center of mass:

$$d\boldsymbol{R}_c = \frac{1}{N_{\text{b}}}\sum_{\alpha}\sum_{\beta}\mathbf{M}_{\alpha\beta}\cdot\left(\boldsymbol{F}_{\text{spring},\beta} + \boldsymbol{F}_{\text{ev},\beta}\right)dt + \sum_{\alpha}\sum_{\beta}\boldsymbol{B}_{\alpha\beta}\cdot d\boldsymbol{W}_{\beta}. \tag{8.55}$$

The short-time diffusivity $D_{\text{s}}$ is defined by the relation

$$\langle d\boldsymbol{R}_c\cdot d\boldsymbol{R}_c\rangle = 6D_s dt. \tag{8.56}$$

From (8.55), we can write that

$$\langle d\boldsymbol{R}_c\cdot d\boldsymbol{R}_c\rangle = \left\langle\frac{1}{N_{\text{b}}^2}\sum_{\alpha}\sum_{\beta}\sum_{\gamma}\sum_{\delta}\left[\boldsymbol{B}_{\alpha\beta}\cdot d\boldsymbol{W}_{\beta}\right]\cdot\left[\boldsymbol{B}_{\gamma\delta}\cdot d\boldsymbol{W}_{\delta}\right]\right\rangle + O(dt\; dW) + O(dt^2). \tag{8.57}$$

The first term is $O(dW^2) = O(dt)$, which is much larger than the other terms, so they can be neglected. Now, denoting $\left(\boldsymbol{B}_{\alpha\beta}\cdot d\boldsymbol{W}_{\beta}\right)_m = B_{\alpha\beta}^{mn}dW_{\beta n}$ (summation implied), we rewrite (8.57) as

$$\begin{aligned}\langle dR_c^2\rangle &= \frac{1}{N_{\text{b}}^2}\sum_{\alpha}\sum_{\gamma}\langle B_{\alpha\beta}^{mn}dW_{\beta n}B_{\gamma\delta}^{mp}dW_{\delta p}\rangle\\ &= \frac{1}{N_{\text{b}}^2}\sum_{\alpha}\sum_{\gamma}\langle B_{\alpha\beta}^{mn}B_{\gamma\delta}^{mp}\rangle\langle dW_{\beta n}dW_{\delta p}\rangle.\end{aligned}$$

Finally, we note that $\langle dW_{\beta n}dW_{\delta p}\rangle = \delta_{\beta\delta}\delta_{np}dt$ and $\langle B_{\alpha\beta}^{mn}B_{\gamma\delta}^{mp}\rangle = 2k_BT\,\mathrm{Tr}\,\mathbf{M}_{\alpha\gamma}$, yielding that

$$\left\langle dR_c^2\right\rangle = \frac{2k_BT}{N_{\text{b}}^2}\sum_{\alpha}\sum_{\gamma}\langle\mathrm{Tr}\,\mathbf{M}_{\alpha\gamma}\rangle dt$$

and thus that

$$D_s = \frac{k_BT}{3N_{\text{b}}^2}\sum_{\alpha}\sum_{\gamma}\langle\mathrm{Tr}\,\mathbf{M}_{\alpha\gamma}\rangle. \tag{8.58}$$

In the polymer dynamics literature, this expression is called the *Kirkwood* diffusivity; we will denote it $D_{\text{K}}$.

Now recall the expression, (2.64), for the mobility of a system of point particles in an unbounded fluid:

$$\mathbf{M}_{\alpha\beta} = \frac{1}{\zeta}\delta\delta_{\alpha\beta} + (1 - \delta_{\alpha\beta})\mathbf{G}(\boldsymbol{R}_{\alpha\beta}).$$

Neglecting the hydrodynamic interactions between beads leads to the so-called "free-draining" approximation, where the mobility becomes

$$\mathbf{M}^{\mathrm{FD}}_{\alpha\beta} = \frac{1}{\zeta}\delta\delta_{\alpha\beta}.$$

In this approximation, **M** retains no dependence on the relative positions of the particles; each bead contributes equally and independently to the resistance to motion, and the resulting expression for the diffusivity is

$$D_{\mathrm{K}}^{\mathrm{FD}} = \frac{k_B T}{3N_{\mathrm{b}}^2}\frac{3N_{\mathrm{b}}}{\zeta} = \frac{k_B T}{N_{\mathrm{b}}\zeta}. \tag{8.59}$$

Notice the scaling with the number of beads, which in turn is proportional to $N_{\mathrm{m}}$: $D_{\mathrm{K}}^{\mathrm{FD}} \sim N_{\mathrm{m}}^{-1}$. This is called "Rouse" scaling; for relatively short polymer chains, it is an adequate representation of some experimental data for short chains (Smith, Perkins & Chu 1996), but it is not a good approximation for $N_{\mathrm{K}} \gg 1$.

A more accurate expression for $D_{\mathrm{K}}$ requires incorporation of the hydrodynamic interactions. In this case,

$$\begin{aligned}\langle \mathrm{Tr}\,\mathbf{M}_{\alpha\beta}\rangle &= \frac{3}{\zeta}\delta_{\alpha\beta} + \mathrm{Tr}\left\langle (1 - \delta_{\alpha\beta})\left(\frac{1}{8\pi\eta r_{\alpha\beta}}\left(\delta + \frac{\boldsymbol{R}_{\alpha\beta}\boldsymbol{R}_{\alpha\beta}}{r_{\alpha\beta}^2}\right)\right)\right\rangle \\ &= \frac{3}{\zeta}\delta_{\alpha\beta} + \frac{(1 - \delta_{\alpha\beta})}{2\pi\eta}\left\langle \frac{1}{R_{\alpha\beta}}\right\rangle.\end{aligned}$$

The resulting expression for diffusivity is

$$D_{\mathrm{K}}^{\mathrm{HI}} = \frac{k_B T}{N_{\mathrm{b}}\zeta} + \frac{k_B T}{6\pi\eta R_H}, \tag{8.60}$$

where

$$R_{\mathrm{H}} = \left(\frac{1}{N_{\mathrm{b}}^2}\sum_{\alpha}\sum_{\beta\neq\alpha}\left\langle\frac{1}{r_{\alpha\beta}}\right\rangle\right)^{-1} \tag{8.61}$$

is called the *hydrodynamic radius* of the polymer. This will be determined by the average conformation of the chain; for a linear ideal chain, $R_{\mathrm{H}} \sim N_{\mathrm{m}}^{1/2}$ due to the random-walk statistics of the chain (Problem 8.6). For sufficiently long chains, the second term in $D_{\mathrm{K}}^{\mathrm{HI}}$ dominates over the first, yielding the result that $D_{\mathrm{K}} \sim N_{\mathrm{m}}^{-1/2}$; this is called "Zimm" scaling, and is a good representation of experimental data for long chains under $\theta$-solvent conditions. For a sufficiently long polymer in a good solvent, the chain expands from the random walk conformation to a self-avoiding random walk (Doi & Edwards 1986), in which case $R_{\mathrm{H}} \sim N_{\mathrm{m}}^{\nu}$, where $\nu \approx 0.588$.

The scalings of the Rouse and Zimm models for diffusion are very different – the Rouse model dramatically overpredicts the resistance of a chain to diffusive motion. The origin of this substantial difference lies in our discussion in Section 2.5 of cooperative and competitive modes of motion in a system of hydrodynamically interacting particles. Diffusion is a cooperative motion of all the segments of the polymer molecule. As it moves, each segment of the chain drags fluid in the same direction, thus making it easier for the other segments to move in the same direction by reducing the relative velocity between the segment and the solvent. In the Rouse model, on the other hand, the fluid motion generated by the segments is neglected; each segment experiences a stationary solvent, resulting in a substantial overprediction of the drag on the chain. There are situations, however, in which the Rouse model may be appropriate, including a chain moving in a porous medium such as a gel or a chain in a concentrated solution or melt. In these cases, the primary resistance to motion is not a freely moving solvent, but rather a background of either elements of the porous medium or the physical contacts with other polymer molecules (Doi & Edwards 1986, Rubinstein & Colby 2003). Similarly, for a chain in solution in a confined geometry such as a slit or tube, we learned in Section 4.5 that hydrodynamic interactions are screened in these geometries – the momentum imparted to the fluid by a point force is absorbed by the walls within a distance on the order of the slit height or tube diameter. Thus polymers with $R_0$ much larger than this scale exhibit Rouse dynamics (Graham 2011).

Similar analyses can be performed to determine the molecular-weight scaling of a number of properties of long-chain polymers in solutions. They are all closely related to our result for diffusion, which indicates that at equilibrium the chain behaves in many ways as a sphere with radius $R_{\mathrm{H}}$. For example, the Stokes's law friction coefficient $\zeta_{\mathrm{p}}$ for a polymer with sufficiently weak forces exerted on it that it does not deform appreciably from its equilibrium conformation is given by the Stokes–Einstein expression $\zeta_{\mathrm{p}} = k_BT/D_{\mathrm{K}}$. If a polymer is exposed to flow or some other field that distorts it from its equilibrium conformation, the time scale $\tau$ for relaxation back to equilibrium can be estimated as the time it takes to diffuse its own size (de Gennes 1979):

$$\tau \sim \frac{R_0^2}{D_{\mathrm{K}}} \sim \frac{R_0^2 \zeta_{\mathrm{p}}}{k_BT}. \tag{8.62}$$

From this result, we can estimate how the relaxation time scales with molecular weight. For an ideal chain with Rouse dynamics, $R_0^2 \sim N_{\mathrm{m}}$ and $\zeta_{\mathrm{p}} \sim N_{\mathrm{m}}$, so

$$\tau \sim N_{\mathrm{m}}^2. \tag{8.63}$$

For Zimm dynamics (the more appropriate case for long chains in dilute solution), $\zeta_{\mathrm{p}} \sim R_{\mathrm{H}} \sim R_0 \sim N_{\mathrm{m}}^{1/2}$, so

$$\tau \sim R_0^3 \sim N_{\mathrm{m}}^{3/2}. \tag{8.64}$$

In a good solvent where $R_0 \sim N_{\mathrm{m}}^{\nu}$, this changes to

$$\tau \sim R_0^3 \sim N_{\mathrm{m}}^{3\nu}. \tag{8.65}$$

## 8.4 Equilibrium Dynamics of Internal Degrees of Freedom: The Relaxation Spectrum

The diffusivity of a polymer is a measure of the cooperative motion of the entire chain. Of course, parts of the chain can move relative to one another as well: i.e., there are internal chain motions with their own time scales. For a so-called *Rouse chain*, a bead–spring chain with Hookean springs and no hydrodynamic or excluded volume interactions, the SDE for motion of bead $\alpha$ simplifies to

$$d\boldsymbol{R}_\alpha = \frac{1}{\zeta}\left(\boldsymbol{F}_{c,\alpha} - \boldsymbol{F}_{c,\alpha-1}\right) + \sqrt{2D_b}\,d\boldsymbol{W}_\alpha, \tag{8.66}$$

where $D_b = k_BT/\zeta$ is the diffusivity of an individual bead and $\boldsymbol{F}_{c,\alpha} = H\boldsymbol{Q}_\alpha$. Since we are concerned here with the dynamics of the internal degrees of freedom, we write the equations of motion for the connector vectors $\boldsymbol{Q}_\alpha$:

$$d\boldsymbol{Q}_\alpha = d\boldsymbol{R}_{\alpha+1} - d\boldsymbol{R}_\alpha = \frac{1}{\zeta}\left(\boldsymbol{F}_{c,\alpha+1} - 2\boldsymbol{F}_{c,\alpha} + \boldsymbol{F}_{c,\alpha-1}\right) + \sqrt{2D_b}\left(d\boldsymbol{W}_{\alpha+1} - d\boldsymbol{W}_\alpha\right) \tag{8.67}$$

$$= \frac{H}{\zeta}\sum_{\beta=1}^{N_b-1} A_{\alpha\beta}\boldsymbol{Q}_\beta + \sqrt{2D_b}\sqrt{2}\,d\boldsymbol{W}'_\alpha, \tag{8.68}$$

where

$$A_{\alpha\beta} = \delta_{\alpha+1,\beta} - 2\delta_{\alpha\beta} + \delta_{\alpha-1,\beta} \tag{8.69}$$

and we have used (7.17) to write

$$d\boldsymbol{W}_{\alpha+1} - d\boldsymbol{W}_\alpha = \sqrt{2}\,d\boldsymbol{W}'_\alpha.$$

We now have a system of linear constant coefficient first-order differential equations of the general form $\dot{\boldsymbol{y}} = \boldsymbol{L}\cdot\boldsymbol{y} + \boldsymbol{f}(t)$; the usual way of solving such an equation is to rewrite it in the basis formed by the eigenvectors of the matrix $\boldsymbol{L}$. In this new basis, the equations become uncoupled and thus trivial to solve. In the present case, we need the eigenvalues and eigenvectors of the $N_b - 1 \times N_b - 1$ matrix $\boldsymbol{A}$, whose rows are $A_{\alpha\beta}$:

$$\boldsymbol{A} = \begin{bmatrix} -2 & 1 & & & & \\ 1 & -2 & 1 & & & \\ & 1 & -2 & 1 & & \\ & & \ddots & \ddots & \ddots & \\ & & & 1 & -2 & 1 \\ & & & & 1 & -2 \end{bmatrix}. \tag{8.70}$$

Ignoring for the moment the top and bottom rows, the eigenvalue problem for this matrix can be written

$$\phi_{\alpha+1} - 2\phi_\alpha + \phi_{\alpha-1} = a\phi_\alpha,$$

where $\phi_\alpha$ is the $\alpha$-component of the eigenvector and $a$ is an eigenvalue. This is a simple constant-coefficient difference equation. Seeking solutions in the form of complex exponentials $\phi_\alpha = e^{i\kappa\alpha} + \text{c.c.}$ and noting that $e^{i\kappa(\alpha\pm 1)} = e^{i\kappa\alpha}e^{\pm i\kappa}$, this equation becomes

$$a = e^{-\kappa} + e^{-i\kappa} - 2 = 2(\cos\kappa - 1).$$

As yet, $\kappa$ is arbitrary; it will be determined by applying the "boundary conditions," that is, the equations corresponding to connector vectors 1 and $N_b - 1$. By rewriting the eigenvector in the form $c\cos\kappa\alpha + d\sin\kappa\alpha$, it can be shown that to satisfy the equation corresponding to the top row of $\boldsymbol{A}$,

$$-2\phi_1 + \phi_2 = 2(\cos\kappa - 1)\phi_1,$$

the eigenvector must have the form

$$\phi_\alpha = \sin\kappa\alpha.$$

Similarly, to satisfy the bottom equation requires that

$$\sin\kappa N_b = 0$$

or

$$\kappa = \frac{n\pi}{N_b}, \quad n = 1, 2, \ldots, N_b - 1. \tag{8.71}$$

(One way to understand these results is to observe that $\boldsymbol{A}$ is a finite difference approximation to the second derivative operator $\frac{d^2}{dx^2}$ with homogeneous Dirichlet boundary conditions in the domain $0 < x < \ell$, the eigenfunctions of which are $\sin n\pi x/\ell$ for $n = 1, 2, 3, \ldots$.) To summarize, the eigenvalues and eigenvectors of $\boldsymbol{A}$ are

$$a_n = 2(\cos\frac{n\pi}{N_b} - 1), \quad n = 1, 2, \ldots, N_b - 1 \tag{8.72}$$

$$\phi_{\alpha n} = \sin\frac{n\pi\alpha}{N_b}, \tag{8.73}$$

where $\phi_{\alpha n}$ denotes the $\alpha$-component of the $n$th eigenvector. Each of these eigenvectors is called a *Rouse mode*. Because $\boldsymbol{A}$ is symmetric, the matrix $\boldsymbol{\phi}$ whose columns are the eigenvectors $\boldsymbol{\phi}_n$ is orthogonal: $\boldsymbol{\phi}^{-1} = \boldsymbol{\phi}^{\mathrm{T}}$. In this case, it is symmetric as well.

Now we can make a change of basis, letting

$$\boldsymbol{X}_n = \sum_{\alpha=1}^{N_b-1} \phi_{\alpha n}\boldsymbol{Q}_\alpha. \tag{8.74}$$

The Rouse mode amplitudes $\boldsymbol{X}_{n-1}$ are the Fourier sine series coefficients of the $\boldsymbol{Q}_\alpha$. Using this coordinate transformation and the eigenvalue relation $\boldsymbol{A}\boldsymbol{\phi}_n = a_n\boldsymbol{\phi}_n$, (8.68) can now be written

$$d\boldsymbol{X}_n = \sigma_n\boldsymbol{X}_n + \sqrt{4D_b}\sum_{\alpha=1}^{N_b-1}\phi_{n\alpha}d\boldsymbol{W}'_\alpha, \tag{8.75}$$

where

$$\sigma_n = \frac{H}{\zeta}a_n = \frac{2H}{\zeta}(\cos\frac{n\pi}{N_b} - 1). \tag{8.76}$$

Although the fluctuating terms here appear to be coupled through the eigenvector matrix $\boldsymbol{\phi}$, they are not, because again we can rewrite a linear combination of Wiener increments as another Wiener increment using (7.17):

$$\sum_{\alpha=1}^{N_b-1} \phi_{n\alpha} d\mathbf{W}'_\alpha = \left( \sum_{\alpha=1}^{N_b-1} \phi_{n\alpha}^2 \right)^{1/2} d\mathbf{W}''_n.$$

Thus by transforming into the Rouse mode basis, we have found a set of $N_b - 1$ independent equations that govern the internal dynamics. Each of the Rouse modes has a relaxation rate $\sigma_n$, or equivalently a relaxation time

$$\tau_n = -\frac{1}{\sigma_n} = \frac{\zeta}{2H(1 - \cos\frac{n\pi}{N_b})}. \tag{8.77}$$

This set of time scales is the *relaxation spectrum* of the Rouse chain. For $N_b \gg 1$ and $n/N_b \ll 1$, i.e., for the most slowly decaying modes of a long chain, we can Taylor-expand the cosine term to find that

$$\tau_n \approx \frac{\zeta}{2H} \frac{N_b^2}{n^2 \pi^2}. \tag{8.78}$$

We can attach a simple physical interpretation to this expression by inserting (8.25) and rewriting it as

$$\tau_n \sim \frac{R_n^2}{D_n}, \tag{8.79}$$

where $R_n^2 = N_{K,s} L_K^2 N_b / n$ is the mean squared end-to-end length of a chain with $N_b/n$ beads and $D_n = k_B T/(\zeta N_b/n)$ its diffusivity (in the absence of hydrodynamic interactions). Thus the $n$th relaxation time can be estimated as the time it takes for a chain of $N_b/n$ beads to diffuse its own size (we anticipated this result in the case of the longest relaxation time at the end of the previous section) and scales with chain length as follows:

$$\tau_n \sim \left( \frac{N_m}{n} \right)^2. \tag{8.80}$$

This dependency of relaxation time on chain length $N_m$ and mode number $n$ is called *Rouse scaling*.

We noted previously that neglecting hydrodynamic interactions is not realistic for long chains in dilute solution. If hydrodynamic interactions are included, then $\mathbf{M}$ depends on the bead positions so the Langevin equations for the bead motions are nonlinear. Thus the analysis just performed for the Rouse chain cannot be accomplished without further approximations. For example, replacing $\mathbf{M}$ with its value averaged over the equilibrium bead distribution (Doi & Edwards 1986, Bird, Armstrong, & Hassager 1987) leads to the *Zimm* model for the polymer relaxation spectrum. Rather than going through the detailed analysis, we simply point out that the scaling relation (8.79) continues to provide a useful approximation as long as the diffusivity $D_n$ is taken to obey *Zimm* scaling: $D_n \sim R_n^{-1} \sim \sqrt{n/N_b}$. In this case,

$$\tau_n \sim \left(\frac{N_m}{n}\right)^{3/2}. \tag{8.81}$$

Finally, under good solvent conditions, this expression becomes

$$\tau_n \sim \left(\frac{N_m}{n}\right)^{3\nu}. \tag{8.82}$$

## 8.5 Polymer Contribution to the Stress Tensor

In Sections 2.3 and 2.4, we learned that the stress exerted by a particle on a fluid is closely related to the force dipole induced by the presence of the particle. This is true as well for a polymer molecule, with an additional kinetic contribution due to the Brownian motion of the polymer segments. For a particle, the force dipole is given by (2.47):

$$\mathbf{D} = \int_{S_P} \boldsymbol{x}'(\hat{\boldsymbol{n}} \cdot \boldsymbol{\sigma})\, dS(\boldsymbol{x}').$$

For a polymer chain comprised of point-particle beads, the analogous expression is

$$\mathbf{D} = \sum_\alpha \boldsymbol{R}_\alpha(\boldsymbol{F}_{\text{drag},\alpha} + \breve{\boldsymbol{F}}_{\text{Brown},\alpha}), \tag{8.83}$$

where $\boldsymbol{F}_{\text{drag},\alpha}$ is the hydrodynamic (drag) force exerted by the fluid on bead $\alpha$. The sum of $\boldsymbol{F}_{\text{drag},\alpha}$ and $\breve{\boldsymbol{F}}_{\text{Brown},\alpha}$ is the total velocity-space averaged force exerted by the fluid on bead $\alpha$. From the force balance on each bead,

$$\boldsymbol{F}_{\text{drag},\alpha} + \breve{\boldsymbol{F}}_{\text{Brown},\alpha} = -(\boldsymbol{F}_{\text{spring},\alpha} + \boldsymbol{F}_{\text{ev},\alpha}) = \frac{\partial \Phi}{\partial \boldsymbol{R}_\alpha},$$

where $\Phi = \Phi_{\text{spring}} + \Phi_{\text{ev}}$ is the potential associated with the spring and excluded volume forces exerted between the beads. It is important to observe that this potential depends only on relative bead positions, not the absolute position of the chain in space.

In a dilute polymer solution, the polymer contribution to the stress, $\boldsymbol{\sigma}_p$, will be the sum of the ensemble-averaged force dipole per polymer times the number density of polymers and the kinetic stress described in Section 6.5:

$$\begin{aligned}
\boldsymbol{\sigma}_p &= n\langle \mathbf{D}\rangle_Q + \boldsymbol{\sigma}_{\text{kin}} \\
&= n\left\langle \sum_\alpha \boldsymbol{R}_\alpha(\boldsymbol{F}_{\text{drag},\alpha} + \breve{\boldsymbol{F}}_{\text{Brown},\alpha})\right\rangle_Q + \boldsymbol{\sigma}_{\text{kin}} \\
&= -n\left\langle \sum_\alpha \boldsymbol{R}_\alpha(\boldsymbol{F}_{\text{spring},\alpha} + \boldsymbol{F}_{\text{ev},\alpha})\right\rangle_Q + \boldsymbol{\sigma}_{\text{kin}} \\
&= n\left\langle \sum_\alpha \boldsymbol{R}_\alpha \frac{\partial \Phi}{\partial \boldsymbol{R}_\alpha}\right\rangle_Q + \boldsymbol{\sigma}_{\text{kin}}.
\end{aligned} \tag{8.84}$$

The total stress will be

$$\sigma = -p\delta + 2\eta \mathbf{E} + \sigma_{\rm p}, \tag{8.85}$$

which is equivalent to the expression (2.57) we found for a suspension under the assumption that ensemble averages and volume averages are equivalent. The kinetic stress arises from the independent contributions of the velocities of all the beads and is given simply by

$$\sigma_{\rm kin} = -nN_{\rm b}k_BT\delta. \tag{8.86}$$

This would be the osmotic stress of the solution if the monomers were not connected – the monomer number density is $Nn$. But the monomers are connected, so we expect the total stress at equilibrium to instead be the conventional dilute solution osmotic stress (which is independent of molecular structure):

$$\sigma_{p,\rm eq} = -nk_BT\delta. \tag{8.87}$$

At equilibrium, (8.84)–(8.86) imply that

$$n\left\langle \sum_\alpha \boldsymbol{R}_\alpha \frac{\partial \Phi}{\partial \boldsymbol{R}_\alpha} \right\rangle_Q = (N_{\rm b} - 1)nk_BT\delta. \tag{8.88}$$

A direct calculation by evaluation of the left-hand side of (8.88) verifies this result; this calculation requires use of integration by parts, the fact that the interbead potential $\Phi$ does not depend on the center-of-mass position of the polymer, and the use of the equilibrium (Boltzmann) distribution function $p_{\rm eq}(\mathbf{R}) = p_0 e^{-\Phi(\mathbf{R})/k_BT}$, which vanishes as $|\boldsymbol{R}_\alpha - \boldsymbol{R}_c| \to \infty$.

Often the stress is written in terms of the deviation from the equilibrium value: i.e.,

$$\tau_{\rm p} = \sigma_{\rm p} - \sigma_{\rm p,eq} = -(N_{\rm b} - 1)nk_BT\delta - n\left\langle \sum_\alpha \boldsymbol{R}_\alpha (\boldsymbol{F}_{{\rm spring},\alpha} + \boldsymbol{F}_{{\rm ev},\alpha}) \right\rangle_Q. \tag{8.89}$$

This expression is sometimes known as the *modified Kramers* form of the stress tensor (Bird et al. 1987). Now we can write

$$\sigma = -(p + nk_BT)\delta + \tau. \tag{8.90}$$

Here

$$\tau = \tau_{\rm s} + \tau_{\rm p}, \tag{8.91}$$

where

$$\tau_{\rm s} = 2\eta \mathbf{E} \tag{8.92}$$

is the viscous stress arising from the solvent and $\tau_{\rm p}$ is given by (8.89). In applications where the polymer number density $n$ is constant, the osmotic term $-nk_BT\delta$ is generally absorbed into the pressure.

The stress tensor can also be written in terms of $\mathbf{Q}$ rather than $\mathbf{R}$ (Problem 8.2) as

$$\tau_{\rm p} = -(N_{\rm b} - 1)nk_BT\delta + n\sum_{\alpha=1}^{N_{\rm b}-1} \langle \boldsymbol{Q}_\alpha \boldsymbol{F}_{{\rm c},\alpha}\rangle_Q - n\sum_{\alpha=1}^{N_{\rm b}-1}\sum_{\beta=1}^{N_{\rm b}} B_{\beta\alpha}\left\langle \boldsymbol{Q}_\alpha \boldsymbol{F}_{{\rm ev},\beta}\right\rangle. \tag{8.93}$$

This is the so-called *Kramers* form of the stress tensor. Finally, returning to (8.83) it can be shown that the direct Brownian contribution to the dipole

$$\sum_{\alpha} \boldsymbol{R}_{\alpha} \breve{\boldsymbol{F}}_{\text{Brown},\alpha}$$

has the same value irrespective of whether the molecule is at equilibrium or not, in which case

$$\boldsymbol{\tau}_{\text{p}} = \boldsymbol{\sigma}_{\text{p}} - \boldsymbol{\sigma}_{\text{p,eq}} = n \sum_{\alpha} \langle \boldsymbol{R}_{\alpha} \boldsymbol{F}_{\text{drag},\alpha} \rangle_{Q}. \tag{8.94}$$

This is the *Kramers–Kirkwood* form of the stress tensor. Although presented here for bead–spring chain models, its derivation (Problem 8.4) makes no reference to the forces between beads, so it also holds for models such as the rigid rod or freely jointed chain, in which the spring forces are replaced by constraint forces that keep the chain segments at constant length. We use this form of the stress tensor in Section 8.7, where we study the rigid rod model.

## 8.6 Bead–Spring Dumbbell Model

The bead–spring chain model has allowed us to understand the molecular-weight dependence of the diffusivity and the presence of internal modes of relaxation of a polymer in solution. For computations of polymer dynamics in flow, however, incorporating a large number of internal degrees of freedom is analytically intractable and computationally cumbersome and comes with some loss of transparency regarding mechanistic understanding of the interaction between polymer dynamics in flow. And in any case, in flow the most important relaxation mode of the molecule is the slowest: as the deformation rate in a flow increases, this will be the first mode to be driven from equilibrium. It also makes the largest contribution to the stress because it involves the largest-scale collective motion of the chain segments – it generates the largest force dipole. Put more simply, the main effect of flow on a flexible polymer molecule is to stretch and orient it. These effects can be qualitatively captured with a bead–spring chain model with only two beads: i.e., a dumbbell (Figure 8.2(a)). The spring constant for this model comes from the force required to separate the two ends of the chain, which near equilibrium is given by Hooke's law with spring constant given by (8.25) with $L_{\text{s}} = L$. Using the ideal chain relation $R_0^2 = L_{\text{K}} L$, this yields

$$H = \frac{3k_B T}{R_0^2}. \tag{8.95}$$

We present here the Smoluchowski equation for the bead–spring dumbbell model as well as a generalized Fick's law expression for the flux. With some further approximations, we then generate an evolution equation for the stress tensor. It will be convenient to change coordinates to explicitly address the evolution of the center of mass of the dumbbell, $\boldsymbol{R}_c = \frac{1}{2}(\boldsymbol{R}_1 + \boldsymbol{R}_2)$ and the end-to-end vector, $\boldsymbol{Q} = \boldsymbol{R}_2 - \boldsymbol{R}_1$. In these coordinates, with $\boldsymbol{J}_{\text{c}}$

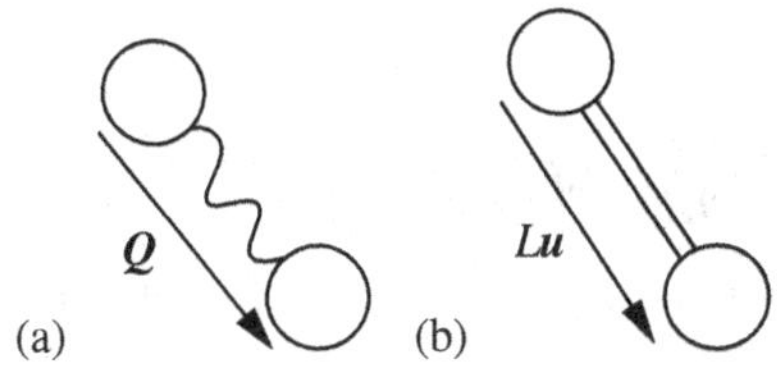


**Figure 8.2** Bead–spring (a) and rigid dumbbell (b) models for flexible and rigid polymers, respectively.

and $\boldsymbol{J}_Q$ representing the probability fluxes for $\boldsymbol{R}_c$ and $\boldsymbol{Q}$, respectively, the Smoluchowski equation has the form

$$\frac{\partial p(\boldsymbol{Q}, \boldsymbol{R}_c)}{\partial t} = -\frac{\partial}{\partial \boldsymbol{R}_c} \cdot \boldsymbol{J}_c - \frac{\partial}{\partial \boldsymbol{Q}} \cdot \boldsymbol{J}_Q. \tag{8.96}$$

The center-of-mass flux is given by

$$\boldsymbol{J}_c = \frac{1}{2}(\boldsymbol{J}_2 + \boldsymbol{J}_1) = \left[\frac{1}{2}\sum_{\alpha=1}^{2} \boldsymbol{v}_\infty(\boldsymbol{R}_\alpha) + \frac{1}{2}\sum_{\alpha=1}^{2}\sum_{\beta=1}^{2} \mathbf{M}_{\alpha\beta} \cdot \check{\boldsymbol{F}}_{\text{Brown},\beta}\right] p(\boldsymbol{R}_c, \boldsymbol{Q}, t). \tag{8.97}$$

The spring forces do not appear in this expression; they cancel out because $\boldsymbol{F}_{\text{spring},2} = -\boldsymbol{F}_{\text{spring},1}$. The connector vector flux is given by

$$\begin{aligned} \boldsymbol{J}_Q &= \boldsymbol{J}_2 - \boldsymbol{J}_1 \\ &= \left[(\boldsymbol{v}_\infty(\boldsymbol{R}_2) - \boldsymbol{v}_\infty(\boldsymbol{R}_1)) + \sum_{\alpha=1}^{2} (\mathbf{M}_{2\alpha} - \mathbf{M}_{1\alpha}) \cdot \left(\boldsymbol{F}_{\text{spring},\alpha} + \check{\boldsymbol{F}}_{\text{Brown},\alpha}\right)\right] p(\boldsymbol{R}_c, \boldsymbol{Q}, t). \end{aligned} \tag{8.98}$$

In these two equations, the terms inside the square brackets are the (velocity-space averaged) center-of-mass velocity and rate of change of connector vector, respectively. Each of these equations has a contribution from the imposed velocity field $\boldsymbol{v}_\infty$. In most situations, velocity gradients occur on much larger scales than the polymer molecules themselves, in which case it is appropriate to replace these terms by their Taylor expansions:

$$\frac{1}{2}\sum_{\alpha=1}^{2} \boldsymbol{v}_\infty(\boldsymbol{R}_i) \approx \boldsymbol{v}_\infty(\boldsymbol{R}_c) + \frac{1}{8}\boldsymbol{Q}\boldsymbol{Q} : \boldsymbol{\nabla}\boldsymbol{\nabla}\boldsymbol{v}_\infty(\boldsymbol{R}_c), \tag{8.99}$$

$$\frac{1}{2}\sum_{\alpha=1}^{2} (\boldsymbol{v}_\infty(\boldsymbol{R}_2) - \boldsymbol{v}_\infty(\boldsymbol{R}_1)) \approx \boldsymbol{Q} \cdot \boldsymbol{\nabla}\boldsymbol{v}_\infty(\boldsymbol{R}_c) = \mathbf{L}_\infty \cdot \boldsymbol{Q}. \tag{8.100}$$

Here $\mathbf{L}_\infty = \boldsymbol{\nabla}\boldsymbol{v}_\infty^{\mathrm{T}}$. Thus to a first approximation, the imposed velocity field convects the molecule with the velocity at the position of the center of mass, but with a correction that arises from gradients in velocity gradient, and it stretches the molecule as if it were a material line (cf. (1.1)).

With these expressions, we can now determine the form of the center-of-mass flux for the dumbbell model. Integrating (8.96) over $\boldsymbol{Q}$ and applying (8.47) gives the governing equation for the number density $n(\boldsymbol{R}_c, t)$,

$$\frac{\partial n}{\partial t} = -\frac{\partial}{\partial \boldsymbol{R}_c} \cdot \langle \boldsymbol{J}_c \rangle_Q .$$

To determine the conformation-averaged flux $\langle \boldsymbol{J}_c \rangle_Q$, it will be useful to first define the *Kirkwood diffusivity tensor*:

$$\boldsymbol{D}_K = \frac{1}{4} k_B T \sum_{\alpha=1}^{2} \sum_{\beta=1}^{2} \mathbf{M}_{\alpha\beta}. \tag{8.101}$$

Taking the trace and ensemble average of (8.101) yields the short-time (Kirkwood) diffusivity result (8.58) for the case $N_b = 2$. To determine $\langle \boldsymbol{J}_c \rangle_Q$, we average (8.97) over $\boldsymbol{Q}$ and apply (8.48), finding that (Ma & Graham 2005)

$$\begin{aligned} \langle \boldsymbol{J}_c \rangle_Q = n\boldsymbol{v}_\infty - \langle \boldsymbol{D}_K \rangle_Q \cdot \frac{\partial n}{\partial \boldsymbol{R}_c} + \frac{n}{8} \langle \boldsymbol{Q}\boldsymbol{Q} \rangle_Q : \boldsymbol{\nabla\nabla v}_\infty \\ - \left\langle \boldsymbol{D}_K \cdot \frac{\partial \ln \psi}{\partial \boldsymbol{R}_c} \right\rangle_Q n. \end{aligned} \tag{8.102}$$

The first two terms on the right-hand side are simply the fluxes due to convective transport and Fickian diffusion, respectively, where the diffusivity $\langle \boldsymbol{D}_K \rangle_Q$ can in general be anisotropic because the polymer conformation and thus the friction on the polymer will change as the polymer is deformed by flow. The third term arises if the polymer is large enough to sample gradients in the velocity gradient, as may happen, for example, for large polymers such as genomic DNA or other biopolymers in a microfluidic geometry, and the last term is a flux that arises due to spatial variations in the diffusivity itself. Again, this term will generally only arise when there are small-scale gradients in the velocity gradient, which will yield gradients in the diffusivity. Finally, for a polymer near solid boundaries, there is an additional contribution to the flux, not included here, that arises from hydrodynamic migration. If the polymer is treated as a hydrodynamic point force dipole, this term is precisely as described in Section 4.4 – see also Section 9.3. For many circumstances, including most macroscopic flows, the first two terms are completely dominant and the others can be neglected. Furthermore, if the dumbbell is considered to be free-draining, then $\langle \boldsymbol{D}_K \rangle_Q = \frac{k_B T}{2\zeta} \boldsymbol{\delta}$. In this case,

$$\langle \boldsymbol{J}_c \rangle_Q = n\boldsymbol{v}_\infty - D \frac{\partial n}{\partial \boldsymbol{R}_c}, \tag{8.103}$$

where

$$D = \frac{k_B T}{2\zeta} \tag{8.104}$$

in agreement with (8.59) when $N_b = 2$.

The other primary object of interest is the stress tensor. To make analytical progress in developing a closed form expression for evolution of the stress, we will make two simplifications: (1) the polymer concentration $n$ is spatially uniform, and (2) hydrodynamic interactions are neglected so $\mathbf{M}_{\alpha\beta} = \frac{1}{\zeta} \boldsymbol{\delta} \delta_{\alpha\beta}$ and we can replace $\boldsymbol{v}_\infty$ with $\boldsymbol{v}$ and $\mathbf{L}_\infty$

with $\mathbf{L}$. Furthermore, since there are only two beads, any excluded volume forces can be included in the expression for the spring force. Now (8.93) reduces to

$$\boldsymbol{\tau}_{\mathrm{p}} = -nk_BT\boldsymbol{\delta} + n\langle \boldsymbol{Q}\boldsymbol{F}_c\rangle_Q . \tag{8.105}$$

Here $\boldsymbol{F}_{\mathrm{c}} = \boldsymbol{F}_{\mathrm{spring},1} = -\boldsymbol{F}_{\mathrm{spring},2}$. For the Hookean dumbbell, where $\boldsymbol{F}_{\mathrm{c}} = H\boldsymbol{Q}$, this simplifies to

$$\boldsymbol{\tau}_{\mathrm{p}} = -nk_BT\boldsymbol{\delta} + nH\langle \boldsymbol{Q}\boldsymbol{Q}\rangle_Q . \tag{8.106}$$

This expression motivates us to seek an evolution equation for a nondimensional *conformation tensor*

$$\boldsymbol{\alpha} = \frac{H}{k_BT}\langle \boldsymbol{Q}\boldsymbol{Q}\rangle_Q .$$

At equilibrium, $\langle Q^2\rangle_Q = R_0^2 = 3k_BT/H$, so $\boldsymbol{\alpha} = \boldsymbol{\delta}$. The stress tensor for the Hookean case is then

$$\boldsymbol{\tau}_{\mathrm{p}} = G_0(\boldsymbol{\alpha} - \boldsymbol{\delta}), \tag{8.107}$$

where $G_0 = nk_BT$ is the *shear modulus* of the solution. To find the evolution equation for $\boldsymbol{\alpha}$ as a function of position $\boldsymbol{R}_{\mathrm{c}}$, we multiply both sides of the Smoluchowski equation, (8.96), by $\boldsymbol{Q}\boldsymbol{Q}$ and take the average over $\boldsymbol{Q}$:

$$\int \boldsymbol{Q}\boldsymbol{Q}\,\frac{\partial p(\boldsymbol{R}_{\mathrm{c}}, \boldsymbol{Q}, t)}{\partial t}\,d\boldsymbol{Q} = -\int \boldsymbol{Q}\boldsymbol{Q}\frac{\partial}{\partial \boldsymbol{R}_{\mathrm{c}}}\cdot \boldsymbol{J}_c\,d\boldsymbol{Q} - \int \boldsymbol{Q}\boldsymbol{Q}\frac{\partial}{\partial \boldsymbol{Q}}\cdot \boldsymbol{J}_Q\,d\boldsymbol{Q}. \tag{8.108}$$

The number density $n$ will drop out because it is constant. Inserting (8.97), (8.98), and (8.100) and the free-draining expression for the mobility, $\mathsf{M}_{\alpha\beta} = \frac{1}{\zeta}\boldsymbol{\delta}\delta_{\alpha\beta}$, results in the equation[1]

$$\frac{\partial \boldsymbol{\alpha}}{\partial t} + \boldsymbol{v}\cdot\nabla\boldsymbol{\alpha} = \mathbf{L}\cdot\boldsymbol{\alpha} + \boldsymbol{\alpha}\cdot\mathbf{L}^{\mathrm{T}} - \frac{1}{G_0\lambda}\boldsymbol{\tau}_{\mathrm{p}} + D\nabla^2\boldsymbol{\alpha}, \tag{8.109}$$

where

$$\lambda = \frac{\zeta}{4H} \tag{8.110}$$

is the *stress relaxation time* for the polymer molecule and $D$ is the diffusivity given by (8.104). In Section 8.4, we introduced the spectrum of relaxation times for a bead–spring chain. For the dumbbell case, there is just one relaxation mode, given by $\boldsymbol{Q}$ itself, with a relaxation time given by (8.77) with $n = 1$ and $N_{\mathrm{b}} = 2$: $\tau_1 = \zeta/2H$. This differs from $\lambda$ by a factor of two, because stress is a quadratic function $\boldsymbol{Q}$ and thus relaxes twice as fast as $\boldsymbol{Q}$ does. Observe that (8.109) can be rewritten

$$\overset{\triangledown}{\boldsymbol{\alpha}} = -\frac{1}{G_0\lambda}\boldsymbol{\tau}_{\mathrm{p}} + D\nabla^2\boldsymbol{\alpha}. \tag{8.111}$$

Recalling the results of Section 1.1.2, the upper convected derivative arises because in the absence of relaxation or Brownian motion, $\boldsymbol{Q}$ would evolve according to the same equation as a material line, so $\boldsymbol{Q}\boldsymbol{Q}$ would evolve as does the Green tensor, the upper

[1] Because $\boldsymbol{\alpha}$ depends only on center of mass position, we simplify notation by replacing $\partial/\partial\boldsymbol{R}_{\mathrm{c}}$ with $\nabla$.

convected derivative of which equals zero. Further discussion of convected derivatives and evolution equations for complex fluids is found in Section 9.4.

For the case of a Hookean dumbbell, we can eliminate $\boldsymbol{\tau}_{\mathrm{p}}$ to yield a closed evolution equation for $\boldsymbol{\alpha}$:

$$\overset{\nabla}{\boldsymbol{\alpha}} = -\frac{1}{\lambda}(\boldsymbol{\alpha} - \boldsymbol{\delta}) + D\nabla^2\boldsymbol{\alpha}. \tag{8.112}$$

Clearly, in the absence of flow, any initial deviation from the equilibrium conformation $\boldsymbol{\alpha} = \boldsymbol{\delta}$ will relax away as $e^{-t/\lambda}$. Alternatively, using (8.107) one can directly write an evolution equation for the polymer stress tensor:

$$\boldsymbol{\tau}_{\mathrm{p}} + \lambda\overset{\nabla}{\boldsymbol{\tau}}_{\mathrm{p}} = 2G_0\lambda\mathbf{E} + D\nabla^2\boldsymbol{\tau}_{\mathrm{p}}. \tag{8.113}$$

The quantity $\sqrt{D\lambda}$ estimates the distance a polymer will diffuse in one relaxation time and is about the size of the polymer at equilibrium; i.e., $D\lambda \sim R_0^2$. Unless stress gradients arise on the scale $R_0$, the stress diffusion term is very small and is often neglected. Equation (8.113) with $D = 0$ is commonly called the "Oldroyd-B" model; it is the starting point for viscoelastic fluid mechanics. The total stress $\boldsymbol{\tau}$ in the fluid will be given by the sum of the solvent and polymer contributions:

$$\boldsymbol{\tau} = 2\eta\mathbf{E} + \boldsymbol{\tau}_{\mathrm{p}}. \tag{8.114}$$

For steady simple shear flow with $\boldsymbol{v} = \dot{\gamma}y\boldsymbol{e}_x$, the shear stress satisfies

$$\tau_{yx} = \eta\dot{\gamma} + G_0\lambda\dot{\gamma},$$

so the quantity $\eta_{\mathrm{p}} = G_0\lambda$ is the polymer contribution to the steady shear viscosity of the fluid and $\beta = \eta/(\eta + \eta_{\mathrm{p}})$ is the ratio of the solvent contribution to the viscosity to the total viscosity $\eta + \eta_{\mathrm{p}}$.

As an aside, we note that when $\eta = 0$, the solvent contribution to the stress vanishes and the Oldroyd-B model becomes the so-called *upper convected Maxwell (UCM) model*. This is sometimes used to describe polymer melts rather than solutions. Additionally, in the limit $1/\lambda \to 0$, the upper convected Maxwell model becomes the *neo-Hookean* model for the stress $\boldsymbol{\tau}$ in an elastic solid:

$$\overset{\nabla}{\boldsymbol{\tau}} = 2G_0\mathbf{E} \tag{8.115}$$

or equivalently

$$\boldsymbol{\tau} = G_0(\mathbf{B} - \boldsymbol{\delta}), \tag{8.116}$$

where **B** is the Green tensor. The former equation is the upper convected derivative of the latter. This a useful model for rubberlike materials under moderate strains. The Hookean solid model is simply a neo-Hookean model in the limit of infinitesimal deformations.

Before leaving the realm of the Hookean spring, we note that the Oldroyd-B equation also governs the stress associated with each mode of a Rouse chain (Bird, Armstrong, & Hassager 1987). The polymer contribution to the stress tensor can be written

$$\boldsymbol{\tau}_{\mathrm{p}} = \sum_{n=1}^{N_{\mathrm{b}}-1} \boldsymbol{\tau}_{\mathrm{p}n}, \tag{8.117}$$

where $\tau_{pn}$ satisfies

$$\tau_{pn} + \lambda_n \overset{\nabla}{\tau}_{pn} = 2G_0\lambda_n \mathbf{E}. \tag{8.118}$$

Here $\lambda_n = 2\tau_n$ with $\tau_n$ given by (8.77). Observe that the viscosity associated with the $n$th mode is $G_0\lambda_n$. Because $\lambda_n$ varies inversely with $n$ (see (8.80)–(8.82)), the internal modes contribute less and less to the stress as $n$ increases. Thus in many fluid dynamical processes the longest relaxation time dominates, and to obtain qualitative understanding of these processes, a dumbbell model, which captures this time scale, suffices.

The Hookean spring model does not take into account the finite length of the polymer molecules. Recall that in Section 8.2, we introduced the FENE spring. Applying this to the dumbbell model results in a stress tensor

$$\boldsymbol{\tau}_{\mathrm{p}} = -nk_BT\boldsymbol{\delta} + nH\left\langle \frac{\mathbf{QQ}}{1-\frac{Q^2}{L^2}} \right\rangle_Q . \tag{8.119}$$

This equation cannot be expressed exactly in terms of $\boldsymbol{\alpha}$, which means that for the FENE dumbbell model there is not an exact closed form evolution equation for $\boldsymbol{\alpha}$. There is, however, an approximation called the *Peterlin* closure that results in a closed equation for $\alpha$ and works well in many circumstances:

$$\left\langle \frac{\mathbf{QQ}}{1-\frac{Q^2}{L^2}} \right\rangle_Q \approx \frac{\langle \mathbf{QQ}\rangle_Q}{1-\frac{\langle Q^2\rangle_Q}{L^2}}.$$

The resulting "FENE-P" stress tensor is

$$\boldsymbol{\tau}_p = nk_BT\left(\frac{\boldsymbol{\alpha}}{1-\frac{\mathrm{tr}\boldsymbol{\alpha}}{b}} - \boldsymbol{\delta}\right), \tag{8.120}$$

where $b = L^2H/k_BT$. Inserting (8.25) for the spring constant of an ideal chain of length $L = N_K L_K$ yields that $b = 3N_K$. Now, (8.109) and (8.120) fully describe the evolution of the stress tensor. This model, usually with $D = 0$, is very widely used in theoretical and computational studies of the fluid dynamics of polymer solutions, because it captures the main qualitative aspects of the polymer dynamics: the stretching and orientation of finitely extensible polymers in flow and the corresponding stresses that those polymers exert on the solvent. It is substantially more realistic than the Oldroyd-B model in a number of respects – in particular, it predicts that the polymer contribution to the shear viscosity decreases with increasing shear rate, while in the Oldroyd-B model the viscosity is constant; and the polymer contribution to the extensional viscosity is bounded at high extension rates, while in the Oldroyd-B model it blows up at a finite extension rate (proportional to $1/\lambda$). We elaborate on these points in Chapter 9.

## 8.7 Rigid Rod Model

At the opposite limit of the flexible chain models described so far is the rigid rod model. This model describes the dynamics of an object whose persistence length is much longer than its contour length. In flow situations, the rigidity assumption also requires that the rod be able to resist flow-induced deformations. We focus on the case of a fore-aft-symmetric and axisymmetric rod, in which case the translational and rotational degrees of freedom are decoupled from one another (see Section 3.7). A helix, for example, does not lie in this category, since application of a torque can lead to translational motion.

In Section 7.7, we addressed the rotational and translational diffusion of a Brownian rigid rod. Here we consider the rotational dynamics and the stress associated with a dilute solution of rigid rods in an unbounded flow. For rods with number density $n$ and length $L$, the solution is dilute if $nL^3 \ll 1$, semidilute if $nL^3 \approx 1$, and concentrated if $nL^3 \gg 1$. We will take the number density of rods to be spatially uniform and neglect the spatial diffusion of rod orientations; as in the previous dumbbell case, this effect is only important if very large gradients in orientation are generated. We also take the rod to be small enough that it experiences a locally linear flow with angular velocity $\boldsymbol{\omega}$ and strain rate $\mathbf{E}$. In contrast to Section 8.6 on the bead–spring dumbbell model, where we started from the Smoluchowski equation, here we begin with the Langevin equation arising from the torque balance on the rod:

$$\boldsymbol{I} \cdot \frac{d\boldsymbol{\Omega}}{dt} = \boldsymbol{T}_{\text{drag}} + \boldsymbol{T}_{\text{ext}} + \boldsymbol{T}_{\text{fluc}}, \tag{8.121}$$

where $\boldsymbol{T}_{\text{fluc}}$ is the fluctuating torque exerted on the rod by the fluid and

$$\boldsymbol{I} = I^{\parallel}\boldsymbol{uu} + I^{\perp}(\boldsymbol{\delta} - \boldsymbol{uu}) \tag{8.122}$$

is the moment of inertia tensor for the rod. Because we assume constant number density and negligible conformational diffusion, here and below the time derivatives are assumed to be taken in a reference frame moving with the rod, which in turn is moving with the fluid velocity $\boldsymbol{v}$. The drag torque is given by (3.83). In the absence of inertia or fluctuations, (8.121) reduces to (3.82). By separately considering the motions along $\boldsymbol{u}$ and orthogonal to it in the absence of flow or external torques, the analysis of Section 6.2 can be used to show (Problem 6.4) that $\boldsymbol{T}_{\text{fluc}}$ satisfies

$$\langle \boldsymbol{T}_{\text{fluc}}(t)\boldsymbol{T}_{\text{fluc}}(t+\tau)\rangle = 2k_BT\delta(\tau)\boldsymbol{\zeta}_{\text{r}}. \tag{8.123}$$

In Section 3.7, we derived the inertialess evolution equation for $\boldsymbol{u}$ in the absence of Brownian motion. Incorporating it yields

$$\frac{d\boldsymbol{u}}{dt} = \boldsymbol{\omega} \times \boldsymbol{u} + B(\boldsymbol{\delta} - \boldsymbol{uu}) \cdot \mathbf{E} \cdot \boldsymbol{u} + \frac{1}{\zeta_{\text{r}}}\boldsymbol{T}_{\text{ext}} \times \boldsymbol{u} + \frac{1}{\zeta_{\text{r}}}\boldsymbol{T}_{\text{fluc}} \times \boldsymbol{u}. \tag{8.124}$$

Application of this equation requires the following result:

$$\langle (\boldsymbol{T}_{\text{fluc}}(t) \times \boldsymbol{u}(t))\,(\boldsymbol{T}_{\text{fluc}}(t+\tau) \times \boldsymbol{u}(t+\tau))\rangle = 2k_BT\zeta_{\text{r}}\delta(\tau)\,(\boldsymbol{\delta} - \boldsymbol{uu}), \tag{8.125}$$

which follows from (8.123). Now we can recast (8.124) as an SDE:

$$d\boldsymbol{u} = \left[\left(\boldsymbol{\omega} + \frac{1}{\zeta_{\text{r}}}\boldsymbol{T}_{\text{ext}}\right) \times \boldsymbol{u} + B(\boldsymbol{\delta} - \boldsymbol{uu}) \cdot \mathbf{E} \cdot \boldsymbol{u} - \frac{2k_BT}{\zeta_{\text{r}}}\boldsymbol{u}\right] dt + \sqrt{\frac{2k_BT}{\zeta_{\text{r}}}}\,(\boldsymbol{\delta} - \boldsymbol{uu}) \cdot d\boldsymbol{W}. \tag{8.126}$$

The last term inside the square brackets is the drift arising from the constraint that $\boldsymbol{u}$ has unit magnitude, as discussed in Section 7.5.2. Finally, taking the rod to move with the local fluid velocity $\boldsymbol{v}$, we can write the evolution equation for the orientational probability density $\psi(\boldsymbol{u}, t)$ that follows directly from (8.126) and (7.105):

$$\frac{D\psi(\boldsymbol{u},t)}{Dt} = -\frac{\partial}{\partial \boldsymbol{u}} \cdot \left( \left( \left( \omega_\infty + \frac{1}{\zeta_\mathrm{r}} \boldsymbol{T}_\mathrm{ext} \right) \times \boldsymbol{u} + B\,(\boldsymbol{\delta} - \boldsymbol{u}\boldsymbol{u}) \cdot \mathbf{E} \cdot \boldsymbol{u} \right) \psi(\boldsymbol{u},t) \right)$$
$$+ D_\mathrm{r} \frac{\partial}{\partial \boldsymbol{u}} \cdot \frac{\partial}{\partial \boldsymbol{u}} \psi(\boldsymbol{u},t), \tag{8.127}$$

where

$$D_\mathrm{r} = \frac{k_B T}{\zeta_\mathrm{r}} \tag{8.128}$$

is the rotational diffusivity. The distribution $\psi(\boldsymbol{u}, t)$ satisfies the normalization condition

$$\int_{||\boldsymbol{u}||=1} \psi(\boldsymbol{u},t)\, d\boldsymbol{u} = \int_0^\pi \int_0^{2\pi} \psi(\theta, \phi, t) \sin\theta \, d\phi \, d\theta = 1 \tag{8.129}$$

and we denote ensemble averages over $\boldsymbol{u}$ with the notation

$$\langle f \rangle_u = \int_{||\boldsymbol{u}||=1} f(\boldsymbol{u}) \psi(\boldsymbol{u},t)\, d\boldsymbol{u}.$$

At equilibrium in the absence of any external torques, every orientation is equally probable, so the equilibrium distribution is simply

$$\psi_\mathrm{eq}(\boldsymbol{u},t) = \frac{1}{4\pi}. \tag{8.130}$$

We saw in Section 7.5.2 that $1/2D_\mathrm{r}$ represents the scale over which an initially oriented ensemble of rods will relax to an isotropic distribution in the absence of imposed flow or torque.

To develop an expression for the polymer contribution to the stress in a dilute solution of rigid rods, we specialize to the case shown in Figure 8.2(b), where the rod is treated as a rigid dumbbell consisting of two point-particle beads with friction coefficient $\zeta$ connected by a rigid massless and frictionless rod of length $L$. Additionally, hydrodynamic interactions between the beads will be neglected, in which case the Bretherton constant $B = 1$ and the rotational friction coefficient

$$\zeta_\mathrm{r} = \frac{L^2 \zeta}{2}. \tag{8.131}$$

We will take the external torque $\boldsymbol{T}_\mathrm{ext}$ to be zero. This situation is slightly different from the bead–spring case we treated in Section 8.5 because here there are no spring forces; instead, a constraint force keeps the two beads a distance $L$ apart (Doi & Edwards 1986). In this case, the Kramers–Kirkwood expression (8.94) for the stress tensor is useful, since it contains only the drag forces on the beads:

$$\boldsymbol{\tau}_\mathrm{p} = n \sum_{\alpha=1}^{2} \left\langle \boldsymbol{R}_\alpha \boldsymbol{F}_{\mathrm{drag},\alpha} \right\rangle_u. \tag{8.132}$$

Taking the origin to be the center of mass of the rod, $\boldsymbol{R}_1 = -\boldsymbol{u}L/2$, $\boldsymbol{R}_2 = \boldsymbol{u}L/2$, and (8.132) become

$$\boldsymbol{\tau}_{\mathrm{p}} = \frac{1}{2} nL \left\langle \boldsymbol{u} \left( \boldsymbol{F}_{\mathrm{drag},2} - \boldsymbol{F}_{\mathrm{drag},1} \right) \right\rangle_u . \tag{8.133}$$

The velocity-space averaged torque balance on the rod is

$$\boldsymbol{u} \times \left( \boldsymbol{F}_{\mathrm{drag},2} - \boldsymbol{F}_{\mathrm{drag},1} + \breve{\boldsymbol{F}}_{\mathrm{Brown},2} - \breve{\boldsymbol{F}}_{\mathrm{Brown},1} \right) = \boldsymbol{0}, \tag{8.134}$$

or equivalently,

$$(\boldsymbol{\delta} - \boldsymbol{uu}) \cdot \left( \boldsymbol{F}_{\mathrm{drag},2} - \boldsymbol{F}_{\mathrm{drag},1} + \breve{\boldsymbol{F}}_{\mathrm{Brown},2} - \breve{\boldsymbol{F}}_{\mathrm{Brown},1} \right) = \boldsymbol{0}. \tag{8.135}$$

Applying Stokes's law, using the fact that

$$\breve{\boldsymbol{U}}_2 - \breve{\boldsymbol{U}}_1 = L \frac{d\breve{\boldsymbol{u}}}{dt}$$

and invoking the assumption that the velocity field is linear on the scale of the molecule leads to the following result:

$$\boldsymbol{F}_{\mathrm{drag},2} - \boldsymbol{F}_{\mathrm{drag},1} = -\zeta L \left( \frac{d\breve{\boldsymbol{u}}}{dt} - \boldsymbol{u} \cdot \boldsymbol{\nabla v} \right). \tag{8.136}$$

Therefore

$$(\boldsymbol{\delta} - \boldsymbol{uu}) \cdot \left( \boldsymbol{F}_{\mathrm{drag},2} - \boldsymbol{F}_{\mathrm{drag},1} \right) = -\zeta L \left( \frac{d\breve{\boldsymbol{u}}}{dt} - (\boldsymbol{\delta} - \boldsymbol{uu}) \cdot \boldsymbol{u} \cdot \boldsymbol{\nabla v} \right). \tag{8.137}$$

The Brownian forces satisfy

$$(\boldsymbol{\delta} - \boldsymbol{uu}) \cdot \breve{\boldsymbol{F}}_{\mathrm{Brown},2} = -(\boldsymbol{\delta} - \boldsymbol{uu}) \cdot \breve{\boldsymbol{F}}_{\mathrm{Brown},1} = -\frac{k_B T}{L} \frac{\partial \ln \psi}{\partial \boldsymbol{u}}$$

so

$$(\boldsymbol{\delta} - \boldsymbol{uu}) \cdot \left( \breve{\boldsymbol{F}}_{\mathrm{Brown},2} - \breve{\boldsymbol{F}}_{\mathrm{Brown},1} \right) = -\frac{2k_B T}{L} \frac{\partial \ln \psi}{\partial \boldsymbol{u}}. \tag{8.138}$$

Inserting (8.137) and (8.138) into the torque balance (8.135) yields an evolution equation for $\breve{\boldsymbol{u}}$:

$$\frac{d\breve{\boldsymbol{u}}}{dt} = (\boldsymbol{\delta} - \boldsymbol{uu}) \cdot \boldsymbol{u} \cdot \boldsymbol{\nabla v} - \frac{2k_B T}{\zeta L^2} \frac{\partial \ln \psi}{\partial \boldsymbol{u}}. \tag{8.139}$$

This expression is the velocity-space averaged analogue of (8.126) for $B = 1, \boldsymbol{T}_{\mathrm{ext}} = \boldsymbol{0}$, and $\zeta_{\mathrm{r}} = L^2\zeta/2$. The last term corresponds to the entropic driving force for diffusion on a sphere. As a brief aside, we note that this equation allows us to derive the Smoluchowski equation using the approach described in Section 6.9. By analogy with (6.90), the equation for conservation of probability on the unit sphere is

$$\frac{\partial \psi(\boldsymbol{u}, t)}{\partial t} = -\frac{\partial}{\partial \boldsymbol{u}} \cdot \boldsymbol{J}_u, \tag{8.140}$$

where $\boldsymbol{J}_u$ is the probability flux for $\boldsymbol{u}$ on the surface of the sphere. By analogy with (6.91), this is given by

$$\boldsymbol{J}_\psi = \psi(\boldsymbol{u}, t) \frac{d\breve{\boldsymbol{u}}}{dt}. \tag{8.141}$$

Using (8.139), these equations combine to yield

$$\frac{\partial \psi(\boldsymbol{u},t)}{\partial t} = -\frac{\partial}{\partial \boldsymbol{u}} \cdot (((\boldsymbol{\delta} - \boldsymbol{uu}) \cdot \boldsymbol{u} \cdot \nabla \boldsymbol{v})\, \psi(\boldsymbol{u},t)) + \frac{2k_B T}{\zeta L^2} \frac{\partial}{\partial \boldsymbol{u}} \cdot \frac{\partial \psi(\boldsymbol{u},t)}{\partial \boldsymbol{u}}. \tag{8.142}$$

This is equivalent to (8.127) for $B = 1$, $\boldsymbol{T}_{\text{ext}} = \boldsymbol{0}$ and $\zeta_r = L^2\zeta/2$.

Returning to the expression for the stress tensor, we substitute (8.136) and (8.139) back into (8.133) to find

$$\boldsymbol{\tau}_\text{p} = \frac{1}{2} n\zeta L^2 \langle \boldsymbol{uu}\,(\boldsymbol{uu} : \nabla \boldsymbol{v})\rangle_u + nk_B T \left\langle \boldsymbol{u} \frac{\partial \ln \psi}{\partial \boldsymbol{u}} \right\rangle_u . \tag{8.143}$$

The last term in this expression can be evaluated using spherical coordinates to yield

$$\boldsymbol{\tau}_\text{p} = \frac{1}{2} n\zeta L^2 \langle \boldsymbol{uu}\,(\boldsymbol{uu} : \nabla \boldsymbol{v})\rangle_u + 3nk_B T \left( \langle \boldsymbol{uu} \rangle_u - \frac{1}{3}\boldsymbol{\delta} \right). \tag{8.144}$$

Finally, because $\boldsymbol{uu}$ is symmetric, the antisymmetric part of $\nabla \boldsymbol{v}$, i.e., the vorticity, makes no contribution to the double dot product, and we can write

$$\boldsymbol{\tau}_\text{p} = \frac{1}{2} n\zeta L^2 \langle \boldsymbol{uuuu}\rangle_u : \mathbf{E} + 3nk_B T \left( \langle \boldsymbol{uu} \rangle_u - \frac{1}{3}\boldsymbol{\delta} \right). \tag{8.145}$$

A physically important feature of this expression is that is contains a term that depends on $\mathbf{E}$ but is independent of $k_B T$, as well as one that is independent of $\mathbf{E}$ but depends on $k_B T$. The first term can be viewed as the viscous contribution of the rod to the stress – it arises directly from the viscous drag between the fluid and the rod and vanishes instantly if the flow is turned off. The second term is entropic and is sometimes called the elastic contribution – it arises as Brownian motion tries to drive the orientation distribution back to its equilibrium state $\langle \boldsymbol{uu} \rangle_u = \frac{1}{3}\boldsymbol{\delta}$. In Section 7.7, we saw that $\langle \boldsymbol{uu} \rangle_u$ relaxes to equilibrium as $\exp(-6D_\text{r} t)$. Here

$$D_\text{r} = \frac{k_B T}{\zeta_\text{r}} = \frac{k_B T}{\frac{1}{2} L^2 \zeta} \tag{8.146}$$

so we can define a stress relaxation time for this model as

$$\lambda = \frac{1}{6D_\text{r}} = \frac{\zeta_\text{r}}{6k_B T} = \frac{L^2\zeta}{12 k_B T}. \tag{8.147}$$

In the limit of small strain rate, the polymer contribution to the viscosity is

$$\eta_{p0} = nk_B T \lambda \tag{8.148}$$

(Problem 8.9). Here the subscript "0" denotes that this is the contribution to the viscosity in the zero-shear rate limit, which is called the *zero-shear* viscosity. Note that this is the same formula as found for Hookean dumbbells. At high shear rates, the formulas do not continue to coincide, as we describe in Section 9.3. Furthermore, even at low shear rates, the transient response of the rigid rod and bead-spring dumbbell models have qualitative differences. In particular, there is no analogue in the bead–spring dumbbell case to the term in (8.145) that contains $\mathbf{E}$; because of this viscous term, there is an instantaneous response of the stress to a change in strain rate that is absent from the bead–spring dumbbell model.

For the bead–spring dumbbell model, it was possible in the Hookean spring case to find an closed form evolution equation for $\langle \boldsymbol{QQ} \rangle_Q$ and in turn the stress tensor. For the FENE spring, the Peterlin closure yields a good approximation. For the rigid rod, the quantity analogous to $\langle \boldsymbol{QQ} \rangle_Q$ is $\langle \boldsymbol{uu} \rangle_u$. We address now the issue of finding a closed form evolution equation the stress in a rigid dumbbell solution. When $B = 1$ and in the absence of external torques (8.126) simplifies to

$$d\boldsymbol{u} = ((\boldsymbol{\delta} - \boldsymbol{uu}) \cdot \mathbf{L} \cdot \boldsymbol{u} - 2D_\mathrm{r}\boldsymbol{u})\ dt + \sqrt{2D_\mathrm{r}}\,(\boldsymbol{\delta} - \boldsymbol{uu}) \cdot d\boldsymbol{W}.$$

Here $\mathbf{L} = (\boldsymbol{\nabla v})^\mathrm{T}$ and we have used the result $\omega \times \boldsymbol{u} = \mathbf{W}^\mathrm{T} \cdot \boldsymbol{u}$, where $\mathbf{W}$ is the vorticity tensor. Applying Itô's formula (7.44) with $f = \boldsymbol{uu}$ yields that

$$d\langle \boldsymbol{uu} \rangle_u = \left(-2D_\mathrm{r}(\boldsymbol{\delta} - 3\langle \boldsymbol{uu} \rangle_u) + \mathbf{L} \cdot \langle \boldsymbol{uu} \rangle_u + (\mathbf{L} \cdot \langle \boldsymbol{uu} \rangle_u)^\mathrm{T} - 2\langle \boldsymbol{uuuu} \rangle_u : \mathbf{E}\right) dt. \tag{8.149}$$

Letting $\mathbf{S} = \langle \boldsymbol{uu} \rangle_u$ and recalling that we are evaluating derivatives as the rod moves with the fluid velocity, we can rewrite this equation as

$$\overset{\nabla}{\mathbf{S}} = -\frac{1}{\lambda}\left(\mathbf{S} - \frac{1}{3}\boldsymbol{\delta}\right) - 2\langle \boldsymbol{uuuu} \rangle_u : \mathbf{E}. \tag{8.150}$$

We cannot solve this equation for $\mathbf{S}$ because of the term containing $\langle \boldsymbol{uuuu} \rangle_u$, which cannot generally be written in terms of $\mathbf{S}$. Furthermore, to obtain an evolution equation for the stress tensor would require an evolution equation for $\langle \boldsymbol{uuuu} \rangle_u$. This equation would depend on $\langle \boldsymbol{uuuuuu} \rangle_u$, which again cannot be written exactly in terms of $\langle \boldsymbol{uu} \rangle_u$ and $\langle \boldsymbol{uuuu} \rangle_u$, and so on for higher moments. Thus no closed form evolution equation exists for the stress in a solution of rigid dumbbells. Nevertheless, as with the FENE dumbbell model, there exist closure approximations that enable accurate approximate computations. A very simple approximation is

$$\langle \boldsymbol{uuuu} \rangle_u \approx \langle \boldsymbol{uu} \rangle_u \langle \boldsymbol{uu} \rangle_u = \mathbf{S} \cdot \mathbf{S}.$$

This relation would be exact if the rods were perfectly aligned with, say $\boldsymbol{u} = \boldsymbol{p}$, in which case $\psi(\boldsymbol{u}) = \delta(\boldsymbol{u} - \boldsymbol{p})$ and

$$\left\langle u_i u_j u_k u_l \right\rangle_u = p_i p_j p_k p_l, \tag{8.151}$$

so it yields reasonable results in strong uniaxial extension (Doi & Edwards 1986). However, it works very poorly in shear, so a more refined approximation is necessary. At equilibrium, $\langle \boldsymbol{uuuu} \rangle_u$ is given by an isotropic fourth-order tensor:

$$\langle u_i u_j u_k u_l \rangle_\mathrm{eq} = \frac{1}{15}\left(\delta_{ij}\delta_{kl} + \delta_{il}\delta_{kj} + \delta_{ik}\delta_{jl}\right). \tag{8.152}$$

In addition to these limiting cases, the following properties are important:

$$\langle \boldsymbol{uuuu} \rangle_u : \mathbf{M} = (\langle \boldsymbol{uuuu} \rangle_u : \mathbf{M})^\mathrm{T}, \tag{8.153}$$

$$\mathrm{Tr}\,\langle \boldsymbol{uuuu} \rangle_u : \mathbf{M} = \mathbf{S} : \mathbf{M}, \tag{8.154}$$

$$\langle \boldsymbol{uuuu} \rangle_u : \mathbf{M} = \langle \boldsymbol{uuuu} \rangle_u : \mathbf{M}^\mathrm{S}, \tag{8.155}$$

where $\mathbf{M}$ is an arbitrary second-order tensor and $\mathbf{M}^{\mathrm{S}}$ its symmetric part. The most general approximation that satisfies these conditions and is quadratic in $\mathbf{S}$ is given by

$$\langle \boldsymbol{uuuu} \rangle_u : \mathbf{M} \approx \frac{1}{5}(\mathbf{S} \cdot \mathbf{M}^{\mathrm{S}} + \mathbf{M}^{\mathrm{S}} \cdot \mathbf{S} - \mathbf{S} \cdot \mathbf{S} \cdot \mathbf{M}^{\mathrm{S}} - \mathbf{M}^{\mathrm{S}} \cdot \mathbf{S} \cdot \mathbf{S} + 2\mathbf{S} \cdot \mathbf{M}^{\mathrm{S}} \cdot \mathbf{S} + 3\mathbf{SS} : \mathbf{M}^{\mathrm{S}}) \tag{8.156}$$

(Dhont & Briels 2003). With this approximation, (8.150) and (8.145) form a closed system of equations for finding the time evolution of $\mathbf{S}$ and $\boldsymbol{\tau}_{\mathrm{p}}$ during flow that yields good agreement with exact solutions of (8.149).

## Problems

**8.1** Following Section 6.9, write a velocity-space averaged force balance on each end of a Gaussian subchain of length $L_{\mathrm{s}}$ to recover the equilibrium distribution (8.21).

**8.2** Use (8.40)–(8.44) to show that the stress tensor expression (8.89) can be rewritten as (8.93).

**8.3** Verify (8.88) for a bead–spring dumbbell. Hints: (1) Rewrite the left-hand side of (8.88) in terms of center of mass and connector vector and recall that the equilibrium probability distribution function is independent of the center of mass. (2) Use the identity

$$\frac{\partial}{\partial \boldsymbol{Q}} e^{-\beta\Phi} = -\beta e^{-\beta\Phi} \frac{\partial \Phi}{\partial \boldsymbol{Q}}.$$

(3) Apply (A.12) and assume that $e^{-\Phi(\boldsymbol{Q})/k_B T}$ decays very rapidly at large $|\boldsymbol{Q}|$.

**8.4** Derive the Kramers–Kirkwood expression for the stress tensor (8.94) for a bead–spring dumbbell. Hint: use (A.12) and assume that $\psi(\boldsymbol{Q}, \boldsymbol{R}_{\mathrm{c}}, t)$ decays rapidly at large $|\boldsymbol{Q}|$.

**8.5** Derive (8.96)–(8.98) from (8.38)–(8.39). Neglect excluded volume or external forces.

**8.6** For a bead–spring model of a Gaussian chain, the quantities required to determine the Kirkwood diffusivity can be explicitly computed for long chains.

(a) For a bead–spring chain with spring constant $H$, write the equilibrium (Gaussian) probability distribution $p(\boldsymbol{R}_{\alpha\beta})$ for an arbitrary pair of beads $\alpha$ and $\beta$.

(b) Use this result to find

$$\left\langle \frac{1}{R_{\alpha\beta}} \right\rangle = \int \frac{1}{R_{\alpha\beta}} p(\boldsymbol{R}_{\alpha\beta})\, d\boldsymbol{R}_{\alpha\beta}.$$

Use spherical coordinates. Answer:

$$\left\langle \frac{1}{R_{\alpha\beta}} \right\rangle = \sqrt{\frac{2H}{|\alpha - \beta|\, \pi k_B T}}.$$

Observe that the answer scales as $|\alpha - \beta|^{-1/2}$. On what physical grounds should we expect this scaling?

(c) To compute the Kirkwood diffusivity of a bead–spring chain, we require the sum

$$\sum_{\alpha=1}^{N_{\mathrm{b}}} \sum_{\beta=1,\beta\neq\alpha}^{N_{\mathrm{b}}} \left\langle \frac{1}{R_{\alpha\beta}} \right\rangle = 2 \sum_{\alpha=1}^{N_{\mathrm{b}}} \sum_{\beta=\alpha+1}^{N_{\mathrm{b}}} \left\langle \frac{1}{R_{\alpha\beta}} \right\rangle.$$

This sum can be approximated by the integral

$$2\sqrt{\frac{2H}{\pi k_B T}}\int_1^{N_b}\int_{\alpha+1}^{N_b}\sqrt{\frac{1}{\beta-\alpha}}\,d\beta\,d\alpha.$$

Estimate this integral for $N_b \gg 1$. From this result, it follows that for large $N_b$ the Kirkwood (short-time) diffusivity of a bead–spring chain scales as $N_b^{-1/2}$.

**8.7** Derive (8.109) from (8.108).

**8.8** Derive (8.125) from (8.123).

**8.9** For the rigid rod model with $\boldsymbol{T}_{\text{ext}} = \mathbf{0}$ and $B = 1$, obtain the zero-shear viscosity as follows:

(a) Derive (8.131).

(b) Use (8.149) and Itô's formula to derive (8.149) .

(c) Approximate $\langle u_i u_j u_k u_l \rangle$ in the resulting expression by its isotropic equilibrium value (8.152) and solve for the steady-state value of $\langle u_i u_j \rangle$. This will give the first (linear) correction of $\langle u_i u_j \rangle$ due to the flow. We do not need to correct $\langle u_i u_j u_k u_l \rangle$ at this level of approximation because to do so would only change the result for the stress at second (quadratic) order in the magnitude of the strain rate.

(d) Use this result and the equilibrium expression for $\langle u_i u_j u_k u_l \rangle$ in (8.145) to find that

$$\boldsymbol{\tau}_p = 2\eta_p \mathbf{E},$$

where $\eta_p = nk_B T\lambda$. Thus at small strain rates, the flow remains Newtonian on time scales longer than $\lambda$.

**8.10** Write an inertialess Brownian dynamics code to simulate ensembles of bead–spring chains (i.e., to solve (8.53)–(8.54)) with Hookean springs and no excluded volume forces. Take each bead to have radius $a$ and incorporate hydrodynamic interactions using one of the regularized Stokeslets described in Section 2.5. Take the spring constant to be $H = 3k_BT/(4a)^2$, which corresponds to an equilibrium distance between beads of $4a$. Perform simulations in the absence of flow over a range of $N_b$, in the absence and presence of hydrodynamic interactions. Compute $R_H$ using the simulation results and (8.61) as well as the mean-squared displacement. Compare the long-time diffusivity to the Kirkwood formula results (8.59) and (8.60). For simplicity, take $k_BT = a = 1$, which corresponds to nondimensionalizing energy and distance with $k_BT$ and $a$, respectively.

# 9 Rheology and Viscoelastic Flow Phenomena

In the previous chapter, we learned that polymers in solution display a spectrum of relaxation times and at low shear rates make a Newtonian contribution to the steady-state viscosity of a fluid. This chapter builds on these results, providing a systematic introduction to the response of a complex fluid to deformation: i.e., the *rheology* of a fluid. We describe the basic framework of the theory of *linear viscoelasticity*, the stress response to small or slow deformations, as well as some results on limiting behavior at very high deformation rates. Observation of Brownian motion of a particle in a complex fluid can reveal information about the rheology of the fluid, so we provide an analysis of this phenomenon. Finally, we introduce some elementary aspects of the fluid dynamics of viscoelastic fluids, including flow instability and connections between continuum mechanics and the basic structure of models of viscoelastic materials.

## 9.1 Fundamentals of Linear Viscoelasticity

For many materials, if the strain is sufficiently small, i.e., if the material is perturbed only slightly from its equilibrium structure, then the stress response depends linearly on the strain and strain rate and the material can be characterized by its *linear viscoelastic properties* (Bird, Armstrong, & Hassager 1987). These properties are a reflection of the dynamic features of the equilibrium structure of the material. In some important cases, as we describe, linear viscoelastic behavior can arise even if the strain is not small, as long as the strain *rate* is small. The theory of linear viscoelasticity is one aspect of a broader framework or understanding dynamic material properties called *linear response theory* (Berne & Pecora 1976, Chaikin & Lubensky 1995, Reichl 1998, Zwanzig 2001).

### 9.1.1 The Relaxation Modulus and the Weissenberg Number

We begin by considering simple shear deformations. Imagine a thin layer of the material of interest placed between parallel horizontal plates. Horizontal and vertical directions are $x$ and $y$, respectively. The top plate is moved horizontally to deform the material; we assume that the no-slip boundary condition holds. Assuming that there are no horizontal gradients and that the layer is thin enough that inertia in the material can be neglected (i.e., that momentum transfer across the thin layer is rapid), then the Cauchy momentum balance yields that $\tau_{yx}$ is independent of position in the material. The velocity gradient at any point in the material is

$$\nabla \boldsymbol{v} = \dot{\gamma}(t)\boldsymbol{e}_y\boldsymbol{e}_x. \tag{9.1}$$

That is, a spatially homogeneous but time-dependent shear rate $\dot{\gamma}(t)$ is imposed. The total shear strain $\gamma(t)$ experienced by any point in the material is then

$$\gamma(t) = \int_0^t \dot{\gamma}(t')\, dt'. \tag{9.2}$$

To begin understanding linear viscoelasticity, imagine that the material of interest satisfies the Oldroyd-B model (8.113). In Cartesian tensor notation, this equation is

$$\tau_{\mathrm{p},ij} + \lambda\left(\frac{\partial \tau_{\mathrm{p},ij}}{\partial t} + \left[v_k\frac{\partial \tau_{\mathrm{p},ij}}{\partial x_k}\right] - \left\{\tau_{\mathrm{p},ik}\frac{\partial v_j}{\partial x_k} + \tau_{\mathrm{p},jk}\frac{\partial v_i}{\partial x_k}\right\}\right) = G_0\lambda\left(\frac{\partial v_i}{\partial x_j} + \frac{\partial v_j}{\partial x_i}\right).$$

Let $\epsilon \ll 1$ represent the magnitude of the total strain and $t_{\mathrm{c}}$ be a characteristic time for the strain rate. As long as $t_{\mathrm{c}}$ is bounded, then both $\gamma(t)$ and $\dot{\gamma}$ are $O(\epsilon)$ as $\epsilon \to 0$. Thus $\boldsymbol{v}$ and $\nabla\boldsymbol{v}$ are $O(\epsilon)$. Expanding $\boldsymbol{\tau}_{\mathrm{p}}$ in powers of $\epsilon$ yields that the shear stresses $\tau_{\mathrm{p},yx}$ and $\tau_{\mathrm{p},xy}$ are also $O(\epsilon)$. The terms in the square and curly brackets only enter at $O(\epsilon^2)$. In particular, the stress components other than $\tau_{\mathrm{p},yx}$ and $\tau_{\mathrm{p},xy}$ arise due to the quadratic interaction between the stress and velocity gradient – the terms in the curly brackets in this equation – and thus are $O(\epsilon^2)$. Since they do not scale linearly with $\epsilon$, they do not play a role in the linear viscoelastic response of a material. In simple shear or any homogeneous flow, the term in square brackets, corresponding to convection of stress, vanishes identically.

By symmetry, $\tau_{\mathrm{p},yx}$ and $\tau_{\mathrm{p},xy}$ evolve identically, with $\tau_{\mathrm{p},yx}$ satisfying

$$\tau_{\mathrm{p},yx} + \lambda\frac{d\tau_{\mathrm{p},yx}}{dt} = G_0\lambda\dot{\gamma}. \tag{9.3}$$

A linear viscoelastic model of this form is called a *Maxwell model*. Taking the initial stress to be zero, this can be integrated to yield

$$\tau_{\mathrm{p},yx}(t) = \int_0^t G_0 \exp\left(-\frac{t-t'}{\lambda}\right)\dot{\gamma}(t')\, dt'. \tag{9.4}$$

Incorporating the viscous stress that is also present in the Oldroyd-B model yields that

$$\tau_{yx}(t) = \eta\dot{\gamma}(t) + \tau_{\mathrm{p},yx}(t) = \eta\dot{\gamma}(t) + \int_0^t G_0 \exp\left(-\frac{t-t'}{\lambda}\right)\dot{\gamma}(t')\, dt'. \tag{9.5}$$

In the nomenclature of linear viscoelasticity, this superposition of Newtonian fluid and Maxwell model behavior is called a *Jeffreys model*.

More generally, for any material whose stress responds linearly to small amplitude strains, we can write

$$\tau_{yx}(t) = \int_{-\infty}^t G(t-t')\dot{\gamma}(t')\, dt', \tag{9.6}$$

where we have now pushed the initial time back infinitely far in the past. The quantity $G(\tau)$, where $\tau = t - t'$, is called the *relaxation modulus*. It represents the memory that the material has for past deformations and as such can only be nonzero for $\tau \geq 0$. Functions that satisfy this condition are said to be *causal*. For the Oldroyd-B model, the total shear stress is comprised of both the polymer contribution and a solvent contribution, so

$$\tau_{yx}(t) = \eta\dot{\gamma}(t) + \int_{-\infty}^{t} G_0 \exp\left(-\frac{t-t'}{\lambda}\right) \dot{\gamma}(t')\, dt' \tag{9.7}$$

$$= \int_{\infty}^{t} \left[\eta\delta(t-t') + G_0 \exp\left(-\frac{t-t'}{\lambda}\right)\right] \dot{\gamma}(t')\, dt' \tag{9.8}$$

so the relaxation modulus is the quantity in the square brackets. From this example, two limiting cases can be identified. If $G_0 = 0$, then the material is a viscous fluid, which has no memory of past deformations. Its relaxation modulus is

$$G(\tau) = \eta\delta(\tau). \tag{9.9}$$

The opposite limit is a purely elastic Hookean solid, whose stress results from the accumulated response to all past deformations. This case is recovered from the Maxwell model when $\lambda \to \infty$. Now the relaxation modulus is

$$G(\tau) = G_0 H(\tau), \tag{9.10}$$

where $H(t)$ is the unit step function. Materials whose response is intermediate between these extremes are viscoelastic. The Jeffreys model corresponds to

$$G(\tau) = \eta\delta(\tau) + G_0 \exp\left(-\frac{\tau}{\lambda}\right) H(\tau), \tag{9.11}$$

with the Maxwell model being the special case $\eta = 0$. A simple description of a viscoelastic solid is the *Kelvin–Voigt* model, for which

$$G(\tau) = G_0 H(\tau) + \eta\delta(\tau). \tag{9.12}$$

The relaxation modulus contains all the information necessary to characterize the linear response of a material. One way to experimentally determine $G(\tau)$ is to perform a so-called *step strain* experiment, where $\gamma(t) = \epsilon H(t)$, so $\dot{\gamma}(t) = \epsilon\delta(t)$. Substituting into (9.6) yields that

$$G(t) = \frac{\tau_{yx}(t)}{\epsilon} \tag{9.13}$$

so the relaxation modulus is determined directly by the stress response.

To better understand the general distinction between viscoelastic liquids and solids, consider a material subjected to a constant infinitesimal shear rate $\dot{\gamma}$. In the linear viscoelastic regime, we can write

$$\tau_{yx}(t) = \left[\int_0^t G(t-t')\, dt'\right] \dot{\gamma}. \tag{9.14}$$

If

$$\int_0^\infty G(\tau)\, d\tau < \infty, \tag{9.15}$$

then $\tau_{yx}$ approaches a constant as $t \to \infty$. That is, at long times the shear stress is proportional to the shear rate and the material is a viscoelastic liquid. Otherwise, the material is a solid. In this case, the stress will eventually build to the point where the linear viscoelasticity approximation breaks down.

If (9.15) holds, as it does, for example, for the Jeffreys model, we can push the initial time in (9.14) infinitely far back in the past and write

$$\begin{aligned}\tau_{yx} &= \int_{-\infty}^{t} G(t-t')\dot{\gamma}\, dt' \\ &= \left[\int_0^{\infty} G(\tau)\, d\tau\right]\dot{\gamma} \\ &\equiv \eta_0\dot{\gamma}, \end{aligned} \tag{9.16}$$

where $s = t - t'$. The quantity $\eta_0$ is called the *zero-shear viscosity* of the liquid. One can define a characteristic relaxation time $\lambda_c$ for the material by the simple relation

$$\lambda_c = \frac{\eta_0}{G(0)}. \tag{9.17}$$

(For a viscoelastic solid, $\lambda_c$ would be infinite.)

From this discussion of viscoelastic liquids and solids, we can better understand the conditions under which the assumption of linear viscoelasticity will be valid. The microstructure of a viscoelastic solid never completely forgets past deformations – its relaxation time $\lambda_c$ is infinite, so the material only displays linear viscoelastic behavior when the total strain it has experienced is small. On the other hand, for a viscoelastic liquid, from (9.14) we see that regardless of the total strain $\gamma(t) = \dot{\gamma}t$, the stress will remain small for all time as long as $\dot{\gamma}$ is sufficiently small. In nondimensional terms, $\tau_{yx}/G(0) \ll 1$ as long as $\lambda_c\dot{\gamma}_c \ll 1$. For a general flow with characteristic deformation rate $\dot{\gamma}_c$, a liquid displays linear viscoelastic behavior as long as $\lambda_c\dot{\gamma}_c \ll 1$, even if the total strain is large. In this case, the liquid only deforms a small amount before losing memory of its earlier deformations, and the microstructure remains very close to equilibrium. It does not matter if the total deformation is large – all but the most recent deformation history is forgotten.

For flexible polymers, $\lambda_c$ is often taken to be the longest relaxation time of the chain and the *Weissenberg number*

$$\mathrm{Wi} = \lambda_c\dot{\gamma}_c \tag{9.18}$$

characterizes the deviation of the polymer from equibrium. For a solution of rigid polymers, relaxation to equilibrium occurs via rotational diffusion, $\lambda_c \sim D_r^{-1}$, and the quantity analogous to Wi is a Péclet number based on rotational diffusivity:

$$\mathrm{P\acute{e}}_r = \frac{\dot{\gamma}_c}{D_r}. \tag{9.19}$$

## 9.1.2 Frequency Response of a Viscoelastic Material

Substantial insight and information can be gained by considering the linear viscoelastic response of a material to oscillatory strains with different frequencies. Imagine applying a sinusoidal strain of the form

$$\gamma(t) = \gamma_0 \sin(\omega t) \tag{9.20}$$

corresponding to a strain rate of

$$\dot{\gamma}(t) = \dot{\gamma}_0 \cos(\omega t) \tag{9.21}$$

where $\dot{\gamma}_0 = \gamma_0 \omega$. The stress response at long times after an initial transient is

$$\begin{aligned}\tau_{yx}(t) &= \dot{\gamma}_0 \int_{-\infty}^{t} G(t-t') \cos \omega t' \, dt' \\ &= \dot{\gamma}_0 \int_{0}^{\infty} G(\tau) \cos \omega (t-\tau) \, d\tau \\ &= \dot{\gamma}_0 \left\{ \left[ \int_0^{\infty} G(\tau) \cos \omega\tau \, d\tau \right] \cos \omega t + \left[ \int_0^{\infty} G(\tau) \sin \omega\tau \, d\tau \right] \sin \omega t \right\} \\ &\equiv \eta'(\omega) \dot{\gamma}_0 \cos \omega t + \eta''(\omega) \dot{\gamma}_0 \sin \omega t. \end{aligned} \tag{9.22}$$

The coefficient $\eta'(\omega)$ measures the component of the stress that is in phase with the strain rate, while $\eta''(\omega)$ measures the out-of-phase component.

The quantities in square brackets can be recognized as being related to the Fourier transform of the relaxation modulus. Recalling the causality condition $G(\tau < 0) = 0$, the *complex viscosity*[1] $\eta^*$ can be defined as

$$\eta^*(\omega) = \eta'(\omega) - i\eta''(\omega) = \int_0^{\infty} G(\tau) e^{-i\omega\tau} \, d\tau = \int_{-\infty}^{\infty} G(\tau) e^{-i\omega\tau} \, d\tau = F\{G(\tau)\}. \tag{9.23}$$

For a Newtonian fluid, $\eta'' = 0$. Now, taking the Fourier transform of (9.6) and applying the Fourier convolution theorem (A.29) yields that

$$\hat{\tau}_{yx}(\omega) = F\left\{ \int_{-\infty}^{t} G(t-t') \dot{\gamma}(t') \, dt' \right\} = \eta^*(\omega) \hat{\dot{\gamma}}(\omega). \tag{9.24}$$

Observing that $\hat{\dot{\gamma}} = i\omega\hat{\gamma}$, we can also write

$$\hat{\tau}_{yx}(\omega) = G^*(\omega) \hat{\gamma}(\omega), \tag{9.25}$$

where

$$G^*(\omega) = G'(\omega) + iG''(\omega) = i\omega\eta^*(\omega) \tag{9.26}$$

is called the *complex modulus*. Thus

$$G'(\omega) = \omega\eta''(\omega), \quad G''(\omega) = \omega\eta'(\omega), \tag{9.27}$$

and we can rewrite (9.22) as

$$\tau_{yx}(t) = G'(\omega)\gamma_0 \sin \omega t + G''(\omega)\gamma_0 \cos \omega t. \tag{9.28}$$

This can be rewritten to explicitly illustrate the phase difference between stress and strain:

$$\tau_{yx}(t) = |G^*(\omega)| \, \gamma_0 \sin(\omega t + \delta(\omega)), \tag{9.29}$$

[1] Thus one might define a complex fluid as any fluid whose viscosity is complex.

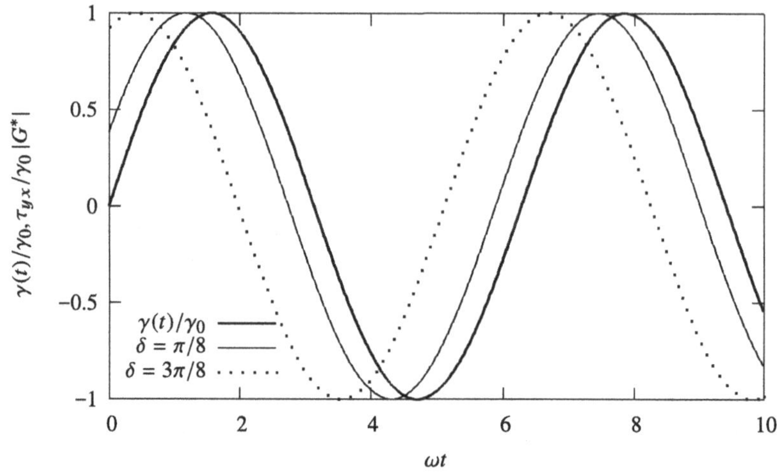


**Figure 9.1** Oscillatory shear strain and the stress response for $\delta = \pi/8$ and $3\pi/2$.

where

$$\tan\delta(\omega) = \frac{G''(\omega)}{G'(\omega)}. \tag{9.30}$$

If $\delta = 0$, then the stress is in phase with the strain, while if $\delta = \pi/2$ it is in phase with the strain rate. Figure 9.1 illustrates the strain and resulting shear stress when $\delta = \pi/8$ and $3\pi/8$.

The quantity $G'(\omega)$ is called the *storage modulus* and $G''(\omega)$ the *loss modulus*. This nomenclature arises from energy considerations (cf. Section 1.3.4). The total work per unit area performed on the material and thus lost to heat over one period of oscillation is

$$W = \int_0^{2\pi/\omega} \tau_{yx}\dot{\gamma}(t)\, dt. \tag{9.31}$$

The strain rate goes as $\cos\omega t$ while the term multiplying $G'(\omega)$ goes as $\sin\omega t$. The product of these integrates to zero over one period, so $G'(\omega)$ plays no role in the net work done on the fluid. It corresponds to reversible storage of mechanical energy in the deformed material. Only the term containing $G''(\omega)$ contributes to the loss of work to heat. For a purely elastic solid, $G'' = 0$. In other words, the component of response that is in phase with the strain rate $\dot{\gamma}(t)$ contributes to energy loss, while the component in phase with the strain amplitude $\gamma(t)$ does not. Accordingly, the quantity $\tan\delta(\omega)$ is called the *loss tangent*.

Some intuition about the physical significance of $\eta^*$ and $G^*$ can be gained from examining the Maxwell model, for which

$$G(\tau) = G_0 \exp\left(-\frac{\tau}{\lambda}\right) H(\tau). \tag{9.32}$$

The zero-shear viscosity $\eta_0$ as determined from (9.16) is $G_0\lambda$. Taking the Fourier transform of $G(\tau)$ to determine $\eta^*$ yields

$$\begin{aligned}\eta^*(\omega) &= \int_0^\infty G_0 \exp\left(-\frac{\tau}{\lambda}\right) e^{-i\omega\tau}\, d\tau \\ &= G_0 \int_0^\infty \exp\left(\left(-\lambda^{-1} - i\omega\right)\tau\right)\, d\tau \\ &= \frac{G_0}{\lambda^{-1} + i\omega} \\ &= \frac{\eta_0}{1+\lambda^2\omega^2} - i\frac{\eta_0\lambda\omega}{1+\lambda^2\omega^2}. \end{aligned} \tag{9.33}$$

Therefore, for this model

$$\eta'(\omega) = \frac{\eta_0}{1+\lambda^2\omega^2} \tag{9.34}$$

$$\eta''(\omega) = \frac{\eta_0\lambda\omega}{1+\lambda^2\omega^2} \tag{9.35}$$

$$G'(\omega) = \frac{G_0\lambda^2\omega^2}{1+\lambda^2\omega^2} \tag{9.36}$$

$$G''(\omega) = \frac{G_0\lambda\omega}{1+\lambda^2\omega^2}. \tag{9.37}$$

The dimensionless quantity $\lambda\omega$ that shows up in these expressions is called the *Deborah number* De. For De $\ll 1$, the rate of relaxation $\lambda^{-1}$ is very fast compared to the frequency of oscillation. In this limit, $\eta' \to \eta_0$ and $\eta'' \to 0$, indicating that the shear stress is in phase with the shear rate, as is the case for a purely viscous material. These results reflect that in this limit the material response is very fast compared to the oscillation frequency – the material has time to relax to its equilibrium structure. In the opposite limit De $\gg 1$, the fluid relaxes very slowly in comparison to the oscillation and the material tends to deform along with the imposed strain. Now $\eta' \to 0$ and $G'' \to 0$, so the viscous response vanishes. On the other hand, $G' \to G_0$, indicating that at high frequency the material behaves as if it were an elastic solid with shear modulus $G_0$.

In Section 8.4, we saw that a polymer chain in solution has a spectrum of relaxation times. The relaxation modulus in this situation has the general *multimode* form

$$G(\tau) = \sum_{n=1}^{N} G_{0n} \exp\left(\frac{\tau}{\lambda_n}\right) H(\tau) \tag{9.38}$$

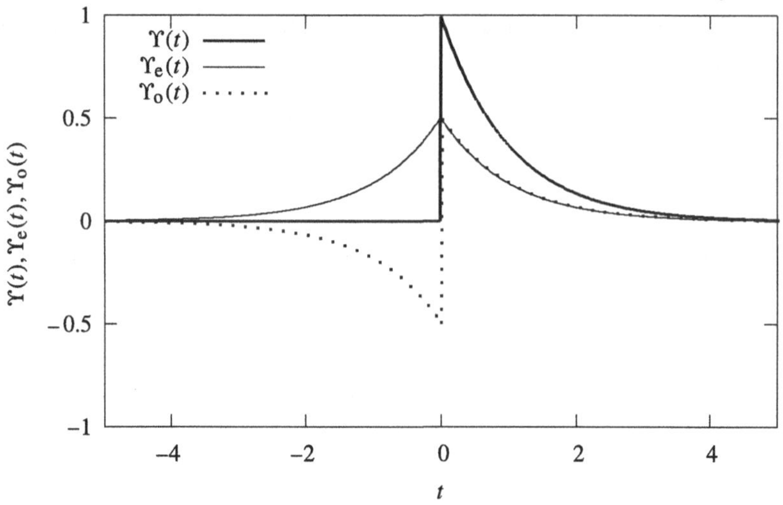


**Figure 9.2** A causal function $\Upsilon(t)$ and its even and odd parts.

and (9.34)–(9.37) generalize to

$$\eta'(\omega) = \sum_{n=1}^{N} \frac{G_{0n}\lambda_n}{1 + \lambda_n^2\omega^2} \tag{9.39}$$

$$\eta''(\omega) = \sum_{n=1}^{N} \frac{G_{0n}\lambda_n(\lambda_n\omega)}{1 + \lambda_n^2\omega^2} \tag{9.40}$$

$$G'(\omega) = \sum_{n=1}^{N} \frac{G_{0n}\lambda_n^2\omega^2}{1 + \lambda_n^2\omega^2} \tag{9.41}$$

$$G''(\omega) = \sum_{n=1}^{N} \frac{G_0\lambda_n\omega}{1 + \lambda_n^2\omega^2}. \tag{9.42}$$

The causality property of the relaxation modulus leads to a set of important results that relate the real and imaginary parts of $\eta^*$ and $G^*$ known as *Kramers–Kronig* relations. These relations are general results of linear response theory and can be derived (Johnson 1975) using an elementary result from Fourier analysis: the Fourier transform of a real even function is real and that of a real odd function is imaginary. Thus the real and imaginary parts of $\hat{f}(\omega) = F\{f(t)\}$ are determined by the Fourier transforms of the even and odd parts of $f(t)$, respectively.

Let $\Upsilon(t)$ be a causal response function that satisfies

$$\int_0^\infty \Upsilon(s)\, ds < \infty.$$

If, for example, the response function consists of a step function plus a part that satisfies this boundedness condition, then the following discussion applies to the latter. The function $\Upsilon(t)$ can be written

$$\Upsilon(t) = \Upsilon_e(t) + \Upsilon_o(t), \tag{9.43}$$

where

$$\Upsilon_e(t) = \frac{1}{2}\left(\Upsilon(t) + \Upsilon(-t)\right)$$

and

$$\Upsilon_o(t) = \frac{1}{2}\left(\Upsilon(t) - \Upsilon(-t)\right)$$

are the even and odd parts of $\Upsilon(t)$, respectively. An example of this decomposition is shown in Figure 9.2. From the figure, the following property is apparent:

$$\Upsilon_e(t) = \Upsilon_o(t)\mathrm{sgn}(t), \tag{9.44}$$

where

$$\mathrm{sgn}(t) = \begin{cases} 1 & t \geq 0, \\ -1 & t < 0. \end{cases} \tag{9.45}$$

The Fourier transform of $\Upsilon(t)$ is $\hat{\Upsilon}(\omega) = \hat{\Upsilon}_r(\omega) + i\hat{\Upsilon}_i(\omega)$, and because $\Upsilon(t)$ is real,

$$\hat{\Upsilon}_r(\omega) = F\{\Upsilon_e(t)\}, \quad i\hat{\Upsilon}_i = F\{\Upsilon_o(t)\}. \tag{9.46}$$

Taking the Fourier transform of (9.44) and applying the Fourier convolution theorem (A.30) and the relations (9.46) yields that

$$\hat{\Upsilon}_r = \frac{1}{2\pi} i\hat{\Upsilon}_i(\omega) * \widehat{\mathrm{sgn}}(\omega).$$

Using the result $F\{\mathrm{sgn}(t)\} = 2/i\omega$ and writing out the convolution integral, this becomes

$$\Upsilon_r(\omega) = \frac{1}{\pi}\int_{-\infty}^{\infty} \frac{\Upsilon_i(\omega')}{\omega - \omega'}\, d\omega'. \tag{9.47}$$

We can also write

$$\Upsilon_o(t) = \Upsilon_e(t)\mathrm{sgn}(t), \tag{9.48}$$

and thus

$$\Upsilon_i(\omega) = -\frac{1}{\pi}\int_{-\infty}^{\infty} \frac{\Upsilon_r(\omega')}{\omega - \omega'}\, d\omega'. \tag{9.49}$$

Equations (9.47) and (9.49) are the Kramers–Kronig relations for a general response function $\Upsilon(t)$.

For $G^*$ and $\eta^*$, these relationships become

$$G'(\omega) = \frac{1}{\pi} \fint_{-\infty}^{\infty} \frac{G''(\omega')}{\omega - \omega'} \, d\omega', \tag{9.50}$$

$$G''(\omega) = -\frac{1}{\pi} \fint_{-\infty}^{\infty} \frac{G'(\omega')}{\omega - \omega'} \, d\omega', \tag{9.51}$$

$$\eta'(\omega) = -\frac{1}{\pi} \fint_{-\infty}^{\infty} \frac{\eta''(\omega')}{\omega - \omega'} \, d\omega', \tag{9.52}$$

$$\eta''(\omega) = \frac{1}{\pi} \fint_{-\infty}^{\infty} \frac{\eta'(\omega')}{\omega - \omega'} \, d\omega'. \tag{9.53}$$

Since these expressions contain principal value integrals, they cannot be directly used for computations. Problem 9.3 addresses this issue.

To conclude this section, we briefly illustrate how linear viscoelasticity generalizes for flows more complex than the previous examples of simple shear flows discussed so far. For a general flow whose deformation rate everywhere satisfies the requirements of the linear viscoelastic approximation, we can generalize (9.6) as

$$\boldsymbol{\tau}(t) = \int_{-\infty}^{t} G(t - t') \left( \nabla \boldsymbol{v}(t') + \nabla \boldsymbol{v}^{\mathrm{T}}(t') \right) \, dt'. \tag{9.54}$$

Inserting this expression into the inertialess (deterministic) Cauchy momentum balance and imposing incompressibility yields

$$0 = -\nabla p + \int_{-\infty}^{t} G(t - t') \nabla^2 \boldsymbol{v}(t') \, dt'. \tag{9.55}$$

Applying the Fourier transform in time and recalling that $\eta^*(\omega) = F\{G(\tau)\}$ results in the following equation:

$$0 = -\nabla \hat{p}(\omega) + \eta^*(\omega) \nabla^2 \hat{\boldsymbol{v}}(\omega). \tag{9.56}$$

This is simply the Stokes equation with a complex viscosity. Everything we learned about solving the Stokes equation thus carries over to the linear viscoelastic case if we take all quantities to be in the Fourier transform domain in time (Zwanzig & Bixon 1970, Xu, Forest & Klapper 2007, Schieber, Córdoba & Indei 2013). For example, the drag force on a sphere in a linear viscoelastic fluid can be written

$$\boldsymbol{F}_{\mathrm{drag}}(\omega) = -\hat{\zeta}(\omega) \hat{\boldsymbol{U}}(\omega), \tag{9.57}$$

where

$$\hat{\zeta}(\omega) = 6\pi\eta^*(\omega) a. \tag{9.58}$$

In the time domain, the drag relation becomes

$$\boldsymbol{F}_{\mathrm{drag}}(t) = -\int_{-\infty}^{t} \zeta(t - t') \boldsymbol{U}(t') \, dt', \tag{9.59}$$

where

$$\zeta(\tau) = 6\pi G(\tau) a. \tag{9.60}$$

## 9.2 Brownian Motion in a Viscoelastic Fluid: Linear Microrheology

By observing the diffusion of a small particle in a Newtonian fluid, it is possible to extract from the Stokes–Einstein formula the viscosity of the fluid. Equation (9.59) suggests that by observing particle motion in a complex fluid we might be able to extract the complex viscosity of a viscoelastic fluid. *Passive particle tracking microrheology* is an experimental technique that builds on this idea (Mason & Weitz 1995, Xu et al. 2007, Squires & Mason 2010, Indei et al. 2012). It is particularly useful for determining properties *in situ* under a microscope as it does not require material to be placed in a simple shear flow device or subjected to an external forcing – the rheology is extracted from the response of the particle to Brownian fluctuations and a so-called *generalized Stokes–Einstein relation* (GSER), which we now derive.

To do so, we first recall (6.71), which relates the mean squared displacement of a particle to its velocity autocorrelation function. Because in an experiment, we will record particle position starting from some initial time (which we take to be zero), it will be convenient to work in the Laplace transform domain. We assume that the dynamics are stationary at $t = 0$. Combining the two integrals in (6.71) and recognizing that this expression has the form of a Laplace convolution, we can find that

$$\left\langle \widetilde{\boldsymbol{RR}}(s) \right\rangle = \frac{2}{s^2} \tilde{\boldsymbol{\phi}}_{\mathrm{v}}(s). \tag{9.61}$$

To determine the velocity autocorrelation function, we return to the momentum balance for a Brownian particle, (6.6), replacing the Stokes's law drag force expression with (9.59) to yield a generalized Langevin equation:

$$m\frac{d\boldsymbol{U}}{dt} = -\int_{-\infty}^{t} \zeta(t-t')\boldsymbol{U}(t')\, dt' + \boldsymbol{F}_{\mathrm{fluc}}(t). \tag{9.62}$$

To take advantage of Laplace transform solution techniques, this can be rewritten

$$m\frac{d\boldsymbol{U}}{dt} = -\int_{0}^{t} \zeta(t-t')\boldsymbol{U}(t')\, dt' + \boldsymbol{F}'_{\mathrm{fluc}}(t), \tag{9.63}$$

where

$$\boldsymbol{F}'_{\mathrm{fluc}}(t) = \boldsymbol{F}_{\mathrm{fluc}}(t) - \int_{-\infty}^{0} \zeta(t-t')\boldsymbol{U}(t')\, dt'. \tag{9.64}$$

Multiplying (9.63) by $\boldsymbol{U}(0)$, ensemble-averaging, and noting that $\boldsymbol{\phi}_{\mathrm{v}}(\tau) = \langle \boldsymbol{U}(0)\boldsymbol{U}(\tau)\rangle$ yields that

$$m\frac{d\boldsymbol{\phi}_{\mathrm{v}}(t)}{dt} = -\int_{0}^{t} \zeta(t-t')\boldsymbol{\phi}_{\mathrm{v}}(t')\, dt' + \langle \boldsymbol{U}(0)\boldsymbol{F}'_{\mathrm{fluc}}(t)\rangle. \tag{9.65}$$

The fluctuating term $\boldsymbol{F}_{\mathrm{fluc}}(t)$ arises from the random motions of the fluid surrounding the particle, which we can take to be independent of the velocity of the particle (Zwanzig 2001). Therefore, $\langle \boldsymbol{U}(0)\boldsymbol{F}_{\mathrm{fluc}}(t)\rangle = \boldsymbol{0}$. However, because $\boldsymbol{U}$ arises explicitly in the expression for $\boldsymbol{F}'_{\mathrm{fluc}}$, we cannot use the same argument for the term $\langle \boldsymbol{U}(0)\boldsymbol{F}'_{\mathrm{fluc}}(t)\rangle$. Nevertheless, it is commonly assumed that this term vanishes (Zwanzig & Bixon 1970, Kubo et al. 1991, Mason & Weitz 1995, Squires & Mason 2010), and for certain models (see, e.g., Zwanzig (2001) and Eqs. (9.71)–(9.74) later in this chapter) it can be shown to do so, in which case (9.65) becomes

$$m\frac{d\boldsymbol{\phi}_{\mathrm{v}}(t)}{dt} = -\int_0^t \zeta(t-t')\boldsymbol{\phi}_{\mathrm{v}}(t')\,dt'. \tag{9.66}$$

This equation can be solved by the Laplace transform:

$$\widetilde{\boldsymbol{\phi}}_{\mathrm{v}}(s) = \frac{m\boldsymbol{\phi}(0)}{ms+\tilde{\zeta}(s)}.$$

Taking the velocity to be stationary at $t = 0$ implies that $\boldsymbol{\phi}(0) = k_BT\boldsymbol{\delta}/m$, so

$$\widetilde{\boldsymbol{\phi}}_{\mathrm{v}}(s) = \frac{k_BT}{ms+\tilde{\zeta}(s)}\boldsymbol{\delta}. \tag{9.67}$$

Combining (9.61) and (9.67) yields the generalized Stokes–Einstein relation

$$\left\langle \widetilde{\boldsymbol{RR}}(s)\right\rangle = \frac{2}{s^2}\frac{k_BT}{ms+\tilde{\zeta}(s)}\boldsymbol{\delta}. \tag{9.68}$$

For a medium in which we can measure the $d$-dimensional mean squared displacement $\left\langle R_d^2(t)\right\rangle$, (9.68) becomes

$$\tilde{\zeta}(s) = \frac{2dk_BT}{s^2\left\langle \widetilde{R_d^2}(s)\right\rangle} - ms. \tag{9.69}$$

We allow for $d$-dimensional measurements because, for example, data from microscope images might consist only of two-dimensional projections of the full three-dimensional displacements. Finally, since $\zeta(t)$ is causal, its Fourier and Laplace tranforms are related: $\hat{\zeta}(\omega) = \tilde{\zeta}(i\omega)$. This result along with the relation $\hat{\zeta}(\omega) = 6\pi\eta^*(\omega)a$ provides a connection between mean squared displacement and complex viscosity:

$$\eta^*(\omega) = -\frac{2dk_BT}{6\pi a\omega^2\left\langle \widetilde{R_d^2}(i\omega)\right\rangle} - \frac{i\omega m}{6\pi a}. \tag{9.70}$$

This equation is widely used, usually with $m$ set to zero, to probe the rheology of materials at small scales and *in situ* using observations under a microscope. For an inertialess particle in a Newtonian fluid, $\left\langle R_d^2(t)\right\rangle = 2dk_BTt/\zeta$ for $t \geq 0$ and thus $\left\langle \widetilde{R_d^2}(i\omega)\right\rangle = -2dk_BT/\zeta\omega^2$. Inserting this expression into (9.70) with $m = 0$, we find that

$$\eta^* = -\frac{2dk_BT}{6\pi a\omega^2(-2dk_BT/\zeta\omega^2)} = \frac{\zeta}{6\pi a} = \eta,$$

thus recovering the correct Newtonian result.

The preceding results have been found for a general form of the relaxation modulus, necessitating analysis of a generalized Langevin equation and an assumption about the fluctuating force. If the memory function can be represented as a sum of exponentials, however, then the generalized Langevin equation can be rewritten (nonuniquely) as a system of ordinary Langevin equations – a Markov process. Now analysis is more straightforward and trajectories can be generated using a standard Euler–Maruyama scheme. For example, consider the system

$$m\frac{d\boldsymbol{U}}{dt} = -\alpha \boldsymbol{Z}, \tag{9.71}$$

$$\frac{d\boldsymbol{Z}}{dt} = \beta(\boldsymbol{U} - \boldsymbol{Z}) + \boldsymbol{G}_{\text{fluc}}(t), \tag{9.72}$$

$$\frac{d\boldsymbol{R}}{dt} = \boldsymbol{U}, \tag{9.73}$$

where

$$\langle \boldsymbol{G}_{\text{fluc}}(t)\boldsymbol{G}_{\text{fluc}}(t')\rangle = \sigma^2\delta(t-t')\boldsymbol{\delta}. \tag{9.74}$$

We have supplemented the equations for the velocity and position with one for an auxiliary variable $\boldsymbol{Z}$, which is driven by a fluctuating term and relaxes exponentially toward $\boldsymbol{U}$. The solution for $\boldsymbol{Z}(t)$ starting at initial time $t_0$ is

$$\boldsymbol{Z}(t) = \boldsymbol{Z}(t_0)e^{-\beta(t-t_0)} + \int_{t_0}^{t} \beta e^{-\beta(t-t')}\boldsymbol{U}(t')\,dt' + \int_{t_0}^{t} e^{-\beta(t-t')}\boldsymbol{G}_{\text{fluc}}(t')\,dt'. \tag{9.75}$$

Substituting (9.75) into (9.71), we find that

$$m\frac{d\boldsymbol{U}}{dt} = -\alpha \boldsymbol{Z}(t_0)e^{-\beta(t-t_0)} - \int_{t_0}^{t} \alpha\beta e^{-\beta(t-t')}\boldsymbol{U}(t')\,dt' + \boldsymbol{F}_{\text{fluc}}(t), \tag{9.76}$$

where

$$\boldsymbol{F}_{\text{fluc}}(t) = -\int_{t_0}^{t} \alpha e^{-\beta(t-t')}\boldsymbol{G}_{\text{fluc}}(t')\,dt'. \tag{9.77}$$

As $t_0 \to -\infty$, the first term of (9.76) vanishes, and this equation has the same form as (9.62) with $\zeta(\tau) = \alpha\beta e^{-\beta\tau}H(\tau)$. From (9.60),

$$G(\tau) = \frac{\zeta(\tau)}{6\pi a} = \frac{\alpha\beta e^{-\beta\tau}}{6\pi a}H(\tau).$$

Comparing with (9.32), we see that by setting $\beta = 1/\lambda$ and $\alpha = 6\pi a\lambda G_0 = 6\pi a\eta_0$, this model corresponds to motion of a sphere in a Maxwell fluid, and $\alpha$ is the friction coefficient based on the zero-shear viscosity $\eta_0 = G_0\lambda$ of the fluid. In the limit $\beta \to \infty$, $\zeta(t) \to 6\pi\eta a\delta(t)$, the memory function for a Newtonian fluid. The system (9.71)–(9.73) is much more straightforward to analyze or simulate than the generalized Langevin equation (9.62) since we can treat it as a standard initial value problem. We will now use this approach to find the velocity autocorrelation function and mean squared displacement for a particle in a Maxwell fluid.

Let $t_0 = 0$ so we can consider the standard initial value problem with initial conditions $\boldsymbol{U}(0)$ and $\boldsymbol{Z}(0)$. We assume as before that the system is stationary by this time. Taking the Laplace transform of (9.71)–(9.72) and rearranging gives

$$\widetilde{\boldsymbol{U}}(s) = \frac{m\boldsymbol{U}(0) - \alpha\widetilde{\boldsymbol{Z}}(s)}{ms},$$

$$\widetilde{\boldsymbol{Z}}(s) = \frac{\beta\widetilde{\boldsymbol{U}}(s) + \widetilde{\boldsymbol{Z}}(0) + \widetilde{\boldsymbol{G}}_{\text{fluc}}(s)}{s+\beta}.$$

Inserting the second equation into the first, we find that

$$\widetilde{U}(s) = \left(mU(0) - \frac{\alpha Z(0) + \alpha \widetilde{G}_{\text{fluc}}(s)}{s+\beta}\right)\left(ms + \frac{\alpha\beta}{s+\beta}\right)^{-1}.$$

Taking the dot product with $U(0)$ and ensemble-averaging yields an expression for the Laplace transform of the trace of the velocity autocorrelation function:

$$\langle U(0)\cdot\widetilde{U}(s)\rangle = \left(m\left\langle|U(0)|^2\right\rangle - \frac{\alpha\langle U(0)\cdot Z(0)\rangle + \alpha\langle U(0)\cdot\widetilde{G}_{\text{fluc}}(s)\rangle}{s+\beta}\right)\left(ms + \frac{\alpha\beta}{s+\beta}\right)^{-1}. \tag{9.78}$$

Comparing to (9.65), we see that the term $\alpha\langle U(0)\cdot Z(0)\rangle + \alpha\langle U(0)\cdot\widetilde{G}_{\text{fluc}}(s)\rangle$ corresponds to the noise term that we assumed earlier to vanish. The average $\langle U(0)\cdot\widetilde{G}_{\text{fluc}}(s)\rangle = 0$ because $U(0)$ is independent of $G_{\text{fluc}}(t)$ for $t \geq 0$. The value of $\langle U(0)\cdot Z(0)\rangle$ can be determined from (9.71). Multiplying this equation by $U(t)$ and ensemble averaging yields that

$$\left\langle U\cdot\frac{dU}{dt}\right\rangle = \frac{1}{2}\frac{d}{dt}\langle U\cdot U\rangle = -\alpha\langle U\cdot Z\rangle. \tag{9.79}$$

Since $U(t)$ is stationary at $t = 0$, $\frac{d}{dt}\langle U\cdot U\rangle = 0$ and thus $\langle U(0)\cdot Z(0)\rangle = 0$. Therefore, we have found that the noise term vanishes from the expression for the evolution of the velocity autocorrelation function, just as we assumed in going from (9.65) to (9.66). Now (9.78) becomes

$$\langle U(0)\cdot\widetilde{U}(s)\rangle = \frac{3k_BT}{ms + \frac{\alpha\beta}{s+\beta}}. \tag{9.80}$$

For the Maxwell model,

$$\tilde{\zeta}(s) = L\left\{6\pi a G_0 \exp\left(\frac{-\tau}{\lambda}\right)\right\} = \frac{6\pi a G_0}{s+1/\lambda} = \frac{\alpha\beta}{s+\beta}$$

so we see that (9.80) agrees with (9.67). In Problem 9.6, we see that this model can be generalized for a modulus comprised of a sum of exponentials of the form (9.38). In the Newtonian limit $\beta \to \infty$, (9.80) reduces to

$$\langle U(0)\cdot\widetilde{U}(s)\rangle = \frac{3k_BT}{ms+\alpha}, \tag{9.81}$$

whose inverse Laplace transform yields

$$\langle U(0)\cdot U(t)\rangle = \frac{3k_BT}{m}e^{-6\pi\eta a t/m} \tag{9.82}$$

(for $t > 0$) in agreement with the results of Section 6.2.

Now we can relate the velocity autocorrelation function to the mean squared displacement using the trace of (9.61):

$$\langle\widetilde{R^2}(s)\rangle = \frac{2}{s^2}\langle U(0)\cdot\widetilde{U}(s)\rangle = \frac{6k_BT}{s^2\left(ms + \frac{\alpha\beta}{s+\beta}\right)}. \tag{9.83}$$

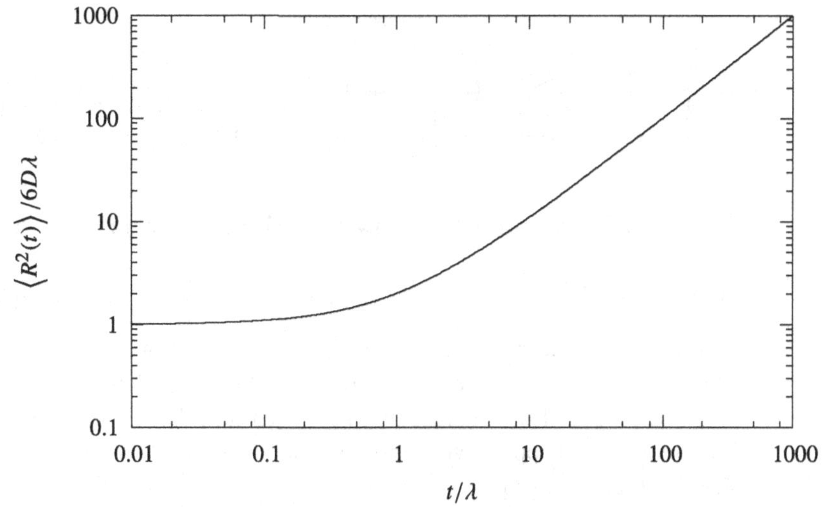


**Figure 9.3** Mean squared displacement of an inertialess Brownian particle in a Maxwell fluid, from (9.84).

Setting $m = 0$ to neglect inertia and then inverting, we find that

$$\langle R^2(t)\rangle = \frac{6k_BT}{\alpha\beta}(1+\beta t) = \frac{6k_BT\lambda}{6\pi\eta_0 a}\left(1+\frac{t}{\lambda}\right) = 6D_P\lambda\left(1+\frac{t}{\lambda}\right), \tag{9.84}$$

where we have recognized the appearance of the Stokes–Einstein diffusivity of the particle, $D_P = k_BT/6\pi\eta_0 a$. This result is illustrated in Figure 9.3. For $t \gg \lambda$, this expression reduces to the simple diffusive result $\langle R^2(t)\rangle/t \to 6D_P$, while for $t \lesssim \lambda$, the mean squared displacement is nearly constant at $D_P\lambda$. This quantity is the diffusive mean squared displacement of the particle in an interval of one fluid relaxation time. For a polymeric fluid, a crude scaling estimate of $\lambda$ would be $\lambda \sim R_0^2/D_K$, and taking $D_P/D_K \sim R_0/a$, the short time plateau value $D_P\lambda \sim R_0^3/a$, which is much less than $a^2$.

One artifact of the neglect of inertia in going from (9.83) to (9.84) is the prediction of a nonzero mean squared displacement at $t = 0$; in reality, the short-time behavior is rather complicated – at very short times, inertia dominates and the particle transport is ballistic. At intermediate times $t \lesssim \lambda$, however, the response oscillates around the plateau value predicted by the inertialess analysis (Indei et al. 2012); (9.83) has complex poles. If a solvent contribution to $G(t)$ is included, however, these oscillations disappear for sufficiently small $m$ while the plateau persists (Problem 9.5).

Finally, it should be noted that the preceding analysis completely neglects the inertia of the fluid, i.e., the Basset and added mass effects described in Section 5.3. These effects can strongly affect short-time dynamics in particle-tracking microrheology (Indei et al. 2012) and even the long-time behavior of the velocity autocorrelation function (Problem 9.7).

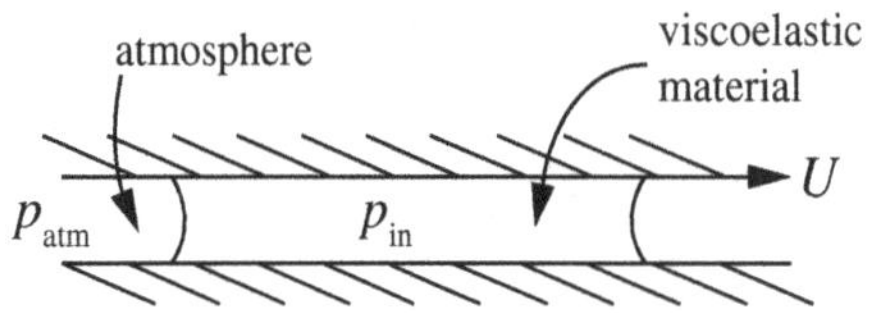


**Figure 9.4** Shear flow of a viscoelastic fluid exposed to the atmosphere.

## 9.3 Nonlinear Viscoelasticity: Shear and Extensional Flows

### 9.3.1 Simple Shear Flow: Normal Stress Differences and Cross-stream Migration

Analysis using linear viscoelasticity is limited to situations where either the total deformation is small or the deformation is sufficiently slow that the material is not perturbed very far from equilibrium: i.e., when the Weissenberg number Wi is small. To examine what happens when this condition is not satisfied, we begin again with the Oldroyd-B model, (8.113).

In simple shear flow, this model can be solved exactly at all shear rates at steady state:

$$\tau_{\mathrm{p},yx} = \tau_{\mathrm{p},xy} = G_0\lambda \frac{\partial v_x}{\partial y} = G_0\lambda\dot{\gamma}, \tag{9.85}$$

$$\tau_{\mathrm{p},xx} = 2\lambda\tau_{\mathrm{p},yx}\frac{\partial v_x}{\partial y} = 2G_0(\lambda\dot{\gamma})^2, \tag{9.86}$$

$$\tau_{\mathrm{p},xz} = \tau_{\mathrm{p},zx} = \tau_{\mathrm{p},yz} = \tau_{\mathrm{p},zy} = \tau_{\mathrm{p},yy} = \tau_{\mathrm{p},zz} = 0. \tag{9.87}$$

We noted earlier that for this model $\eta_0 = G_0\lambda$, so we see here that $\tau_{\mathrm{p},yx} = \eta_0\dot{\gamma}$ for all values of $\dot{\gamma}$: the shear viscosity for an Oldroyd-B fluid is independent of shear rate. Defining a Weissenberg number $\mathrm{Wi} = \lambda\dot{\gamma}$, we can write the expressions for $\tau_{\mathrm{p},yx}$ and $\tau_{\mathrm{p},xx}$ nondimensionally as

$$\frac{\tau_{\mathrm{p},yx}}{G_0} = \mathrm{Wi}, \tag{9.88}$$

$$\frac{\tau_{\mathrm{p},xx}}{G_0} = 2\mathrm{Wi}^2. \tag{9.89}$$

The quadratic dependence of $\tau_{\mathrm{p},xx}$ on Wi reflects its origin in the nonlinear response of the material to flow.

Consider simple shear flow of a sample of fluid confined between parallel plates and subjected to simple shear as in the previous examples, but now we account for the fact that the sample is not infinite and its edges are exposed to the atmosphere, which has pressure $p_{\mathrm{atm}}$. This situation is shown in Figure 9.4. Neglecting the geometrical complexities (meniscus, contact line) of the edge region, the $x$-component of the interfacial stress balance is simply

$$\sigma_{xx,\mathrm{out}} = \sigma_{xx,\mathrm{in}},$$

where subscripts "out" and "in" refer to the regions outside and inside the fluid sample, respectively. Outside, in the surrounding air, $\sigma_{xx} = \sigma_{xx,\mathrm{out}} = -p_{\mathrm{atm}}$. Within the fluid, $\sigma_{xx} = -p_{\mathrm{in}} + \tau_{xx}$ and the interfacial balance shows that

$$p_{\mathrm{in}} - p_{\mathrm{atm}} = \tau_{xx}. \tag{9.90}$$

Now consider the $y$-component of the stress that the fluid exerts on the top plate: $t_y = \hat{\boldsymbol{n}} \cdot \boldsymbol{\sigma} \cdot \boldsymbol{e}_y - p_{\text{atm}} = -\sigma_{yy} - p_{\text{atm}}$. The outside air exerts a downward stress $p_{\text{atm}}$, so the net stress on the plate $t_{y,\text{net}}$ is given by $-\sigma_{yy} - p_{\text{atm}}$. Therefore,

$$\begin{aligned} t_{y,\text{net}} &= -\sigma_{yy} - p_{\text{atm}} \\ &= p_{\text{in}} - \tau_{yy} - p_{\text{atm}} \\ &= p_{\text{atm}} + \tau_{xx} - \tau_{yy} - p_{\text{atm}} \\ &= \tau_{xx} - \tau_{yy}. \end{aligned} \tag{9.91}$$

The quantity

$$N_1 = \tau_{xx} - \tau_{yy} \tag{9.92}$$

is called the *first normal stress difference*. For the Oldroyd-B model, $N_1 = \tau_{\text{p},xx} = 2G_0(\lambda\dot{\gamma})^2 > 0$, so as the fluid is sheared, it pushes upward on the top plate and downward on the bottom, and accordingly the device must be designed to resist these forces. Similarly,

$$N_2 = \tau_{yy} - \tau_{zz} \tag{9.93}$$

is the *second normal stress difference*. For the Oldroyd-B model, this is zero; in general for polymer melts $N_2 < 0$ with $|N_2| \ll |N_1|$. The *first* and *second normal stress coefficients* $\Psi_1$ and $\Psi_2$ are defined by the relations

$$N_1 = \Psi_1(\dot{\gamma})\dot{\gamma}^2 \tag{9.94}$$

$$N_2 = \Psi_2(\dot{\gamma})\dot{\gamma}^2. \tag{9.95}$$

In general, these quantities depend on $\dot{\gamma}$. For the Oldroyd-B model, however, they are constant just as the viscosity is: $\Psi_1 = 2G_0\lambda^2$ and $\Psi_2 = 0$.

Sometimes the Weissenberg number is defined for a shear flow as the ratio between the first normal stress difference and the shear stress: i.e., $N_1/\tau_{yx}$ (Poole 2012). For an upper convected Maxwell model in simple shear, where $N_1 = 2G_0(\lambda\dot{\gamma})^2$ and $\tau_{yx} = \tau_{\text{p},yx} = G_0\lambda\dot{\gamma}$, this definition of Wi reduces to $2\lambda\dot{\gamma}$: i.e., in the case where solvent viscosity is negligible, it coincides with the definition in terms of time scale ratios described in Section 9.1.1.

The normal stresses that develop during shear flow of a polymeric fluid are a reflection of the underlying stretching and alignment of the polymer molecules with the flow direction. In dilute solution, stretched polymer molecules in the vicinity of a wall will tend to migrate as described in Section 4.4. Recalling that the stress tensor is just the number density $n$ times the ensemble-average force dipole of the polymer molecules, we observe that (4.13) can be written as

$$\left\langle v_y^{\text{W}} \right\rangle = \frac{3}{64\pi\eta y_0^2}\frac{N_1 - N_2}{n}. \tag{9.96}$$

During flow, this migration velocity will be balanced by molecular diffusion, leading to a region near solid surfaces that is depleted of polymer molecules (Jendrejack et al. 2004, Ma & Graham 2005, Graham 2011). For a flexible chain,

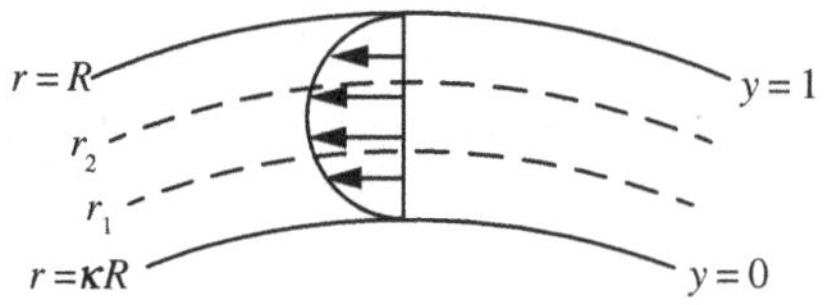


**Figure 9.5** Pressure-driven flow in a curved channel.

when Wi ≫ 1, this region can be much larger than the equilibrium size of a polymer molecule.

## 9.3.2 Flow in a Curved Channel: Hoop Stress and Viscoelastic Flow Instability

Normal stresses in shear flows have particularly important consequences in flows with curved streamlines; a simple model flow for illustrating these is the pressure-driven flow in a curved channel, also known as *Dean flow* (Figure 9.5). The inner and outer walls of the channel are at radial positions $r = \kappa R$ and $r = R$, respectively. Given a constant azimuthal pressure gradient $\boldsymbol{e}_\theta \cdot \boldsymbol{\nabla} p = -\Gamma R$, working in cylindrical coordinates and assuming a unidirectional flow in the $\theta$-direction, the steady-state $r$- and $\theta$-momentum balances are

$$-\rho\frac{v_\theta^2}{r} = -\frac{\tau_{\theta\theta}}{r} - \frac{\partial p}{\partial r}, \tag{9.97}$$

$$0 = \frac{1}{r^2}\frac{\partial}{\partial r}r^2\tau_{r\theta} + \Gamma\frac{R}{r}. \tag{9.98}$$

The only nonzero solvent (viscous) stress is the shear stress,

$$\tau_{s,r\theta} = \eta\left(\frac{\partial v_\theta}{\partial r} - \frac{v_\theta}{r}\right), \tag{9.99}$$

and for the Oldroyd-B model the steady-state expressions for the nonzero polymer stresses are

$$\tau_{p,r\theta} = \eta_p\left(\frac{\partial v_\theta}{\partial r} - \frac{v_\theta}{r}\right), \tag{9.100}$$

$$\tau_{p,\theta\theta} = 2\tau_{p,r\theta}\lambda\left(\frac{\partial v_\theta}{\partial r} - \frac{v_\theta}{r}\right) = \frac{2\tau_{p,r\theta}^2}{G_0}. \tag{9.101}$$

All of the other polymer stress components vanish.

Before proceeding to the specific case of Dean flow, let us make some general observations from the radial momentum balance, which we can rewrite

$$\frac{\partial p}{\partial r} = \rho\frac{v_\theta^2}{r} - \frac{\tau_{\theta\theta}}{r}.$$

Note that the centrifugal force term $v_\theta^2/r$ arising from fluid inertia results in a positive radial pressure gradient and thus an increase in pressure with increasing radial position. In the viscoelastic case, the stretching in the $\theta$-direction, arising in the term $\tau_{\theta\theta}/r$, also contributes to the $r$-momentum balance. This term, sometimes referred to as a "hoop stress," because it arises in an azimuthally stretched hoop of material, is analogous to the radial force exerted by the curved surface of an inflated bubble or balloon – tensions along a concave surface generate a resultant force that is normal to the surface and proportional to the product of tension and curvature. From (9.101), we can see that $\tau_{\theta\theta}$ is nonnegative everywhere in the channel. Therefore, viscoelasticity leads to the opposite effect as inertia: a decreasing pressure with increasing radial position – the hoop stress tends to squeeze material radially inward. In general, shear flows with curved streamlines will result in pressure gradients across the streamlines that must be taken into account. A dramatic example of this phenomenon is the *rod-climbing* phenomenon, also known as the *Weissenberg effect*. If a rotating rod is lowered into a container of a sufficiently viscoelastic liquid, the liquid is observed to climb the rod due to the hoop stress generated as the rod shears the fluid. The opposite happens in the case of a Newtonian liquid, where the centrifugal force tends to fling the liquid away from the rod, lowering the liquid level near it (Bird, Armstrong, & Hassager 1987).

Returning to the specific case of Dean flow, given no-slip boundary conditions at the inner and outer walls $r = \kappa R$ and $r = R$, an exact solution for $v_\theta$ can easily be obtained. It will be instructive, however, to consider the limiting case when the gap $d = (1 - \kappa)R$ between the channel walls is much less than $R$. Let $\epsilon = d/R = 1 - \kappa$ and define a dimensionless radial variable $y = (r - \kappa R)/\epsilon R$, so the inner wall is at $y = 0$ and the outer at $y = 1$. The quantity $\epsilon$ has the physical interpretation of a dimensionless curvature: as $\epsilon \to 0$ with $d$ held constant, the channel becomes straight. To leading order in $\epsilon$, the $r-$ and $\theta$-momentum balances become

$$\frac{\partial p}{\partial y} = \epsilon \left(\rho v_\theta^2 - \tau_{\theta\theta}\right), \tag{9.102}$$

$$\frac{d^2 v_\theta}{dy^2} + \frac{\Gamma d^2}{\eta + \eta_{\mathrm{p}}} = 0. \tag{9.103}$$

As $\epsilon \to 0$, the radial pressure gradient vanishes. The solution to (9.103) is a simple parabolic velocity profile:

$$v_\theta = \frac{\Gamma d^2}{2(\eta + \eta_{\mathrm{p}})} y(1 - y). \tag{9.104}$$

The corresponding polymer stresses are

$$\tau_{\mathrm{p},r\theta} = \frac{1}{2}(1 - \beta)\Gamma d(1 - 2y), \tag{9.105}$$

$$\tau_{\mathrm{p},\theta\theta} = \frac{1}{2G_0}\left((1 - \beta)\Gamma d(1 - 2y)\right)^2. \tag{9.106}$$

Using the shear rate at the bottom wall, $\dot{\gamma}_w = \Gamma d/(\eta + \eta_{\mathrm{p}})$, we can define a Weissenberg number $\mathrm{Wi} = \lambda \dot{\gamma}_w$ and rewrite the polymer stresses as

$$\frac{\tau_{\mathrm{p},r\theta}}{G_0} = \mathrm{Wi}(1 - 2y), \tag{9.107}$$

$$\frac{\tau_{\mathrm{p},\theta\theta}}{G_0} = 2\mathrm{Wi}^2(1 - 2y)^2. \tag{9.108}$$

We noted previously the radial pressure gradient associated with centrifugal force

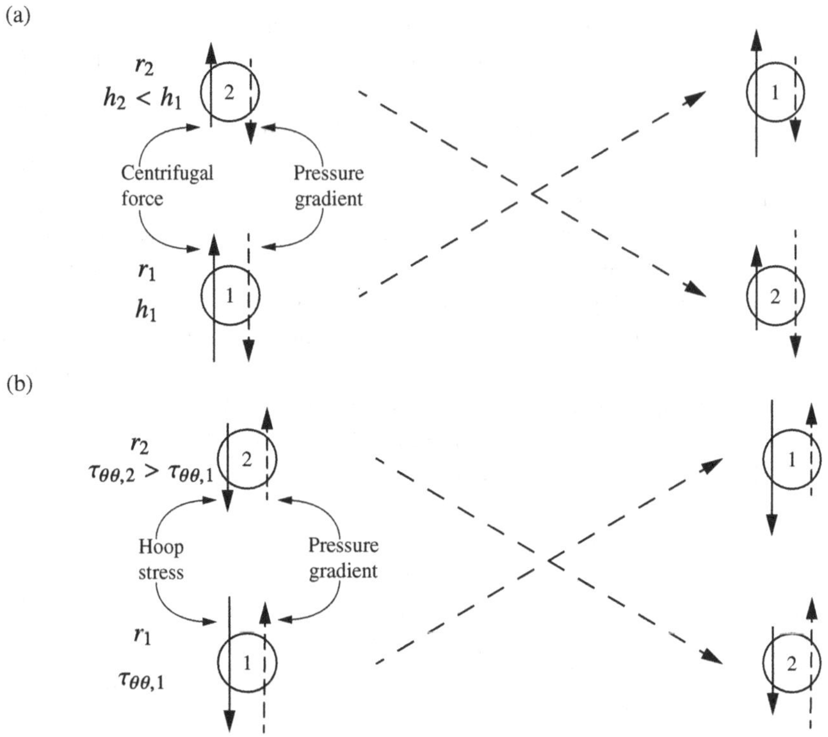


**Figure 9.6** Schematic of instability mechanisms in curved streamline flows: (a) centrifugal, (b) viscoelastic. The base flow is out of the page; each drawing is a cross-section at constant $\theta$ showing rings each with a small cross section and the indicated radius $r_1$ or $r_2$.

and hoop stresses in the presence of curved streamlines in a shear flow. These can lead to hydrodynamic instability: small perturbations to a nominally simple flow field such as circular Couette or Dean flow can grow dramatically, resulting in substantial changes to the flow (Drazin & Reid 1981, Shaqfeh 1996). Detailed stability analyses are often rather involved, but in the Dean flow case, simple physical arguments can be given to rationalize the presence of instability and the circumstances under which it can occur.

Consider first the Newtonian flow case, where in the undisturbed unidirectional flow the centrifugal force and radial pressure gradient terms balance:

$$\frac{\partial p}{\partial r} = \rho \frac{v_\theta^2}{r}.$$

Now imagine perturbing the flow by taking an axisymmetric ring-shaped fluid element of radius $r_1$, which has azimuthal velocity $v_{\theta,1}$, and exchanging it with a ring of equal volume, radius $r_2 > r_1$ and velocity $v_{\theta,2}$, while maintaining the angular momentum per unit mass $h = rv_\theta$ of each fluid element. This process is illustrated in Figure 9.6(a). The fluid element that was initially at $r_1$ now has velocity $v_{\theta,1}r_1/r_2$. At its new position $r_2$, it experiences an inward force $\rho v_{\theta,2}^2/r_2$ due to the pressure gradient there, while exerting an outward centrifugal force $\rho\left(v_{\theta,1}r_1/r_2\right)^2$. If the latter is larger than the former, then the fluid element will continue to move outward and the unidirectional flow will be unstable with respect to the imposed perturbation. Thus the instability condition is

$$\left(r_1 v_{\theta,1}\right)^2 > \left(r_2 v_{\theta,2}\right)^2,$$

which can be rewritten

$$h_1^2 > h_2^2. \tag{9.109}$$

Therefore this analysis predicts instability if the square of the angular momentum of the unidirectional flow profile decreases with increasing radial position:

$$\frac{dh^2}{dr} < 0. \tag{9.110}$$

This instability result is known as the *Rayleigh criterion* (Drazin & Reid 1981). The simple argument presented here is due to von Kármán (1934). For Dean flow, it holds in the outer half of the channel while for circular Couette flow with the inner cylinder rotating and the outer stationary it holds throughout the domain. Viscosity stabilizes the flow: even when the angular momentum criterion is satisfied, the product $\epsilon\mathrm{Re}^2$ must exceed a critical value for instability to occur, where Re is defined as usual based on the maximum velocity, viscosity, and gap width (Drazin & Reid 1981). As the dimensionless curvature $\epsilon$ vanishes, so does the instability.

The preceding arguments can be adapted to the viscoelastic case (Joo & Shaqfeh 1992). In the absence of inertia, the undisturbed unidirectional flow satisfies

$$\frac{\partial p}{\partial r} = -\frac{\tau_{\theta\theta}}{r}.$$

Imagine making the same exchange of fluid elements as described earlier, but now holding the elastic stress $\tau_{\theta\theta}$ of each element constant. At its new position $r_2$, the element that was originally at $r_1$ experiences an outward force $\tau_{\theta\theta,2}/r_2$ due to the pressure gradient, while exerting an inward force $\tau_{\theta\theta,1}/r_2$ due to its hoop stress. If the former term exceeds the latter, then the element will continue to move outward and the unidirectional flow will again be unstable. This will occur if $\tau_{\theta\theta,2} > \tau_{\theta\theta,1}$ or

$$\frac{d\tau_{\theta\theta}}{dr} > 0. \tag{9.111}$$

By this criterion, the flow in the outer half of the curved channel flow is viscoelastically unstable, a result that is confirmed by detailed analysis (Joo & Shaqfeh 1992). Relaxation of the polymer molecules stabilizes the flow, so in fact for an Oldroyd-B model, the product $\epsilon\mathrm{Wi}^2$ must exceed a critical value for instability to occur – note that this is the dimensionless product of curvature and normal stress. There is also a viscoelasticity-driven instability in circular Couette flow, but in contrast to the case of centrifugal instability, where the mechanism is identical in Dean or circular Couette flow, the viscoelastic instability mechanism for Couette flow is somewhat different, though still associated with hoop stresses (Larson, Shaqfeh & Muller 1990). Viscoelastic instabilities due to normal stresses and curved streamlines are observed in a variety of flow geometries (Shaqfeh 1996) and can also occur in viscoelastic flows with curved free surfaces (Graham 2003). These instabilities are suppressed by the presence of second normal stress differences and shear-thinning and are thus primarily observed in highly elastic dilute polymer solutions rather than concentrated solutions or melts, where these effects are more pronounced.

### 9.3.3 Uniaxial Extension: Extensional Viscosity and the Trouton Ratio

Another dramatic viscoelastic effect arises in extensional flows. Consider uniaxial extension as shown in Figure 1.2(a), with uniform velocity gradient

$$\nabla \boldsymbol{v} = \dot{\epsilon}\left(\boldsymbol{e}_z\boldsymbol{e}_z - \frac{1}{2}\left(\boldsymbol{e}_x\boldsymbol{e}_x + \boldsymbol{e}_y\boldsymbol{e}_y\right)\right) = \dot{\epsilon}\left(\boldsymbol{e}_z\boldsymbol{e}_z - \frac{1}{2}\boldsymbol{e}_r\boldsymbol{e}_r\right)$$

and $\dot{\epsilon} > 0$. A good approximation to this flow can be generated by placing material between parallel plates and then pulling the plates apart at an exponentially increasing rate, where now $z$ is the direction normal to the plates. Taking the sample of material to be axisymmetric, we use cylindrical coordinates. We assume that the stress in the material is spatially homogeneous. Neglecting surface tension, the normal stress balance across the interface of the material with the surrounding atmosphere is

$$\sigma_{rr,\text{in}} = -p_{\text{in}} + \tau_{rr} = -p_{\text{atm}}$$

or

$$p_{\text{in}} = p_{\text{atm}} + \tau_{rr}.$$

The total stress exerted on the upper plate by the fluid is $t_{z,\text{net}} = -\sigma_{zz} - p_{\text{atm}}$, so the stress required to pull the plate upward is $-t_{z,\text{net}}$. Using the previous result, this becomes

$$\begin{aligned} -t_{z,\text{net}} &= \sigma_{zz} + p_{\text{atm}} \\ &= \tau_{zz} - p_{\text{in}} + p_{\text{atm}} \\ &= \tau_{zz} - \tau_{rr}. \end{aligned} \tag{9.112}$$

We define

$$\eta_{\text{e}} = \frac{-t_{z,\text{net}}}{\dot{\epsilon}} = \frac{\tau_{zz} - \tau_{rr}}{\dot{\epsilon}} \tag{9.113}$$

as the *extensional* or *elongational* viscosity $\eta_e$ of the fluid. The *Trouton ratio* Tr is the ratio between the extensional and shear viscosities of a fluid:

$$\text{Tr} = \frac{\eta_{\text{e}}}{\eta + \eta_{\text{p}}}. \tag{9.114}$$

For a Newtonian fluid, $\tau_{rr} = 2\eta\frac{\partial v_r}{\partial r} = -\eta\dot{\epsilon}$ and $\tau_{zz} = 2\eta\frac{\partial v_z}{\partial z} = 2\eta\dot{\epsilon}$, so

$$\tau_{zz} - \tau_{rr} = 3\eta\dot{\epsilon}. \tag{9.115}$$

We see from (9.115) that for a Newtonian fluid Tr = 3. For a truly dilute polymer solution $\eta_{\rm p} \ll \eta$, so

$$\mathrm{Tr} \approx \frac{\eta_{\rm e}}{\eta}.$$

Now let us consider the extensional behavior of an Oldroyd-B fluid. The evolution equations for $\tau_{\mathrm{p},rr}$, $\tau_{\mathrm{p},\theta\theta}$ and $\tau_{\mathrm{p},zz}$ become

$$\tau_{\mathrm{p},rr} + \lambda\left(\frac{d\tau_{\mathrm{p},rr}}{dt} + \dot{\epsilon}\tau_{\mathrm{p},rr}\right) = -\dot{\epsilon}G_0\lambda,$$

$$\tau_{\mathrm{p},\theta\theta} + \lambda\left(\frac{d\tau_{\mathrm{p},\theta\theta}}{dt} + \dot{\epsilon}\tau_{\mathrm{p},\theta\theta}\right) = -\dot{\epsilon}G_0\lambda,$$

$$\tau_{\mathrm{p},zz} + \lambda\left(\frac{d\tau_{\mathrm{p},zz}}{dt} - 2\dot{\epsilon}\tau_{\mathrm{p},zz}\right) = 2\dot{\epsilon}G_0\lambda.$$

Introducing a Weissenberg number for extensional flow $\mathrm{Wi} = \lambda\dot{\epsilon}$, these equations have steady-state solutions

$$\frac{\tau_{\mathrm{p},rr}}{G_0} = \frac{\tau_{\mathrm{p},\theta\theta}}{G_0} = \frac{-\mathrm{Wi}}{1+\mathrm{Wi}}$$

$$\frac{\tau_{\mathrm{p},zz}}{G_0} = \frac{2\mathrm{Wi}}{1-2\mathrm{Wi}}.$$

Observe that the steady-state value of $\tau_{\mathrm{p},zz}$ diverges at $\mathrm{Wi} = 1/2$ and is negative for $\mathrm{Wi} > 1/2$. To understand the origin of this divergence, we return to the transient problem for $\tau_{\mathrm{p},zz}$, rewriting it as

$$\lambda\frac{d\tau_{\mathrm{p},zz}}{dt} = 2G_0\mathrm{Wi} + (2\mathrm{Wi} - 1)\,\tau_{\mathrm{p},zz}.$$

Starting from the equilibrium initial condition $\tau_{\mathrm{p},zz}(0) = 0$, if $\mathrm{Wi} > 1/2$ the stress will grow exponentially without bound, never reaching a steady state, so the negative steady state found earlier is not physically accessible. Therefore, for the Oldroyd-B model, the polymer contribution to the steady-state extensional viscosity is given by

$$\eta_{\rm pe}^{\rm ss} = \frac{\tau_{\mathrm{p},zz} - \tau_{\mathrm{p},rr}}{\dot{\epsilon}} = \eta_{\rm p0}\left(\frac{2}{1-2\mathrm{Wi}} + \frac{1}{1+\mathrm{Wi}}\right)$$

when $\mathrm{Wi} < 1/2$, and does not exist when $\mathrm{Wi} > 1/2$. In the latter case, $\eta_{\rm e}$ increases exponentially with time. Returning to the molecular origin of the Oldroyd-B model, recall that for this model $\tau_{\mathrm{p},zz} \sim Q_z^2$. Thus the exponential divergence of $\tau_{zz}$ when $\mathrm{Wi} > 1/2$ implies that that polymer molecules will stretch forever, a result that is unphysical because a real polymer molecule cannot be extended beyond its contour length. This pathology arises from the Hookean spring law that underlies the model. Once the polymer chain is sufficiently extended, this model breaks down.

### 9.3.4 Effects of Finite Extensibility on Shear and Extensional Flows: Scaling Arguments

Although the Oldroyd-B model yields insight into the origins of normal stresses in shear flows and large polymer stretching in extensional flows, it does not predict the shear-thinning that is experimentally observed in polymer solutions, nor does it yield physically relevant predictions in strong extensional flows. These problems originate in the neglect of the finite extensibility of real polymer molecules. The FENE spring model introduced previously incorporates this effect, but even for this simple model, available analytical results are very limited (cf. Bird, Curtiss, Armstrong & Hassager 1987). On the other hand, substantial insight and some quantitative results can be obtained based on physical intuition and simple scaling arguments (Hinch & Leal 1972, Doyle, Shaqfeh & Gast 1997, Larson 1999). These have the additional advantage of being independent of the detailed model for the spring force.

In simple shear flow at very high Weissenberg number, we expect that a dumbbell will spend most of its time stretched to nearly its contour length $L$ and oriented very nearly in the flow direction, as illustrated in Figure 9.7(a). If the chain orients perfectly in the flow direction, the drag forces on the beads vanish and they will relax back toward equilibrium. However, Brownian motion will perturb the bead positions, exposing them again to flow, and the dumbbell will again stretch out. The orientation angle of the connector vector $\boldsymbol{Q}$ with respect to the flow direction will be determined by a balance of the hydrodynamic drag, which is tending to stretch the connector along the flow direction, and Brownian motion, which is tending to push the orientation distribution back to isotropic. Consider the rightmost bead in Figure 9.7(a), which we take to be an as yet unknown distance $\delta_y$ above the $x$-axis. (Equal and opposite forces are exerted on the other bead.) This length scale is the width of the probability distribution function for $\boldsymbol{Q}$ in the $y$-direction; most molecules will be stretched and oriented along the flow direction $x$. The magnitude of the hydrodynamic force on this bead is $F_x = \zeta\dot{\gamma}\delta_y$, and the component tending to rotate the connector down toward the $x$-axis is $\zeta\dot{\gamma}\delta_y(\delta_y/(L/2))$ (see Figure 9.7(a)). (Hereinafter we neglect numerical factors.) This force is counteracted by the $y$-component of the Brownian force; the length scale over which the probability distribution varies is $\delta_y$, so this force $F_y$ scales as $k_B T/\delta_y$. At steady state, these forces balance:

$$\frac{\zeta\dot{\gamma}\delta_y^2}{L} \sim \frac{k_B T}{\delta_y}.$$

Solving for $\delta_y$ yields

$$\delta_y \sim \left(\frac{k_B T L}{\zeta\dot{\gamma}}\right)^{1/3}. \tag{9.116}$$

This is the "boundary layer thickness" over which the probability distribution for $\boldsymbol{Q}$ decays. This result is closely analogous to the Graetz–Leveque result for the thickness of a concentration boundary layer on a semi-infinite flat plate exposed to simple shear, which results from the balance of convective flux in the $x$-direction and molecular diffusion in $y$ (Leal 2007). For a solute of diffusivity $D$ dissolving from the plate into the fluid, the boundary layer thickness $\delta_y(x) \sim (Dx/\dot{\gamma})^{1/3}$.

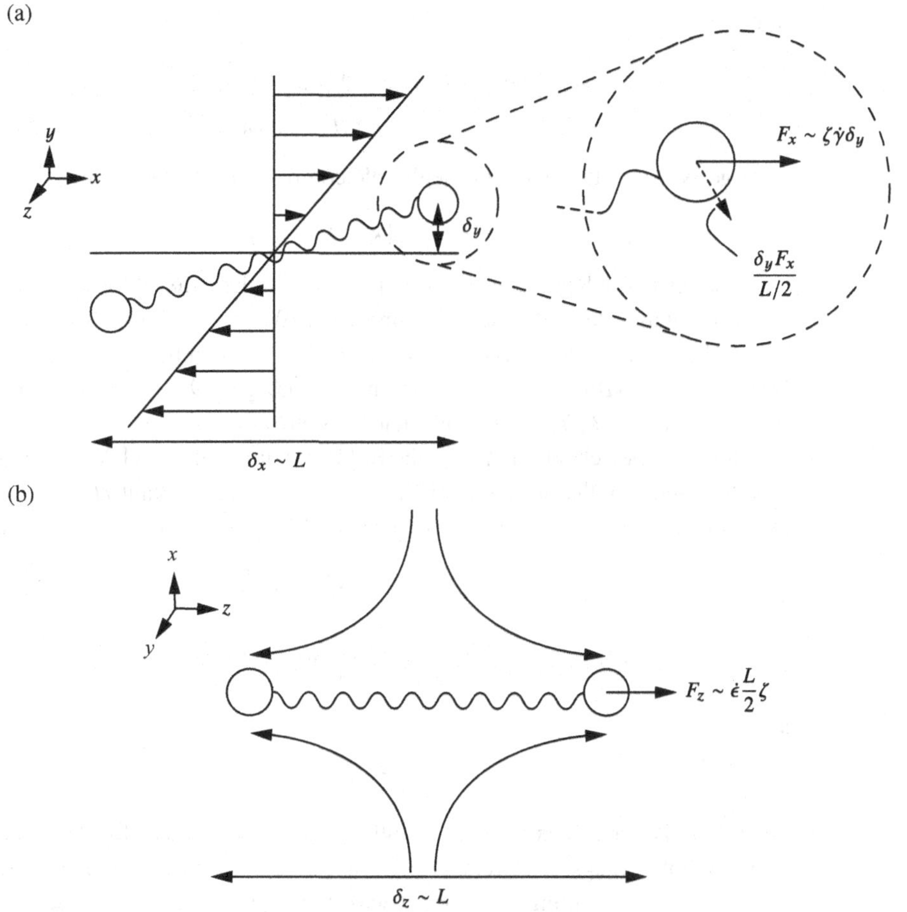


**Figure 9.7** Sketches to illustrate drag force scalings for a finitely extensible dumbbell in (a) shear and (b) uniaxial extension.

We can estimate that the $yx$ component of the force dipole exerted by the polymer molecule on the solvent scales as $\delta_y F_x$, so for a solution with number density $n$, the polymer contribution to the shear stress obeys

$$\tau_{\mathrm{p},yx} \sim n\delta_y F_x \sim n\delta_y^2\zeta\dot{\gamma} \sim n\left(\frac{k_BTL}{\zeta\dot{\gamma}}\right)^{2/3}\zeta\dot{\gamma}. \tag{9.117}$$

(As must happen, the same result arises from taking the product $n\delta_x F_y$, with $\delta_x \sim L$ because the molecule is highly stretched.) Accordingly, the polymer contribution to the shear viscosity $\eta_\mathrm{p}$ satisfies

$$\eta_\mathrm{p} \sim n\delta_y^2\zeta \sim n\left(\frac{k_BTL}{\zeta\dot{\gamma}}\right)^{2/3}\zeta. \tag{9.118}$$

Similarly,

$$\tau_{\mathrm{p},xx} \sim n\delta_x F_x \sim n\left(k_B T L^4 \zeta^2 \dot{\gamma}^2\right)^{1/3}, \tag{9.119}$$

$$\tau_{\mathrm{p},yy} \sim n\delta_y F_y \sim n k_B T. \tag{9.120}$$

At large Wi, the latter scaling is negligible compared to the former so

$$N_1 \sim n\left(k_B T L^4 \zeta^2 \dot{\gamma}^2\right)^{1/3}. \tag{9.121}$$

These results indicate that the combination of finite extensibility and shear-induced chain orientation leads to shear-thinning when Wi $\gg 1$; $\eta_\mathrm{p}$ decreases as $\dot{\gamma}$ increases. The scaling of $\eta_\mathrm{p}$ as $\dot{\gamma}^{-2/3}$ (and $\Psi_1 = N_1/\dot{\gamma}^2 \sim \dot{\gamma}^{-4/3}$) can be found analytically for the FENE-P model (Bird, Curtiss, Armstrong & Hassager 1987). Furthermore, estimating $\delta_z \sim \delta_y$ and $F_z \sim k_B T/\delta_z$, we have that $N_2 \sim n(\delta_y F_y - \delta_z F_z) \sim k_B T$, so $N_2 \ll N_1$ for a highly stretched chain; in fact for the FENE-P dumbbell model $N_2$ is identically zero. Finally, observing that for a dumbbell model the spring constant $H \sim k_B T/N_\mathrm{K} L_K^2$ and the relaxation time $\lambda \sim \zeta/H$, we can rewrite (9.117)- (9.121), respectively, as

$$\frac{\tau_{\mathrm{p},yx}}{G_0} \sim N_\mathrm{K}^{1/3}\mathrm{Wi}^{1/3}, \tag{9.122}$$

$$\frac{\eta_\mathrm{p}}{G_0\lambda} \sim N_\mathrm{K}^{1/3}\mathrm{Wi}^{-2/3}, \tag{9.123}$$

and

$$\frac{N_1}{G_0} \sim N_\mathrm{K}^{2/3}\mathrm{Wi}^{2/3}. \tag{9.124}$$

For real polymer chains in dilute solution the $-2/3$ exponent for the viscosity is too steep except perhaps at extremely high Wi, as the internal degrees of freedom keep the chain from orienting so strongly with flow (Doyle et al. 1997, Jendrejack et al. 2002). Nevertheless the simple argument presented here sheds substantial light into the molecular origin of the shear-thinning phenomenon in dilute solutions.

For rigid rods in shear at high Pé, similar ideas can be applied, but two complications arise (Hinch & Leal 1972, Öttinger 1988). Rods with finite aspect ratio $\ell$ will tumble (undergoing Jeffery orbits) even in the absence of Brownian motion, and as Pé $\to \infty$, the stresses simply become linear functions of $\dot{\gamma}$. In the intermediate regime, $1 \ll$ Pé $\ll \ell^3$, although the orientational distribution function is strongly peaked in a region of thickness $\delta_y \sim \dot{\gamma}^{-1/3}$ as before, orientations that are not highly aligned also make a substantial and indeed dominant contribution to $\tau_{\mathrm{p},yx}$. Roughly speaking, this effect arises because the $y$-component of these orientations makes a contribution $L \gg \delta_y$ to the $yx$ component of the force dipole. The resulting stress scalings are

$$\tau_{\mathrm{p},yx} \sim G_0\mathrm{Pé}^{2/3}, \tag{9.125}$$

$$N_1 \sim G_0\mathrm{Pé}^{2/3}, \tag{9.126}$$

$$N_2 \sim G_0\mathrm{Pé}^{-1/3}. \tag{9.127}$$

(Rods that are not flow-aligned make no contribution to $N_1$, so its shear-rate dependence remains the same as for a flexible chain, which we found earlier.)

For uniaxial extensional flow at high Wi (or Pé), related arguments can be applied, as shown in Figure 9.7(b). In this case, the polymer will be almost fully extended in the $z$-direction so $\tau_{p,rr}$ is negligible and

$$\tau_{p,zz} \sim nLF_z.$$

For a dumbbell model with no HI, $F_z \sim \zeta L\dot{\epsilon}$, so

$$\eta_e = \frac{\tau_{p,zz}}{\dot{\epsilon}} \sim nL^2\zeta. \quad (9.128)$$

At high Wi the extensional viscosity now reaches a plateau rather than diverging as in the Oldroyd-B model. Including the prefactors and using the common notation for FENE models (though this result is insensitive to the specific spring model), we find that at steady state

$$\mathrm{Tr} = \frac{2b}{3}\frac{1-\beta}{\beta}. \quad (9.129)$$

A refinement of this result can be obtained if we approximate the extended polymer as a slender body with aspect ratio $\ell$. If $\bar{F}_z$ is the $z$-component of the drag force per unit length along the polymer, then we can write $\tau_{p,zz}$ as the integral of the force dipole along the chain:

$$\tau_{p,zz} \sim n\int_{-L/2}^{L/2} z\bar{F}_z\, dz. \quad (9.130)$$

Using slender body theory (Section 3.9),

$$\bar{F}_z = \bar{\zeta}^{\|}v_z = \bar{\zeta}^{\|}\dot{\epsilon}z, \quad (9.131)$$

where $\bar{\zeta}^{\|}$ is given by (3.99). Evaluating the integral and inserting (3.99) yields that

$$\mathrm{Tr} \sim \frac{nL^3}{\ln \ell}. \quad (9.132)$$

This expression is valid for both rigid and flexible chains, once Pé or Wi $\gg 1$ as the case may be. For a flexible polymer, we can estimate $\ell$ as $L/L_K = N_K$. The logarithmic term accounts for hydrodynamic interactions along the extended chain backbone. For dilute solutions of long chains, the Trouton ratio can be much larger than unity even when the shear viscosity is barely changed from the solvent viscosity $(1-\beta) \ll 1$. This contrasting behavior in shear and extension is the origin of many of the dramatic fluid dynamical effects that are observed with dilute polymer solutions, including the suppression of drop breakup (Bergeron et al. 2000) and the reduction of drag in turbulent flows (White & Mungal 2008, Graham 2014).

## 9.4 Material Frame Indifference and Models of Viscoelastic Fluids

Up to this point, we have considered models of complex fluids from the point of view of a stationary reference frame. Sometimes it is natural, however, to consider models as viewed in a rotating reference frame. For example, the natural frame for studying

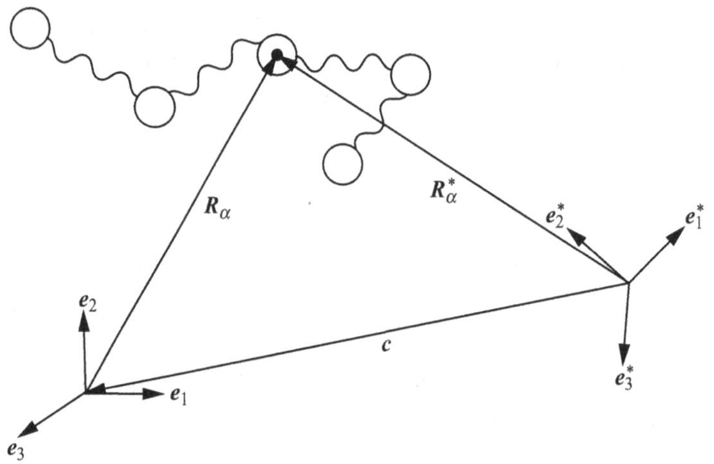


**Figure 9.8** A position vector $\boldsymbol{R}_\alpha$ for a bead of a polymer molecule as represented in frames $F$ and $F^*$.

atmospheric and oceanic dynamics is one rotating with the earth. Furthermore, in a general deforming material, some parts of the material are translating and rotating with respect to other parts, so a reference frame that is stationary with respect to one part of the material will necessarily be translating and rotating with respect to others. Thus a natural expectation would be that the behavior of the material should appear identical whether viewed in a stationary or moving frame, a property called material frame indifference (Malvern 1969, Bird, Armstrong, & Hassager 1987, Gonzalez & Stuart 2008). For concreteness, we will address this issue in the context of a familiar model, the Langevin equation for particle $\alpha$ of a system of $N$ particles: e.g., a bead in a coarse-grained polymer chain in the absence of any external fields.

First, we define a stationary reference frame $F$ with Cartesian basis vectors $\boldsymbol{e}_i$. An arbitrary position vector represented in this frame and basis set will be denoted $\boldsymbol{R}$. Now consider a second reference frame $F^*$ that is rotating and translating relative to the stationary frame. The Cartesian basis vectors in the accelerating frame will be denoted $\boldsymbol{e}_i^*$. Figure 9.8 shows a polymer molecule as seen from $F$ and $F^*$.

At time $t$, an orthogonal rotation matrix[2] $\mathbf{Q}(t)$ relates the two sets of basis vectors, rotating frame $F^*$ to the orientation of $F$:

$$\boldsymbol{e}_i = \mathbf{Q} \cdot \boldsymbol{e}_i^* \tag{9.133}$$

$$\boldsymbol{e}_i^* = \mathbf{Q}^{\mathrm{T}} \cdot \boldsymbol{e}_i. \tag{9.134}$$

By taking the dot product of (9.134) with $\boldsymbol{e}_k$, we find that the components of $\mathbf{Q}$ expressed in the $F$ frame are

$$Q_{ik} = \boldsymbol{e}_i^* \cdot \boldsymbol{e}_k. \tag{9.135}$$

[2] The columns of an orthogonal matrix $\mathbf{Q}$ are orthogonal vectors, so the matrix satisfies $\mathbf{Q}^{\mathrm{T}} \cdot \mathbf{Q} = \mathbf{Q} \cdot \mathbf{Q}^{\mathrm{T}} = \boldsymbol{\delta}$. Thus $\mathbf{Q}^{\mathrm{T}} = \mathbf{Q}^{-1}$.

The $i$th row of $\mathbf{Q}$ contains the components of $\boldsymbol{e}_i^*$. The antisymmetric matrix

$$\mathbf{A}(t) = \frac{d\mathbf{Q}}{dt} \cdot \mathbf{Q}^{\mathrm{T}} = -\mathbf{Q} \cdot \frac{d\mathbf{Q}^{\mathrm{T}}}{dt} \tag{9.136}$$

is the angular velocity tensor of $F$ relative to $F^*$.

Additionally, the origin of frame $F^*$ translates relative to that of $F$; the vector from the former to the latter is $\boldsymbol{c}(t)$. Thus the position vector $\boldsymbol{R}$, written in $F^*$, becomes

$$\boldsymbol{R}^* = \mathbf{Q}(t) \cdot \boldsymbol{R} + \boldsymbol{c}(t). \tag{9.137}$$

Equivalently,

$$\boldsymbol{R} = \mathbf{Q}^{\mathrm{T}}(t) \cdot (\boldsymbol{R}^* - \boldsymbol{c}(t)) . \tag{9.138}$$

Note that $\boldsymbol{R}$ and $\boldsymbol{R}^*$ are the *same vector*, just expressed in different reference frames. Thus what we mean by (9.137) is that it relates the *components* of the same vector in different coordinate systems to one another: $R_i^* = Q_{ij} R_j + c_i$. Now consider an arbitrary vector function of position and time $\boldsymbol{w}(\boldsymbol{R}, t)$ whose respresentation in $F^*$ is $\boldsymbol{w}^*(\boldsymbol{R}^*, t)$. If $\boldsymbol{w}^*(\boldsymbol{R}^*, t)$ satisfies

$$\boldsymbol{w}^*(\boldsymbol{R}^*, t) = \mathbf{Q}(t) \cdot \boldsymbol{w}(\boldsymbol{R}, t), \tag{9.139}$$

then $\boldsymbol{w}$ is said to be *material frame indifferent*. Similarly, any second-order tensor $\mathbf{S}(\boldsymbol{R}, t)$ that maps an indifferent vector onto an indifferent vector is frame indifferent and satisfies

$$\mathbf{S}^*(\boldsymbol{R}^*(t), t) = \mathbf{Q}(t) \cdot \mathbf{S}(\boldsymbol{R}, t) \cdot \mathbf{Q}^{\mathrm{T}}(t). \tag{9.140}$$

A frame indifferent scalar $\phi(\boldsymbol{R}, t)$ is unchanged upon a change of frame:

$$\phi(\boldsymbol{R}^*, t) = \phi(\boldsymbol{R}, t). \tag{9.141}$$

As an important example of a second-order tensor that is *not* frame indifferent, consider the deformation gradient tensor $\mathbf{F}$ introduced in Section 1.1.2. Recall from (1.10) that this quantity characterizes the time evolution of basis vectors $\boldsymbol{g}_i$ that move and deform with the material:

$$\boldsymbol{g}_i(t) = \mathbf{F}(t) \cdot \boldsymbol{e}_i .$$

(Note that for simplity of notation we have suppressed the dependence of $\boldsymbol{g}_i$ and $\mathbf{F}$ on the material element to which they are attached.) Since the basis vectors are indifferent, they transform as $\boldsymbol{g}_i^* = \mathbf{Q} \cdot \boldsymbol{g}_i$, so left-multiplying (1.10) by $\mathbf{Q}$ yields

$$\boldsymbol{g}_i^*(t) = \mathbf{Q}(t) \cdot \mathbf{F}(t) \cdot \boldsymbol{e}_i = \mathbf{F}^*(t) \cdot \boldsymbol{e}_i,$$

where we have defined

$$\mathbf{F}^*(t) = \mathbf{Q}(t) \cdot \mathbf{F}(t). \tag{9.142}$$

Thus the deformation gradient tensor transforms in the same way as an indifferent vector, *not* an indifferent tensor. This result can be understood simply by recalling that the columns of $\mathbf{F}$ are simply the basis vectors $\boldsymbol{g}_i$. Since these transform as indifferent vectors, so must $\mathbf{F}$.

With this background, we turn to the Langevin equation for the position $\boldsymbol{R}_\alpha$ of bead $\alpha$. In the stationary frame, this reads

$$m\ddot{\boldsymbol{R}}_\alpha + \zeta\left(\dot{\boldsymbol{R}}_\alpha - \boldsymbol{v}(\boldsymbol{R}_\alpha)\right) = \sum_{\beta\neq\alpha} \boldsymbol{F}_{\alpha\beta}(\boldsymbol{R}_{\alpha\beta}) + \boldsymbol{F}_{\text{fluc},\alpha}. \tag{9.143}$$

For simplicity, we neglect hydrodynamic interactions, though none of our conclusions will depend on this assumption. The interbead forces $\boldsymbol{F}_{\alpha\beta}$ are material frame indifferent because they depend only on the relative positions of the beads, and since the position vector $\boldsymbol{R}_\alpha$ is also indifferent, it follows that (8.89) for the stress tensor is also indifferent. It remains to be seen, however, whether the evolution of the bead positions themselves is frame indifferent. This issue is addressed by rewriting the Langevin equation in the accelerating frame.

To do so, we must express the time derivatives in this frame (Malvern 1969):

$$\begin{aligned}\frac{d\boldsymbol{R}^*_\alpha}{dt} &= \frac{d}{dt}\left(\mathbf{Q}\cdot\boldsymbol{R}_\alpha + \boldsymbol{c}\right)\\ &= \mathbf{Q}\cdot\frac{d\boldsymbol{R}_\alpha}{dt} + \frac{d\mathbf{Q}}{dt}\cdot\boldsymbol{R}_\alpha + \frac{d\boldsymbol{c}}{dt}.\end{aligned}$$

Rearranging and applying (9.136), we can write that

$$\begin{aligned}\frac{d\boldsymbol{R}_\alpha}{dt} &= \mathbf{Q}^{\mathrm{T}}\cdot\left(\frac{d\boldsymbol{R}^*_\alpha}{dt} - \frac{d\boldsymbol{c}}{dt}\right) + \mathbf{Q}^{\mathrm{T}}\cdot\frac{d\mathbf{Q}}{dt}\cdot\mathbf{Q}^{\mathrm{T}}\cdot\left(\boldsymbol{R}^*_\alpha - \boldsymbol{c}\right)\\ &= \mathbf{Q}^{\mathrm{T}}\cdot\frac{d\boldsymbol{R}^*_\alpha}{dt} - \mathbf{Q}^{\mathrm{T}}\cdot\frac{d\boldsymbol{c}}{dt} - \mathbf{Q}^{\mathrm{T}}\cdot\mathbf{A}\cdot\left(\boldsymbol{R}^*_\alpha - \boldsymbol{c}\right).\end{aligned} \tag{9.144}$$

The presence of the second and third terms on the right-hand side indicates that the particle velocity is *not* frame indifferent – its components do not transform according to (9.137). Repeating the same process for the second derivative yields

$$\frac{d^2\boldsymbol{R}_\alpha}{dt^2} = \mathbf{Q}^{\mathrm{T}}\cdot\left(\frac{d^2\boldsymbol{R}^*_\alpha}{dt^2} - \frac{d^2\boldsymbol{c}}{dt^2} - \frac{d\mathbf{A}}{dt}\cdot\left(\boldsymbol{R}^*_\alpha - \boldsymbol{c}\right) - \mathbf{A}\cdot\left(\frac{d\boldsymbol{R}^*_\alpha}{dt} - \frac{d\boldsymbol{c}}{dt}\right)\right). \tag{9.145}$$

The acceleration is also not frame indifferent.

Because the velocity $\boldsymbol{v}(\boldsymbol{R}_\alpha)$ is the rate of change of the position of the fluid element at position $\boldsymbol{R}_\alpha$, it obeys an equation similar to (9.144):

$$\boldsymbol{v}(\boldsymbol{R}_\alpha) = \mathbf{Q}^{\mathrm{T}}\cdot\boldsymbol{v}^*(\boldsymbol{R}^*) - \mathbf{Q}^{\mathrm{T}}\cdot\frac{d\boldsymbol{c}}{dt} - \mathbf{Q}^{\mathrm{T}}\cdot\mathbf{A}\cdot\left(\boldsymbol{R}^*_\alpha - \boldsymbol{c}\right). \tag{9.146}$$

The second two terms will vanish when $\boldsymbol{v}$ is subtracted from $\frac{d\boldsymbol{R}_\alpha}{dt}$ so the *relative* velocity of the bead and the surrounding fluid *is* frame indifferent. The bead–bead forces are frame indifferent because they only depend on the relative positions of the beads, and we will assume that the fluctuating forces are also frame indifferent. Inserting the preceding expressions into the Langevin equation and multiplying by $\mathbf{Q}$ on the left yields

$$\begin{aligned}m\ddot{\boldsymbol{R}}^*_\alpha + \zeta\left(\dot{\boldsymbol{R}}^*_\alpha - \boldsymbol{v}^*(\boldsymbol{R}_\alpha)\right) &= \sum_{\beta\neq\alpha}\boldsymbol{F}^*_{\alpha\beta}(\boldsymbol{R}^*_{\alpha\beta}) + \boldsymbol{F}^*_{\text{fluc},\alpha}\\ &\quad + m\ddot{\boldsymbol{c}} + m\dot{\mathbf{A}}\cdot\left(\boldsymbol{R}^*_\alpha - \boldsymbol{c}\right) + m\mathbf{A}\cdot\left(\dot{\boldsymbol{R}}^*_\alpha - \dot{\boldsymbol{c}}\right).\end{aligned} \tag{9.147}$$

The terms multiplying $m$ on the right-hand side of this equation are not present in

(9.143), so the Langevin equation is not invariant under a change of reference frame: In general, an observer in that frame will see different behavior than will an observer in a stationary frame. Specifically, the last two terms on the right-hand side are the well-known centrifugal and Coriolis forces experienced by a mass in a rotating reference frame[3] (Goldstein 1980).

On the other hand, we point out in Section 6.1 that for the systems of interest here, the mass of the particles is very small and their momentum is lost to the fluid in a very short time $\lambda_{\mathrm{v}}^{-1}$ so that under most circumstances we can neglect inertia entirely. In this limit, the terms proportional to $m$ vanish from the Langevin equation and the representation in the accelerating frame, (9.147), becomes identical to that in the stationary frame, (9.143). Now not only the expression for the stress tensor, but also the evolution of the particle positions (and thus evolution of the stress tensor), will be material frame indifferent.

We can illustrate this point explicitly with the Hookean dumbbell model with $D = 0$, where we have a evolution equation for the stress tensor, the Oldroyd-B equation

$$\boldsymbol{\tau}_{\mathrm{p}} + \lambda \overset{\nabla}{\boldsymbol{\tau}}_{\mathrm{p}} = 2G_0\lambda \mathbf{E}. \tag{9.148}$$

Indeed, this equation was first presented using continuum mechanics arguments, without reference to a microstructural model (Oldroyd 1950). For the evolution of the stress to be frame indifferent, this equation must be invariant with respect to a change in frame: i.e., we must show that

$$\boldsymbol{\tau}_{\mathrm{p}}^* + \lambda \overset{\nabla}{\boldsymbol{\tau}_{\mathrm{p}}^*} = 2G_0\lambda \mathbf{E}^*. \tag{9.149}$$

The first term is indifferent. The second and third terms depend on $\nabla \boldsymbol{v}$, so we need to determine how this transforms under a change of frame. The following result, which follows from (9.138), will be useful:

$$\frac{\partial x_i}{\partial x_j^*} = Q_{ji}. \tag{9.150}$$

By the same process that leads to (9.144), in $F^*$ the fluid velocity $\boldsymbol{v}^*$ at position $\boldsymbol{x}^*$ is given by

$$\boldsymbol{v}^* = \dot{\boldsymbol{c}} + \mathbf{A} \cdot (\boldsymbol{x}^* - \boldsymbol{c}) + \mathbf{Q} \cdot \boldsymbol{v}. \tag{9.151}$$

[3] The term *fictitious force* is sometimes used to denote the apparent force experienced by an object in an accelerating reference frame. Of course, in that frame there is nothing fictitious about the effects that such apparent forces produce.

Recalling that $\boldsymbol{c}$ and $\mathbf{Q}$ are independent of position, we can write

$$\begin{aligned}(\boldsymbol{\nabla v})^*_{ji} &= \frac{\partial v^*_i}{\partial x^*_j} \\ &= A_{ij} + Q_{ik}\frac{\partial v_k}{\partial x^*_j} \\ &= A_{ij} + Q_{ik}\frac{\partial v_k}{\partial x_l}\frac{\partial x_l}{\partial x^*_j} \\ &= A_{ij} + Q_{ik}\frac{\partial v_k}{\partial x_l}Q^{\mathrm{T}}_{lj}, \\ &= A_{ij} + Q_{jl}\frac{\partial v_k}{\partial x_l}Q^{\mathrm{T}}_{ki},\end{aligned}$$

or

$$(\boldsymbol{\nabla v})^* = \mathbf{A}^{\mathrm{T}} + \mathbf{Q}\cdot\boldsymbol{\nabla v}\cdot\mathbf{Q}^{\mathrm{T}}. \tag{9.152}$$

Thus $\boldsymbol{\nabla v}$ is not frame indifferent. Using this result and the antisymmetry of $\mathbf{A}$, we can write an expression for the vorticity tensor $\mathbf{W}$ in $F^*$:

$$\begin{aligned}2\mathbf{W}^* = (\boldsymbol{\nabla v})^* - (\boldsymbol{\nabla v})^{*\mathrm{T}} &= \mathbf{A}^{\mathrm{T}} - \mathbf{A} + \mathbf{Q}\cdot\boldsymbol{\nabla v}\cdot\mathbf{Q}^{\mathrm{T}} - \mathbf{Q}\cdot\boldsymbol{\nabla v}^{\mathrm{T}}\cdot\mathbf{Q}^{\mathrm{T}} \\ &= 2\left(\mathbf{A}^{\mathrm{T}} + \mathbf{Q}\cdot\mathbf{W}\cdot\mathbf{Q}^{\mathrm{T}}\right).\end{aligned} \tag{9.153}$$

The presence of $\mathbf{A}$ shows that the vorticity is also not frame indifferent. For the rate of strain $\mathbf{E}$, we have that

$$\begin{aligned}2\mathbf{E}^* = (\boldsymbol{\nabla v})^* + (\boldsymbol{\nabla v})^{*\mathrm{T}} &= \mathbf{A}^{\mathrm{T}} + \mathbf{A} + \mathbf{Q}\cdot\boldsymbol{\nabla v}\cdot\mathbf{Q}^{\mathrm{T}} + \mathbf{Q}\cdot\boldsymbol{\nabla v}^{\mathrm{T}}\cdot\mathbf{Q}^{\mathrm{T}} \\ &= 2\mathbf{Q}\cdot\mathbf{E}\cdot\mathbf{Q}^{\mathrm{T}}.\end{aligned} \tag{9.154}$$

Thus the deformation rate *is* material frame indifferent, a result that we can understand on physical grounds because it is independent of the local rotation rate (vorticity) of the fluid.

Having established that the first and third terms in (9.148) are frame indifferent, it remains to analyze the upper convected derivative

$$\overset{\nabla}{\boldsymbol{\tau}}_{\mathrm{p}} \equiv \frac{D\boldsymbol{\tau}_{\mathrm{p}}}{Dt} - \left(\boldsymbol{\tau}_{\mathrm{p}}\cdot\boldsymbol{\nabla v} + \left(\boldsymbol{\tau}_{\mathrm{p}}\cdot\boldsymbol{\nabla v}\right)^{\mathrm{T}}\right).$$

The substantial derivative is simply a time derivative following the motion of a material element, so

$$\frac{D\boldsymbol{\tau}_{\mathrm{p}}}{Dt} = \frac{D}{Dt}\left(\mathbf{Q}\cdot\boldsymbol{\tau}^*_{\mathrm{p}}\cdot\mathbf{Q}\right) = \mathbf{Q}^{\mathrm{T}}\cdot\frac{D\boldsymbol{\tau}^*_{\mathrm{p}}}{Dt}\cdot\mathbf{Q} + \dot{\mathbf{Q}}^{\mathrm{T}}\cdot\boldsymbol{\tau}^*_{\mathrm{p}}\cdot\mathbf{Q} + \mathbf{Q}^{\mathrm{T}}\cdot\boldsymbol{\tau}^*_{\mathrm{p}}\cdot\dot{\mathbf{Q}}.$$

The presence of the second and third terms here indicates that the substantial derivative of an indifferent tensor is not indifferent. Using (9.152), we find that

$$\boldsymbol{\tau}_{\mathrm{p}}\cdot\boldsymbol{\nabla v} = \mathbf{Q}^{\mathrm{T}}\cdot\boldsymbol{\tau}^*_{\mathrm{p}}\cdot\boldsymbol{\nabla v}^*\cdot\mathbf{Q} - \mathbf{Q}^{\mathrm{T}}\cdot\boldsymbol{\tau}^*_{\mathrm{p}}\cdot\mathbf{A}^{\mathrm{T}}\cdot\mathbf{Q}$$

and similarly for $\left(\boldsymbol{\tau}_{\mathrm{p}}\cdot\boldsymbol{\nabla v}\right)^{\mathrm{T}}$. This term is also not frame indifferent. Combining these results, we can write that

$$\overset{\nabla}{\boldsymbol{\tau}}_{\mathrm{p}} = \mathbf{Q}^{\mathrm{T}} \cdot \overset{\nabla}{\boldsymbol{\tau}^*}_{\mathrm{p}} \cdot \mathbf{Q}$$
$$+ \left[ \dot{\mathbf{Q}}^{\mathrm{T}} \cdot \boldsymbol{\tau}_{\mathrm{p}}^* \cdot \mathbf{Q} + \mathbf{Q}^{\mathrm{T}} \cdot \boldsymbol{\tau}_{\mathrm{p}}^* \cdot \dot{\mathbf{Q}} + \mathbf{Q}^{\mathrm{T}} \cdot \boldsymbol{\tau}_{\mathrm{p}}^* \cdot \mathbf{A}^{\mathrm{T}} \cdot \mathbf{Q} + \left( \mathbf{Q}^{\mathrm{T}} \cdot \boldsymbol{\tau}_{\mathrm{p}}^* \cdot \mathbf{A}^{\mathrm{T}} \cdot \mathbf{Q} \right)^{\mathrm{T}} \right].$$

Finally, applying (9.136) for the relation between **A** and **Q**, the term in square brackets is found to vanish. Therefore, the upper convected derivative of a frame indifferent tensor is also frame indifferent:

$$\overset{\nabla}{\boldsymbol{\tau}}_{\mathrm{p}} = \mathbf{Q}^{\mathrm{T}} \cdot \overset{\nabla}{\boldsymbol{\tau}^*}_{\mathrm{p}} \cdot \mathbf{Q}. \tag{9.155}$$

We have now shown that each term in the Oldroyd-B model (9.148) is frame indifferent and thus that (9.149) is valid; the Oldroyd-B model is frame indifferent.

A similar analysis shows that the lower convected derivative is also frame indifferent and therefore so is any linear combination of upper and lower convected derivatives. Specifically, the frame indifferent quantity

$$\frac{1+a}{2} \overset{\nabla}{\mathbf{S}} + \frac{1-a}{2} \overset{\Delta}{\mathbf{S}} \tag{9.156}$$

is called the *Gordon–Schowalter* derivative of **S**. The cases $a = 1, -1$ and $0$ correspond to the upper, lower, and corotational or Jaumann convected derivatives, respectively. This derivative appears in molecular models of polymer dynamics (Larson 1988). The corotational derivative is the rate of change in a frame that rotates with the local angular velocity of the material. Because an inertialess rigid sphere in flow also rotates with the local angular velocity of the fluid (see Section 3.3.3), this derivative arises naturally, for example, in describing the dynamics of suspended drops or capsules in flow (Barthes-Biesel & Rallison 1981).

## Problems

**9.1** Starting with (9.6), derive this equation for the shear strain $\gamma(t)$ as a function of the history of $d\tau_{yx}/dt$:

$$\gamma(t) = \int_{-\infty}^{t} J(t-t') \frac{d\tau_{yx}}{dt}(t')\, dt'. \tag{9.157}$$

How is $J(t)$, which is called the *creep compliance*, related to $G(t)$? Observe that $J(t)$ can be determined by imposing a step function shear stress, then measuring $\gamma(t)$.

**9.2** Find the evolution equation for the shear stress response of a rigid rod suspension ((8.150)) to a shear flow $\dot{\gamma}(t)$ in the linear viscoelastic limit.

**9.3** Show that the loss modulus $G''(\omega)$ is odd. Use this fact to rewrite (9.50) as

$$G'(\omega) = \frac{2}{\pi} \int_0^{\infty} \frac{G''(\omega')\omega'}{\omega^2 - \omega'^2}\, d\omega'.$$

This expression is still a principal value integral. Apply the singularity subtraction trick used in Section 4.6 to rewrite the integral as

$$G'(\omega) = \frac{2}{\pi} \int_0^{\infty} \frac{G''(\omega')\omega' - G''(\omega)\omega}{\omega^2 - \omega'^2}\, d\omega'.$$

Now use the fact that $G'(\omega)$ is even to rewrite (9.51) as

$$G''(\omega) = -\frac{2\omega}{\pi}\int_0^\infty \frac{G'(\omega') - G'(\omega)}{\omega^2 - \omega'^2}\, d\omega'.$$

**9.4** In (9.74), what must the value of $\sigma^2$ be so that the particle has the correct kinetic energy? How would this model be modified to include a solvent component to the fluid behavior (i.e., for a Jeffreys rather than a Maxwell fluid model)?

**9.5** Consider the mean squared displacement of a particle in an Oldroyd-B (Jeffreys) fluid, which has both a viscous and a viscoelastic response:

$$\zeta(t) = 6\pi\eta a\delta(t) + 6\pi\eta_{\mathrm{p}} a\lambda^{-1}e^{-t/\lambda}.$$

Find an expression for the mean squared displacement in the Laplace transform domain. Show that for sufficiently small $m$ and nonzero $\eta$, all the poles of this function are real, implying that there is no oscillatory component in the time domain. Find an analytical expression for the mean squared displacement in the time domain in the inertialess limit and show that there is a parameter range in which the plateau in the mean squared displacement exhibited by Figure 9.3 continues to exist.

**9.6** Show that the system

$$\begin{aligned} m\frac{d\boldsymbol{U}}{dt} &= -\sum_{n=1}^{N} \alpha_n \boldsymbol{Z}_n, \\ \frac{d\boldsymbol{Z}_n}{dt} &= \beta_n(\boldsymbol{U} - \boldsymbol{Z}_n) + \boldsymbol{G}_{\mathrm{fluc},n}(t) \\ \frac{d\boldsymbol{R}}{dt} &= \boldsymbol{U}, \end{aligned}$$

where

$$\langle \boldsymbol{G}_{\mathrm{fluc},n}(t)\boldsymbol{G}_{\mathrm{fluc},m}(t')\rangle = \sigma_i^2\delta(t-t')\boldsymbol{\delta}\delta_{nm}$$

models the dynamics of a particle in a fluid with a relaxation modulus given by a sum of exponentials, (9.38). How do the parameters $\alpha_n, \beta_n, \sigma_n$ relate to the parameters $\lambda_n$ and $G_{0n}$ in (9.38)?

**9.7** Consider the evolution of a Brownian particle in a Newtonian fluid, taking into account the inertia of the fluid. Under the same assumptions used in Section 9.2, find $\widetilde{\boldsymbol{\phi}}_{\mathrm{v}}(s)$ for a sphere whose drag force is given by (5.6). From the final value theorem of Laplace transforms, the long-time behavior of $\boldsymbol{\phi}_{\mathrm{v}}(t)$ will be given by $\lim_{s\to 0} s\widetilde{\boldsymbol{\phi}}_{\mathrm{v}}(s)$. Taylor-expand $s\widetilde{\boldsymbol{\phi}}_{\mathrm{v}}(s)$ and invert to show that $\boldsymbol{\phi}_{\mathrm{v}}(t) \sim t^{-3/2}$ at long times.

**9.8** Repeat the analysis of Section 9.3.3, but for *biaxial extension*, where

$$\nabla\boldsymbol{v} = \dot{\epsilon}\left(\left(\boldsymbol{e}_x\boldsymbol{e}_x + \boldsymbol{e}_y\boldsymbol{e}_y\right) - 2\boldsymbol{e}_z\boldsymbol{e}_z\right) = \dot{\epsilon}\left(\boldsymbol{e}_r\boldsymbol{e}_r - 2\boldsymbol{e}_z\boldsymbol{e}_z\right).$$

Show that as in uniaxial extension, there is a Weissenberg number beyond which the stretching diverges for the Oldroyd-B model. Also, repeat the scaling analysis of Section 9.3.4 to estimate the normal stress components $\tau_{rr}$, $\tau_{\theta\theta}$ and $\tau_{zz}$ at high Weissenberg number. How will the polymers be oriented in the $x - y$ plane?

**9.9** For a Brownian rigid rod undergoing shear flow $\boldsymbol{v} = \dot{\gamma} y \boldsymbol{e}_x$, the Smoluchowski equation (8.149) for the orientation distribution of a Brownian rigid dumbbell (or rod with $B = 1$) becomes

$$\frac{\partial \psi}{\partial t} = \frac{\sin\phi\cos\phi}{\cos\alpha}\frac{\partial}{\partial\alpha}\left(\psi\cos^2\alpha\sin\alpha\right) + \frac{\partial}{\partial\phi}\left(\psi\sin^2\phi\right)$$
$$+\,\epsilon\left\{\frac{1}{\cos\alpha}\frac{\partial}{\partial\alpha}\left(\frac{\partial\psi}{\partial\alpha}\cos\alpha\right) + \frac{1}{\cos^2\alpha}\frac{\partial^2\psi}{\partial\phi^2}\right\},$$

where $\epsilon = \text{Pé}^{-1}$. Here spherical coordinates have been used, but the polar angle $\theta$ has been replaced by $\alpha = \pi/2 - \theta$. As discussed in Section 9.3.4, when $\epsilon \ll 1$ we expect that the molecules are mostly aligned in the flow direction: the probability density $\psi(\alpha, \phi)$ is highly peaked in an orientational boundary layer near the $x$-axis, i.e., $\theta = \pi/2, \phi = 0$ or $\alpha = \phi = 0$ (thus the use of nonstandard spherical coordinates here). To determine the boundary layer scaling, let $\alpha = \delta_\alpha(\epsilon)a$ and $\phi = \delta_\phi(\epsilon)b$, where $\delta_\alpha(\epsilon) \ll 1, \delta_\phi(\epsilon) \ll 1$ and $a$ and $b$ are rescaled angular variables. Let $\delta_\phi = \delta_\alpha = \epsilon^p$ and determine what value of $p$ leads to a balance of terms at steady state. (You should be able to guess a value of $p$ based on the analysis of Section 9.3.4.) Hint: Use the Taylor expansions for the trigonometric functions and the handy relationship $(1+z)^{-n} = 1 - nz + O(z^2)$ for $z \ll 1$.

**9.10** Derive (9.129) for the steady-state Trouton ratio of a fully extended dumbbell with no HI.

**9.11** Show that the lower convected derivative of an indifferent tensor is indifferent.

**9.12** Show that the Green tensor is indifferent and that the Cauchy tensor is *invariant*: i.e., its components are the same in frames $F$ and $F^*$.

# Appendix Mathematical Background

This appendix first reviews some basic mathematical results from vector calculus and Fourier and Laplace transforms. These sections are brief, based on the assumption that the reader has some familiarity with these topics. The delta function and its properties are then addressed in some detail, in particular because it is important to understand how to work with its derivatives. The final sections address fundamentals of random variables and stochastic processes.

## A.1 Definitions and Identities of Vector Algebra and Calculus

### A.1.1 Notation

The identity tensor in $\mathbb{R}^3$ will be denoted $\boldsymbol{\delta}$; for any vector $\boldsymbol{a}$,

$$\boldsymbol{\delta} \cdot \boldsymbol{a} = \boldsymbol{a} \cdot \boldsymbol{\delta} = \boldsymbol{a}. \tag{A.1}$$

Unless otherwise noted, we will always work in a fixed Cartesian coordinate system in $\mathbb{R}^3$ with orthonormal basis vectors $\boldsymbol{e}_i$, $i = 1, 2, 3$, or equivalently $\boldsymbol{e}_x, \boldsymbol{e}_y, \boldsymbol{e}_z$. The orthonormality condition can be written $\boldsymbol{e}_i \cdot \boldsymbol{e}_j = \delta_{ij}$, where

$$\delta_{ij} = \begin{cases} 1, & i = j \\ 0, & i \neq j \end{cases}. \tag{A.2}$$

The quantity $\delta_{ij}$ is called the *Kronecker delta*. Now a general vector $\boldsymbol{a}$ is written

$$\boldsymbol{a} = \sum_{i=1}^{3} a_i \boldsymbol{e}_i.$$

The *dyadic product* $\boldsymbol{e}_i\boldsymbol{e}_j$ between $\boldsymbol{e}_i$ and $\boldsymbol{e}_j$ is called a *dyad* and satisfies the properties

$$(\boldsymbol{e}_i\boldsymbol{e}_j) \cdot \boldsymbol{e}_k = \boldsymbol{e}_i(\boldsymbol{e}_j \cdot \boldsymbol{e}_k) = \delta_{jk}\boldsymbol{e}_i, \quad \boldsymbol{e}_k \cdot (\boldsymbol{e}_i\boldsymbol{e}_j) = (\boldsymbol{e}_k \cdot \boldsymbol{e}_i)\boldsymbol{e}_j = \delta_{ki}\boldsymbol{e}_j.$$

A general dyad $\boldsymbol{ab}$ satisfies $\boldsymbol{ab} \cdot \boldsymbol{c} = \boldsymbol{a}(\boldsymbol{b} \cdot \boldsymbol{c})$ and $\boldsymbol{c} \cdot \boldsymbol{ab} = (\boldsymbol{c} \cdot \boldsymbol{a})\boldsymbol{b}$. The identity $\boldsymbol{\delta}$ can be written

$$\boldsymbol{\delta} = \boldsymbol{e}_1\boldsymbol{e}_1 + \boldsymbol{e}_2\boldsymbol{e}_2 + \boldsymbol{e}_3\boldsymbol{e}_3 = \sum_{i=1}^{3}\sum_{j=1}^{3} \delta_{ij}\boldsymbol{e}_i\boldsymbol{e}_j. \tag{A.3}$$

When using a fixed Cartesian basis, it will not be necessary to distinguish between a second-order tensor and the matrix containing its Cartesian components. Thus we will often write a tensor

$$\mathbf{S} = \sum_{i=1}^{3}\sum_{j=1}^{3} S_{ij}\boldsymbol{e}_i\boldsymbol{e}_j$$

in matrix form

$$\mathbf{S} = \begin{bmatrix} S_{11} & S_{12} & S_{13} \\ S_{21} & S_{22} & S_{23} \\ S_{31} & S_{32} & S_{33} \end{bmatrix}.$$

Similarly,

$$\mathbf{S}^{\mathrm{T}} = \sum_{i=1}^{3}\sum_{j=1}^{3} S_{ji}\boldsymbol{e}_i\boldsymbol{e}_j = \begin{bmatrix} S_{11} & S_{21} & S_{31} \\ S_{12} & S_{22} & S_{32} \\ S_{13} & S_{23} & S_{33} \end{bmatrix}.$$

The gradient operator is given by

$$\boldsymbol{\nabla} = \sum_{i=1}^{3} \boldsymbol{e}_i \frac{\partial}{\partial x_i},$$

where $x_i$ is position along the $i$th coordinate direction. Therefore, the gradient of a vector is

$$\boldsymbol{\nabla v} = \sum_{i=1}^{3} \boldsymbol{e}_i \frac{\partial}{\partial x_i} \sum_{j=1}^{3} v_j \boldsymbol{e}_j = \sum_{i=1}^{3}\sum_{j=1}^{3} \frac{\partial v_j}{\partial x_i}\boldsymbol{e}_i\boldsymbol{e}_j = \begin{bmatrix} \frac{\partial v_1}{\partial x_1} & \frac{\partial v_2}{\partial x_1} & \frac{\partial v_3}{\partial x_1} \\ \frac{\partial v_1}{\partial x_2} & \frac{\partial v_2}{\partial x_2} & \frac{\partial v_3}{\partial x_2} \\ \frac{\partial v_1}{\partial x_3} & \frac{\partial v_2}{\partial x_3} & \frac{\partial v_3}{\partial x_3} \end{bmatrix}.$$

We will sometimes write a vector $\boldsymbol{a}$ as either a matrix with a single column or a single row containing its cartesan coordinates: i.e.,

$$\boldsymbol{a} = \begin{bmatrix} a_1 \\ a_2 \\ a_3 \end{bmatrix} \quad \text{or} \quad \boldsymbol{x} = \begin{bmatrix} a_1 & a_2 & a_3 \end{bmatrix}.$$

Using the present notation, there is no mathematical distinction between a row vector and a column vector, in contrast to the case of second-order tensors: $\mathbf{S} \neq \mathbf{S}^{\mathrm{T}}$. If one wishes to treat the dot product between a tensor and a vector as a matrix-vector product, then one would write

$$\mathbf{S}\cdot\boldsymbol{a} = \begin{bmatrix} S_{11} & S_{12} & S_{13} \\ S_{21} & S_{22} & S_{23} \\ S_{31} & S_{32} & S_{33} \end{bmatrix} \begin{bmatrix} a_1 \\ a_2 \\ a_3 \end{bmatrix}$$

and

$$\boldsymbol{a}\cdot\mathbf{S} = \begin{bmatrix} a_1 & a_2 & a_3 \end{bmatrix} \begin{bmatrix} S_{11} & S_{12} & S_{13} \\ S_{21} & S_{22} & S_{23} \\ S_{31} & S_{32} & S_{33} \end{bmatrix}.$$

Note that $\mathbf{S}\cdot\boldsymbol{a} = \boldsymbol{a}\cdot\mathbf{S}^{\mathrm{T}} \neq \boldsymbol{a}\cdot\mathbf{S}$.

We will often denote vectors and tensors in *index notation*, representing them by their

Cartesian components (e.g., $a_i$ for $\boldsymbol{a}$, $S_{ij}$ for $\mathbf{S}$, $a_i b_j$ for $\boldsymbol{ab}$). The Einstein summation convention will be used unless otherwise noted: the use of repeated indices in expressions involving Cartesian components implies summation over those indices. For example, the $i$th Cartesian component of the quantity $\mathbf{S}\cdot\boldsymbol{a}$ is given by

$$\begin{aligned}(\mathbf{S}\cdot\boldsymbol{a})_i &= \boldsymbol{e}_i\cdot(\mathbf{S}\cdot\boldsymbol{a})\\ &= \boldsymbol{e}_i\cdot\left(\left(\sum_{j=1}^{3}\sum_{k=1}^{3}S_{jk}\boldsymbol{e}_j\boldsymbol{e}_k\right)\cdot\sum_{l=1}^{3}a_l\boldsymbol{e}_l\right)\\ &= \sum_{j=1}^{3}\sum_{k=1}^{3}\sum_{l=1}^{3}S_{jk}a_l\boldsymbol{e}_i\cdot\boldsymbol{e}_j\boldsymbol{e}_k\cdot\boldsymbol{e}_l\\ &= \sum_{j=1}^{3}\sum_{k=1}^{3}\sum_{l=1}^{3}S_{jk}a_l\delta_{ij}\delta_{kl}\\ &= \sum_{k=1}^{3}S_{ik}a_k,\end{aligned}$$

which we will write

$$(\mathbf{S}\cdot\boldsymbol{a})_i = S_{ik}a_k.$$

Similarly, $(\boldsymbol{a}\cdot\mathbf{S})_i = a_k S_{ki}$. We will encounter tensors of order higher than two as well; a third-order tensor $\mathbf{T}$ is written

$$\mathbf{T} = \sum_{i=1}^{3}\sum_{j=1}^{3}\sum_{k=1}^{3}T_{ijk}\boldsymbol{e}_i\boldsymbol{e}_j\boldsymbol{e}_k,$$

where $(\boldsymbol{e}_i\boldsymbol{e}_j\boldsymbol{e}_k)\cdot\boldsymbol{e}_l = (\boldsymbol{e}_i\boldsymbol{e}_j)(\boldsymbol{e}_k\cdot\boldsymbol{e}_l) = \boldsymbol{e}_i\boldsymbol{e}_j\delta_{kl}$ and $\boldsymbol{e}_l\cdot(\boldsymbol{e}_i\boldsymbol{e}_j\boldsymbol{e}_k) = (\boldsymbol{e}_j\boldsymbol{e}_k)(\boldsymbol{e}_l\cdot\boldsymbol{e}_i) = \boldsymbol{e}_j\boldsymbol{e}_k\delta_{ij}$. Using index notation and the summation convention, we can write dot products between $\mathbf{T}$ and a vector $\boldsymbol{a}$ and a second-order tensor $\mathbf{S}$, respectively, as

$$(\mathbf{T}\cdot\boldsymbol{a})_{ij} = T_{ijk}a_k, \quad (\mathbf{T}\cdot\mathbf{S})_i = T_{ijk}S_{jk}.$$

Unless otherwise noted, vectors in three spatial dimensions (positions, velocities, etc.) will be denoted in bold italic characters, and second-order tensors in bold greek or bold roman characters. Quantities involving $3N$ dimensions (such as the vector $\mathsf{R}$ containing the positions of $N$ particles) will be bold sans serif.

The *double dot product* between two second-order tensors $\mathbf{A}$ and $\mathbf{B}$ is defined as

$$\mathbf{A}:\mathbf{B} = \mathrm{tr}(\mathbf{A}\cdot\mathbf{B}^{\mathrm{T}}) = A_{ij}B_{ij}. \tag{A.4}$$

Note that some authors (e.g., Bird, Armstrong, & Hassager 1987) define this differently: $\mathbf{A}:\mathbf{B} = A_{ij}B_{ji}$. Fortunately, the difference does not matter if $\mathbf{A}$ or $\mathbf{B}$ is symmetric, which is often the case in applications. The definition used here has the advantageous property that it serves as a valid inner product between real matrices. Furthermore, the quantity $||\mathbf{A}|| = \sqrt{\mathbf{A}:\mathbf{A}}$ is the *Frobenius norm* of real matrix $\mathbf{A}$.

If $\mathbf{A}$ is antisymmetric and $\mathbf{B}$ is symmetric or vice versa, then

$$\mathbf{A}:\mathbf{B} = \mathbf{0}.$$

### A.1.2 Antisymmetric Tensors and Pseudovectors

Given an antisymmetric tensor $\mathbf{A}$, we can define a *pseudovector* $\boldsymbol{b}$ such that

$$\mathbf{A} = \boldsymbol{\epsilon} \cdot \boldsymbol{b} \quad \text{or} \quad A_{ij} = \epsilon_{ijk} b_k,$$

where $\boldsymbol{\epsilon}$ is called the Levi–Civita symbol or alternating unit tensor or permutation tensor. In Cartesian coordinates,

$$\epsilon_{ijk} = \frac{1}{2}(i-j)(j-k)(k-i) = \begin{cases} 1, & ijk = 123, 231, 312 \\ -1, & ijk = 132, 321, 213\ . \\ 0 & \text{otherwise} \end{cases} \tag{A.5}$$

(A pseudovector is not a true vector because its sign depends on the handedness of the coordinate system. The Levi–Civita symbol is a pseudotensor because it operates differently in right- and left-handed coordinate systems (Aris 1989).) Therefore

$$\mathbf{A} \cdot \boldsymbol{c} = \boldsymbol{c} \times \boldsymbol{b}, \tag{A.6}$$

$$\boldsymbol{c} \cdot \mathbf{A} = \boldsymbol{b} \times \boldsymbol{c}. \tag{A.7}$$

Similarly, if $\mathbf{A}$ is an antisymmetric tensor field, then

$$\boldsymbol{\nabla} \cdot \mathbf{A} = -\boldsymbol{\nabla} \times \boldsymbol{b}.$$

Additional important properties of $\epsilon_{ijk}$ are that

$$\epsilon_{ijk}\epsilon_{klm} = \delta_{il}\delta_{jm} - \delta_{im}\delta_{jl} \tag{A.8}$$

and thus

$$\epsilon_{ijk}\epsilon_{klj} = \delta_{il}\delta_{jj} - \delta_{ij}\delta_{jl} = 2\delta_{il}. \tag{A.9}$$

## A.2 Theorems of Vector Calculus

The *divergence theorem* is central to vector calculus and to all the field theories of physics. Given an $n$-dimensional vector field $\boldsymbol{v}(\boldsymbol{x})$ and a volume $V$ bounded by a surface $S$ with outward unit normal $\boldsymbol{n}$, it states that

$$\int_V \boldsymbol{\nabla} \cdot \boldsymbol{v}\, dV = \int_S \boldsymbol{n} \cdot \boldsymbol{v}\, dS. \tag{A.10}$$

Here $\boldsymbol{\nabla}$ is the gradient operator in $\mathbb{R}^n$ and $\boldsymbol{\nabla}\cdot$ the divergence operator. A number of other important results follow from this, including

$$\int_V \boldsymbol{\nabla}\phi\, dV = \int_S \boldsymbol{n}\phi\, dS, \tag{A.11}$$

$$\int_V \boldsymbol{\nabla}\boldsymbol{v}\, dV = \int_S \boldsymbol{n}\boldsymbol{v}\, dS, \tag{A.12}$$

$$\int_V \boldsymbol{\nabla} \cdot \boldsymbol{\tau}\, dV = \int_S \boldsymbol{n} \cdot \boldsymbol{\tau}\, dS, \tag{A.13}$$

where $\phi(\boldsymbol{x})$ and $\boldsymbol{\tau}(\boldsymbol{x})$ are scalar and tensor fields, respectively. Furthermore, the divergence theorem in $\mathbb{R}^3$ can be used to show that

$$\int_V \boldsymbol{\nabla} \times \boldsymbol{v}\, dV = \int_S \boldsymbol{n} \times \boldsymbol{v}\, dS. \tag{A.14}$$

Another important result of vector calculus is the multidimensional version of *Leibniz's rule*. Consider the time derivative of an integral over a volume that is moving with time, e.g., a region of moving fluid. If a point on the boundary of the region is moving with a velocity $\boldsymbol{q}(\boldsymbol{x})$, then Leibniz's rule states that

$$\frac{d}{dt}\int_{V(t)} m(\boldsymbol{x},t)\, dV = \int_V \frac{\partial m}{\partial t}\, dV + \int_S m(\boldsymbol{x},t)\boldsymbol{n}\cdot\boldsymbol{q}\, dS. \tag{A.15}$$

The second term in this formula appears only if the volume is moving or changing shape with time and represents the net amount of the quantity of interest that is "engulfed" into $V$ because of the motion of its boundaries. In the fluid mechanics literature, (A.15) is often called the *Reynolds transport theorem*, especially in the special case where the boundary of the domain is moving with the material so $\boldsymbol{q} = \boldsymbol{v}$.

*Green's identities* are special cases of the divergence theorem that are useful for working with integrals over quantities involving differential operators other than the divergence. Green's first identity is the divergence theorem for the case where $\boldsymbol{v}$ is replaced by $u\boldsymbol{\nabla}v$, where $u$ and $v$ are now scalars:

$$\int_V (\boldsymbol{\nabla}u\cdot\boldsymbol{\nabla}v + u\nabla^2 v)\, dV = \int_S u\boldsymbol{n}\cdot\boldsymbol{\nabla}v\, dS. \tag{A.16}$$

Green's second identity comes from writing Green's first identity with $u$ and $v$ exchanged and subtracting this expression from Green's first identity as written previously:

$$\int_V (u\nabla^2 v - v\nabla^2 u)\, dV = \int_S \boldsymbol{n}\cdot(u\boldsymbol{\nabla}v - v\boldsymbol{\nabla}u)\, dS. \tag{A.17}$$

Finally, *Green's formula* comes from replacing $\boldsymbol{v}$ in the original expression by $u\boldsymbol{v}$, where $u$ is a scalar and $\boldsymbol{v}$ a vector:

$$\int_V (\boldsymbol{\nabla}u\cdot\boldsymbol{v} + u\boldsymbol{\nabla}\cdot\boldsymbol{v})\, dV = \int_S u\boldsymbol{n}\cdot\boldsymbol{v}\, dS. \tag{A.18}$$

In one dimension, Green's formula reduces to the standard integration by parts formula.

## A.3 Basic Properties of Fourier and Laplace Transforms

### A.3.1 Fourier Transform

For a function $f(t)$, where $t \in \mathbb{R}$, we define the *Fourier transform* $\hat{f}(\omega) = F\{f(t)\}$ as

$$\hat{f}(\omega) = F\{f(x)\} = \int_{-\infty}^{\infty} f(t)e^{-i\omega t}\, dt, \tag{A.19}$$

where $\omega \in \mathbb{R}$. The inverse Fourier transform is then given by

$$f(t) = F^{-1}\{\hat{f}(\omega)\} = \frac{1}{2\pi}\int_{-\infty}^{\infty} f(\omega)e^{i\omega t}\, d\omega. \tag{A.20}$$

These definitions are commonly but not universally used, so it is important to be aware of the definition used by a particular author. For example, sometimes the terms $e^{-i\omega t}$ and $e^{i\omega t}$ used in (A.19) and (A.20), respectively, are interchanged (see, e.g., Chaikin & Lubensky 1995). With that definition, $\hat{f}(\omega)$ is the complex conjugate of that computed with (A.19).

For the Fourier transform of a function of position $x$, the preceding formulas will be used but with $t$ and $\omega$ replaced by $x$ and $k$, respectively. In three dimensions, the spatial Fourier transform becomes

$$\begin{aligned}\hat{f}(\boldsymbol{k}) = F\{f(\boldsymbol{x})\} &= \int_{-\infty}^{\infty}\int_{-\infty}^{\infty}\int_{-\infty}^{\infty} f(\boldsymbol{x})e^{-ik_x x}e^{-ik_y y}e^{-ik_z z}\, dx\, dy\, dz \\ &= \int f(\boldsymbol{x})e^{-i\boldsymbol{k}\cdot\boldsymbol{x}}\, d\boldsymbol{x}.\end{aligned} \tag{A.21}$$

The Fourier transforms of derivatives satisfy

$$F\left\{\frac{df}{dt}\right\} = i\omega\hat{f}(\omega), \tag{A.22}$$

$$F\{\nabla f(\boldsymbol{x})\} = i\boldsymbol{k}\hat{f}(\boldsymbol{k}), \tag{A.23}$$

$$F\{\nabla\cdot\boldsymbol{v}(\boldsymbol{x})\} = i\boldsymbol{k}\cdot\hat{\boldsymbol{v}}(\boldsymbol{k}). \tag{A.24}$$

The following *Fourier shift identities* are often useful:

$$F\{f(t-\tau)\} = e^{-i\omega\tau}\hat{f}(\omega), \tag{A.25}$$

$$F\{f(\boldsymbol{x}-\boldsymbol{\chi}\} = e^{-i\boldsymbol{k}\cdot\boldsymbol{\chi}}\hat{f}(\boldsymbol{k}), \tag{A.26}$$

$$F\left\{e^{i\alpha t}f(t)\right\} = \hat{f}(\omega-\alpha), \tag{A.27}$$

$$F\left\{e^{i\boldsymbol{\kappa}\cdot\boldsymbol{x}}f(\boldsymbol{x})\right\} = \hat{f}(\boldsymbol{k}-\boldsymbol{\kappa}). \tag{A.28}$$

The *Fourier convolution* of two functions $f(t)$ and $g(t)$, denoted $f * g$, is defined as follows:

$$f(t) * g(t) = \int_{-\infty}^{\infty} f(t-\tau)g(\tau)\, d\tau = \int_{-\infty}^{\infty} f(\tau)g(t-\tau)\, d\tau.$$

The *Fourier convolution theorem* states that

$$F\{f(t) * g(t)\} = \hat{f}(\omega)\hat{g}(\omega) \tag{A.29}$$

and

$$F\{f(t)g(t)\} = \frac{1}{2\pi}\hat{f}(\omega) * \hat{g}(\omega) = \frac{1}{2\pi}\int_{-\infty}^{\infty}\hat{f}(\omega-\sigma)\hat{g}(\sigma)\, d\sigma. \tag{A.30}$$

### A.3.2 Laplace Transform

Consider a real function

$$g(t) = f(t)e^{-ct}, \tag{A.31}$$

where $f(t) = 0$ for $t < 0$ (i.e., $f(t)$ is causal) and $c > 0$ is sufficiently large that the Fourier transform of $g(t)$ exists, but is otherwise arbitrary. Thus $f(t)$ can grow with $t$ as long as it does not do so faster than exponentially. The Fourier transform of $g(t)$ is

$$\hat{g}(\omega) = \int_0^\infty f(t)e^{-ct}e^{-i\omega t}\,dt.$$

Defining a new variable $s = c + i\omega$, this becomes

$$\hat{g}(\omega) = \int_0^\infty f(t)e^{-st}\,dt.$$

This expression motivates the definition of the *Laplace transform* of a causal function $f(t)$:

$$L\{f(t)\} = \tilde{f}(s) = \int_0^\infty f(t)e^{-st}\,dt, \tag{A.32}$$

where $s \in \mathbb{C}$.

The inverse Fourier transform of $\hat{g}(\omega)$ is

$$g(t) = f(t)e^{-ct} = \frac{1}{2\pi}\int_{-\infty}^\infty \hat{g}(\omega)e^{i\omega t}\,d\omega.$$

Multiplying both sides by $e^{ct}$ and eliminating $\omega$ and $\hat{g}(\omega)$ in favor of $s$ and $\tilde{f}(s)$ yields the inverse Laplace transform of $\tilde{f}(s)$:

$$f(t) = L^{-1}\left\{\tilde{f}(s)\right\} = \frac{1}{2\pi i}\int_{c-i\infty}^{c+i\infty} \tilde{f}(s)e^{st}\,ds. \tag{A.33}$$

Because the Laplace transform of a function is unique, one rarely needs the inversion formula; Laplace transforms and their inverses can be found from tables. Thus the value of $c$ need not be specified. From this development, it can be seen that for a causal $f(t)$ whose Fourier transform exists (allowing us to take $c = 0$), the Laplace and Fourier transforms are equivalent upon setting $s = i\omega$:

$$\tilde{f}(i\omega) = \hat{f}(\omega). \tag{A.34}$$

The Laplace transform satisfies

$$L\left\{\frac{df}{dt}\right\} = \tilde{f}(s) - sf(0), \tag{A.35}$$

$$L\left\{\int_0^t f(\tau)\,d\tau\right\} = \frac{1}{s}\tilde{f}(s). \tag{A.36}$$

The *Laplace convolution theorem* states that

$$L\left\{\int_0^t f(t)g(t-\tau)\,d\tau\right\} = \tilde{f}(s)\tilde{g}(s). \tag{A.37}$$

## A.4 The Delta Function

The delta function $\delta(x)$ is an idealization of a highly localized function. It can be considered as the limiting case of a tall narrow spike as its height diverges and its width vanishes. Consider the two functions

$$\delta_\sigma(x) = \frac{\sigma}{\sqrt{2\pi}} \exp\left(-\frac{\sigma^2 x^2}{2}\right) \tag{A.38}$$

and

$$\delta_\sigma(x) = \begin{cases} 0, & x < 0, \\ \sigma e^{-\sigma x}, & x \geq 0. \end{cases} \tag{A.39}$$

In the limit $\sigma \to \infty$, these functions have infinite height and zero width and unit area. These are the key properties of the delta function. More generally, we can define a *delta family* as a family of functions $\delta_\sigma(x)$ parameterized by $\sigma$ that satisfy

$$\delta_\sigma(x) = \sigma f(\sigma x), \tag{A.40}$$

$$\int_{-\infty}^{\infty} f(z)\, dz = 1. \tag{A.41}$$

Aside from having unit area, $f(z)$ is arbitrary. Now the delta function is given by

$$\delta(x) = \lim_{\sigma \to \infty} \delta_\sigma(x). \tag{A.42}$$

Operationally, the delta function is defined not by (A.42) but by how it acts under integration, making it, strictly speaking, a *distribution* rather than a classical function. Defining the inner product in one dimension ($\mathbb{R}^1$),

$$(f(x), g(x)) = \int_{-\infty}^{\infty} f(x) g(x)\, dx \tag{A.43}$$

(where $f$ and $g$ are such that this integral exists), the delta function is defined by the following property:

$$(\delta(x), \phi(x)) = \int_{-\infty}^{\infty} \delta(x)\phi(x)\, dx = \phi(0). \tag{A.44}$$

Equivalently,

$$(\delta(x - x_0), \phi(x)) = \phi(x_0). \tag{A.45}$$

More generally,

$$\int_a^b (\delta(x - x_0)\phi(x)\, dx = \begin{cases} \phi(x_0) & a < x_0 < b, \\ 0 & x_0 < a \text{ or } x_0 > b. \end{cases} \tag{A.46}$$

The case where $x_0$ is at one of the boundaries of the domain, i.e., $x_0 = a$ or $x_0 = b$ in the preceding expression, deserves special mention. Observe that the function $\delta_\sigma(x)$ in (A.38) is even, with a nonzero value for positive and negative values of $x$, while that in (A.39) is only nonzero for $x \geq 0$. The only way that both of these functions can be regarded as delta functions as $\sigma \to \infty$ is if they are localized precisely *at* $x = 0$. Thus, (A.46) generalizes to

$$\int_a^b (\delta(x - x_0)\phi(x)\, dx = \begin{cases} \phi(x_0) & a \leq x_0 \leq b, \\ 0 & \text{otherwise.} \end{cases} \tag{A.47}$$

That said, there are situations like those encountered in Sections 6.7 and 6.8 where $\delta(x)$ is specifically used to represent a highly localized *even* function, such as a time correlation function, whose area is divided equally between positive and negative values of $x$. In these special situations, the following results are used:

$$\int_a^0 \delta(x)\phi(x)\, ds = \int_0^b \delta(x)\phi(x)\, ds = \frac{1}{2}\phi(0), \tag{A.48}$$

where $a < 0 < b$.

In $n$ dimensions, the delta function can be represented in Cartesian coordinates as $\delta(\boldsymbol{x}) = \delta(x_1)\delta(x_2)\ldots\delta(x_n)$. Defining the $n$-dimensional inner product

$$(f(\boldsymbol{x}), g(\boldsymbol{x})) = \int_{-\infty}^{\infty}\int_{-\infty}^{\infty}\cdots\int_{-\infty}^{\infty} f(\boldsymbol{x})g(\boldsymbol{x})\, dx_1dx_2\ldots dx_n, \tag{A.49}$$

we have that

$$(\delta(\boldsymbol{x} - \boldsymbol{x}_0), \phi(\boldsymbol{x})) = \phi(\boldsymbol{x}_0). \tag{A.50}$$

Additionally, because the delta function $\delta(\boldsymbol{x} - \boldsymbol{x}_0)$ is only nonzero at the source position $\boldsymbol{x}_0$, all of the preceding results hold over any domain $V$ containing $\boldsymbol{x}_0$, whether bounded or unbounded. If $V$ does not contain $\boldsymbol{x}_0$, the integral in question is zero.

Delta functions are also encountered in cylindrical or spherical coordinates. In polar coordinates $(r, \theta)$ with $\boldsymbol{x} \in \mathbb{R}^2$,

$$\delta(\boldsymbol{x}) = \frac{1}{2\pi r}\delta(r). \tag{A.51}$$

In cylindrical coordinates $(r, \theta, z)$ in $\mathbb{R}^3$,

$$\delta(\boldsymbol{x}) = \frac{1}{2\pi r}\delta(r)\delta(z) \tag{A.52}$$

while in spherical coordinates $(r, \theta, \phi)$ in $\mathbb{R}^3$,

$$\delta(\boldsymbol{x}) = \frac{1}{4\pi r^2}\delta(r). \tag{A.53}$$

Derivatives of delta functions are defined using integration by parts. For a function $f(x)$ that vanishes at the boundaries of the domain, this yields that

$$(f'(x), g(x)) = -(f(x), g'(x)),$$

where $'$ indicates differentiation with respect to $x$. Accordingly, we can define the derivative of the delta function by the result

$$(\delta'(x - x_0), \phi(x)) = -(\delta(x - x_0), \phi'(x)) = -\phi'(x_0). \tag{A.54}$$

With reference to Section 2.4, if a delta function is a source, its derivative is a dipole. Higher derivatives can be defined by repeated application of integration by parts:

$$\left(\frac{d^k}{dx^k}\delta(x-x_0),\phi(x)\right) = (-1)^k\frac{d^k\phi}{dx^k}(x_0). \tag{A.55}$$

Turning to the $n$-dimensional case, a general $k$th derivative operator can be defined as

$$D^k = \frac{\partial^{|k|}}{\partial x_1^{k_1} x_2^{k_2}\dots x_N^{k_N}}.$$

Here $k$ is a *multi-index*, $k = (k_1, k_2, \dots, k_N)$, with $|k| = k_1 + k_2 + \dots + k_N$ and $k_i \geq 0$. Again using integration by parts, (A.55) can be generalized:

$$\left(D^k\delta(\boldsymbol{x}-\boldsymbol{x}_0),\phi(\boldsymbol{x})\right) = (-1)^{|k|}\left(\delta(\boldsymbol{x}-\boldsymbol{x}_0),D^k\phi(\boldsymbol{x})\right) = (-1)^{|k|}D^k\phi(\boldsymbol{x}_0). \tag{A.56}$$

By linearity, this result generalizes to integrals involving the gradient operator acting on a delta function. For example,

$$\int \phi(\boldsymbol{x})\nabla^2\delta(\boldsymbol{x}-\boldsymbol{x}_0)\,d\boldsymbol{x} = \int \nabla^2\phi(\boldsymbol{x})\delta(\boldsymbol{x}-\boldsymbol{x}_0)\,d\boldsymbol{x} = \nabla^2\phi(\boldsymbol{x}_0), \tag{A.57}$$

and

$$\int \boldsymbol{u}(\boldsymbol{x})\cdot\boldsymbol{\nabla}\delta(\boldsymbol{x}-\boldsymbol{x}_0)\,d\boldsymbol{x} = \int -\boldsymbol{\nabla}\cdot\boldsymbol{u}(\boldsymbol{x})\delta(\boldsymbol{x}-\boldsymbol{x}_0)\,d\boldsymbol{x} = -\boldsymbol{\nabla}\cdot\boldsymbol{u}(\boldsymbol{x}_0). \tag{A.58}$$

Similarly,

$$\begin{aligned}\int\int f(x)g(y)\frac{\partial}{\partial x}\frac{\partial}{\partial y}\delta(x-y)\,dxdy &= -\int\int f(x)\frac{dg}{dy}\frac{\partial}{\partial x}\delta(x-y)\,dxdy\\ &= \int \frac{df}{dx}\frac{dg}{dy}\delta(x-y)\,dxdy\\ &= \int \frac{df}{dx}\frac{dg}{dx}\,dx.\end{aligned} \tag{A.59}$$

One further case of interest is that of differentiation with respect to the source position rather than the independent variable. For example,

$$\begin{aligned}\frac{d}{dx_0}\int \phi(x)\delta(x-x_0)\,dx &= \int \phi(x)\frac{d}{dx_0}\delta(x-x_0)\,dx\\ &= -\int \phi(x)\frac{d}{dx}\delta(x-x_0)\,dx\\ &= \frac{d\phi}{dx}(x_0).\end{aligned} \tag{A.60}$$

The integral of the delta function is the Heaviside unit step function

$$H(x) = \int_{-\infty}^{x}\delta(s)\,ds = \begin{cases}0, & x<0,\\ 1, & x\geq 0.\end{cases} \tag{A.61}$$

Equivalently, $H(x)$ solves the differential equation

$$\frac{d}{dx}H(x) = \delta(x). \tag{A.62}$$

Related results include the free-space Green's functions for the Laplacian operator in various numbers of dimensions:

$$\frac{d^2}{dx^2}\frac{|x|}{2} = \delta(x), \tag{A.63}$$

$$\nabla^2\left(\frac{1}{2\pi}\ln r\right) = -\delta(\boldsymbol{x}), \boldsymbol{x} \in \mathbb{R}^2, \tag{A.64}$$

$$\nabla^2\left(\frac{1}{4\pi r}\right) = -\delta(\boldsymbol{x}), \boldsymbol{x} \in \mathbb{R}^3. \tag{A.65}$$

The Fourier transform of the $n$-dimensional delta function, $F\{\delta(\boldsymbol{x})\}$, is given by

$$F\{\delta(\boldsymbol{x})\} = \int \delta(\boldsymbol{x})e^{-i\boldsymbol{k}\cdot\boldsymbol{x}}\, d\boldsymbol{x} = 1. \tag{A.66}$$

This result along with (A.63)–(A.65) provides some useful basic results regarding Fourier transforms of Green's functions:

$$F\{|x|\} = -\frac{2}{k^2}, \tag{A.67}$$

$$F\{\ln r\} = -\frac{2\pi}{k^2}, \boldsymbol{x} \in \mathbb{R}^2, \tag{A.68}$$

$$F\left\{\frac{1}{r}\right\} = -\frac{4\pi}{k^2}, \boldsymbol{x} \in \mathbb{R}^3. \tag{A.69}$$

Other useful results for $n$-dimensional Fourier transforms involving delta functions include the following:

$$F\{1\} = (2\pi)^n\,\delta(\boldsymbol{k}), \tag{A.70}$$

$$F\{\delta(\boldsymbol{x}-\boldsymbol{\chi})\} = e^{-i\boldsymbol{k}\cdot\boldsymbol{\chi}}, \tag{A.71}$$

$$F\left\{e^{i\boldsymbol{\kappa}\cdot\boldsymbol{x}}\right\} = (2\pi)^n\,\delta(\boldsymbol{k}-\boldsymbol{\kappa}). \tag{A.72}$$

A more comprehensive and mathematically thorough treatment of the delta function and the theory of distributions can be found in Stakgold (1998).

## A.5 Random Variables

### A.5.1 Random Variables, Probability Distributions, and Ensemble Averages

A *random variable* $\boldsymbol{X}$ is a quantity that can take on values chosen randomly from a given set $\Omega$. The probability of $\boldsymbol{X}$ taking on a value within the differential element $d\Omega(\boldsymbol{X})$ is $p(\boldsymbol{X})\, d\Omega(\boldsymbol{X})$, where $p(\boldsymbol{X})$ is the *probability density function* or *probability distribution function* of $\boldsymbol{X}$. This function satisfies

$$\int_\Omega p(\boldsymbol{X})\, d\Omega(\boldsymbol{X}) = 1. \tag{A.73}$$

The probability $\Pr(A)$ that a random variable will take on a value in some subset $A \in \Omega$ is given by

$$\Pr(A) = \int_A p(\boldsymbol{X})\, d\Omega(\boldsymbol{X}). \tag{A.74}$$

Let $\boldsymbol{X}^{(m)}, m = 1, 2, \ldots N_s$ be specific samples of $\boldsymbol{X}$ drawn from the set $\Omega$ according to the probability density $p(\boldsymbol{X})$; it is useful to think of these as the outcomes of $N_s$ independent but identical experiments. A *stochastic process* $\boldsymbol{X}(t)$ is a family of random variables parameterized by a scalar $t$, which is usually time. Now $\boldsymbol{X}^{(m)}(t)$ is the outcome of the $m$th $t$-dependent experiment, and the probability density and ensemble averages become functions of $t$.

Given some function $f(\boldsymbol{X})$, the *expected value* or *ensemble average* of $f$ is the average over independent realizations of the experiment and is given by

$$\langle f(\boldsymbol{X})\rangle = \lim_{N_s\to\infty} \frac{1}{N_s} \sum_{m=1}^{N_s} f(\boldsymbol{X}^{(m)}). \tag{A.75}$$

Equivalently,

$$\langle f(\boldsymbol{X})\rangle = \int_\Omega f(\boldsymbol{X}) p(\boldsymbol{X})\, d\Omega(\boldsymbol{X}). \tag{A.76}$$

Sometimes a slight modification of the preceding formula is desirable. For example, if $\Omega = \mathbb{R}^2$ it may be convenient to represent $\boldsymbol{X}$ in either Cartesian coordinates $(x, y)$ or polar coordinates $(r, \theta)$. In the former case, it is natural to take $d\Omega = dx\, dy$. In the latter case, one could choose either $d\Omega = r\, d\theta\, dr$ or $d\Omega = d\theta\, dr$. Here the definitions of the probability distribution functions will differ by a factor of $r$ as illustrated in Section 7.5.1 and Problem 7.7. In this book, we will always choose the former definition. Thus, in spherical coordinates in $\mathbb{R}^3$, we take $d\Omega = r^2 \sin\theta\, dr\, d\phi\, d\theta$, where $\theta$ is the polar angle. Unless we are specifically considering cases in spherical or cylindrical coordinates, we will often simplify notation by writing $d\Omega(\boldsymbol{X})$ as $d\boldsymbol{X}$.

The *mean* $\mu$ and *variance* $\sigma^2$ of a scalar valued random variable $X$ are given, respectively, by

$$\mu = \langle X\rangle = \lim_{N_s\to\infty} \frac{1}{N_s} \sum_{m=1}^{N_s} X^{(m)} = \int_\Omega X p(X)\, dX, \tag{A.77}$$

$$\sigma^2 = \left\langle (X - \langle X\rangle)^2 \right\rangle = \lim_{N_s\to\infty} \frac{1}{N_s} \sum_{m=1}^{N_s} \left(X^{(m)} - \langle X\rangle\right)^2 = \int_\Omega (X - \langle X\rangle)^2 p(X)\, dX. \tag{A.78}$$

The variance is sometimes denoted $\mathrm{var}(X)$. For a process $Y = \alpha X$, these expressions show that that the mean and variance become $\alpha\mu$ and $\alpha^2\sigma^2$, respectively. The $k$th *central moment* of $p(X)$ is $\left\langle (X - \mu)^k \right\rangle$.

If we have two random variables $\boldsymbol{X}$ and $\boldsymbol{Y}$ in sets $\Omega_X$ and $\Omega_Y$, respectively, then we define the *joint probability density* $p(\boldsymbol{X}, \boldsymbol{Y})$, where $p(\boldsymbol{X}, \boldsymbol{Y})\, d\Omega_X, d\Omega_Y$ is the probability of $\boldsymbol{X}$ and $\boldsymbol{Y}$ taking on values within the differential element $d\Omega_X\, d\Omega_Y$. Ensemble averages have the form

$$\langle f(\boldsymbol{X},\boldsymbol{Y})\rangle = \lim_{N_s\to\infty}\frac{1}{N_s}\sum_{m=1}^{N_s} f\left(\boldsymbol{X}^{(m)},\boldsymbol{Y}^{(m)}\right) \tag{A.79}$$

or

$$\langle f(\boldsymbol{X},\boldsymbol{Y})\rangle = \int_{\Omega_X}\int_{\Omega_Y} f(\boldsymbol{X},\boldsymbol{Y})p(\boldsymbol{X},\boldsymbol{Y})\, d\Omega_X(\boldsymbol{X})d\Omega_Y(\boldsymbol{Y}). \tag{A.80}$$

If $\boldsymbol{X}^{(m)}$ is unaffected by the value of $\boldsymbol{Y}^{(m)}$ and vice versa for all realizations $m$, then $\boldsymbol{X}$ and $\boldsymbol{Y}$ are *independent* random variables and

$$p(\boldsymbol{X},\boldsymbol{Y}) = p_X(\boldsymbol{X})p_Y(\boldsymbol{Y}), \tag{A.81}$$

where $p_X$ and $p_Y$ are the probability densities for $X$ and $Y$, respectively.

The *covariance* $\mathrm{cov}(X,Y)$ of two scalar-valued random variables $X$ and $Y$ is given by

$$\mathrm{cov}(X,Y) = \langle (X-\langle X\rangle)(Y-\langle Y\rangle)\rangle\,. \tag{A.82}$$

For vector-valued random variables $\boldsymbol{X}$ and $\boldsymbol{Y}$, the covariance is a matrix $\mathbf{C}$ whose elements are given by

$$C_{ij} = \mathrm{cov}(X_i,Y_j) = \langle (X_i-\langle X_i\rangle)(Y_i-\langle Y_i\rangle)\rangle\,. \tag{A.83}$$

In some situations, we will be interested in quantities averaged over one random variable, but not another, for example, functions of position averaged over the velocity distribution. For example, for functions of random variables $\boldsymbol{X}$ and $\boldsymbol{Y}$, we can construct the *marginal density*

$$\psi(\boldsymbol{X}) = \int_{\Omega_Y} p(\boldsymbol{X},\boldsymbol{Y})\, d\Omega_Y. \tag{A.84}$$

This is also called a *contracted* distribution function. Now the ensemble average of a function $f$ that depends only on $\boldsymbol{X}$ can be written

$$\langle f(\boldsymbol{X})\rangle = \int_{\Omega_X} f(\boldsymbol{X})\psi(\boldsymbol{X})\, d\Omega_X. \tag{A.85}$$

### A.5.2 Sums of Independent Random Variables

Let $X_1$ and $X_2$ be independent scalar-valued random variables that can take on any real value; they are governed by probability density functions $p_{X_1}(X_1)$ and $p_{X_2}(X_2)$, respectively. Now let $Y = X_1 + X_2$; we would like to know the probability density for $Y$, given $p_{X_1}(X_1)$ and $p_{X_2}(X_2)$. The joint probability density for $X_1$ and $X_2$ is simply $p_{X_1}(X_1)p_{X_2}(X_2)$. Now let $X_2 = Y - X_1$ so the joint probability density for $X_1$ and $Y$ is $p_{X_1}(X_1)p_{X_2}(Y-X_1)$. The probability density $p_Y(Y)$ is then determined by integrating the joint density over all possible values of $X_1$:

$$p_Y(Y) = \int_{-\infty}^{\infty} p_{X_1}(X_1)p_{X_2}(Y-X_1)\, dX_1. \tag{A.86}$$

The right-hand side is a convolution integral; taking the Fourier transform in $Y$ yields

$$\hat{p}_Y(K) = \hat{p}_{X_1}(K)\hat{p}_{X_2}(K), \tag{A.87}$$

with $K$ as the Fourier-domain independent variable. This is a simple formula, showing that the transform domain is convenient for working with linear combinations of random variables.

This result is easily generalized. By definition,

$$\hat{p}_Y(K) = \left\langle e^{-iKY} \right\rangle.$$

If $Y = \alpha_1 X_1 + \alpha_2 X_2$, where $X_1$ and $X_2$ are independent and $\alpha_1$ and $\alpha_2$ are constants, then

$$\begin{aligned}
\hat{p}_Y(Y) &= \left\langle e^{-iK(\alpha_1 X_1 + \alpha_2 X_2)} \right\rangle \\
&= \int_{-\infty}^{\infty} \int_{-\infty}^{\infty} e^{-iK\alpha_1 X_1} e^{-iK\alpha_2 X_2} p_{X_1}(X_1) p_{X_2}(X_2)\, dX_1 dX_2 \\
&= \hat{p}_{X_1}(\alpha_1 K)\, \hat{p}_{X_2}(\alpha_2 K).
\end{aligned}$$

More broadly, if

$$Y = \sum_{j=1}^{N} \alpha_j X_j, \tag{A.88}$$

then

$$\hat{p}_Y(Y) = \prod_{j=1}^{N} \hat{p}_{X_j}\left(\alpha_j K\right). \tag{A.89}$$

The complex conjugate of $\hat{p}(K)$ is often used in the statistics literature, where it is called the *characteristic function* for $p(X)$.

### A.5.3 The Central Limit Theorem and Sums of Gaussian Random Variables

A *Gaussian* or *normal* distribution $p(X)$ with mean $\mu$ and variance $\sigma^2$ has the form

$$p(X) = \frac{1}{\sqrt{2\pi\sigma^2}} e^{-\frac{(x-\mu)^2}{2\sigma^2}}. \tag{A.90}$$

For a random variable $X$ with this distribution, it is common notation to write that

$$X \sim \mathcal{N}(\mu, \sigma^2).$$

The Fourier transform of a Gaussian is

$$\hat{p}(K) = e^{-iK\mu - \frac{1}{2}K^2\sigma^2}. \tag{A.91}$$

Gaussian distributions are observed widely in nature and are commonly used to model natural phenomena. The *central limit theorem* helps us understand why this is so.

Consider the case of $N$ independent, identically distributed random variables $X_i$, each with mean $\mu$ and variance $\sigma^2$. We will be interested in the probability density for the sum of these random variables in the limit $N \to \infty$.

In this situation,

$$p(X_1, X_2, \ldots) = \prod_{j=1}^{N} p_{X_j}(X_j) = \prod_{j=1}^{N} p_X(X_j). \tag{A.92}$$

The latter equation indicates that all the $X_j$ have the same probability density $p_X$. The sum $Y$ of these variables satisfies

$$\langle Y\rangle = \left\langle \sum_{j=1}^{N} X_j \right\rangle = \sum_{j=1}^{N} \left\langle X_j \right\rangle = N\mu,$$

$$\mathrm{var}(Y) = \sum_{j=1}^{N} \mathrm{cov}(X_j, X_j) = N\sigma^2.$$

Now we define new random variables $\chi_j$ and $Z$ and their corresponding densities $P_Z(Z)$ and $P_\chi(\chi_j)$, which are normalized to have zero mean and unit variance:

$$\chi_j = \frac{1}{\sigma}(X_j - \mu), \tag{A.93}$$

$$Z = \frac{1}{\sigma\sqrt{N}}(Y - N\mu) = \sum_{j=1}^{N} \frac{1}{\sqrt{N}}\chi_j. \tag{A.94}$$

(The variables $\chi_j$ are independent and identically distributed because the $X_j$ are.)

With these preliminaries, we now derive the main result. Using (A.89) and (A.94) yields that

$$\hat{p}_Z(Z) = \prod_{j=1}^{N} \hat{p}_X\left(\frac{K}{\sqrt{N}}\right) = \hat{p}_X\left(\frac{K}{\sqrt{N}}\right)^N. \tag{A.95}$$

When $N$ is large, the argument $K/\sqrt{N}$ becomes small for any finite $K$, motivating the expansion of $\hat{p}_\chi$ in a Taylor series:

$$\begin{aligned}\hat{p}_X\left(\frac{K}{\sqrt{N}}\right) &= \left\langle e^{-iK\chi/\sqrt{N}} \right\rangle \\ &= 1 - i\frac{K}{\sqrt{N}}\langle\chi\rangle - \frac{1}{2}\left(\frac{K}{\sqrt{N}}\right)^2 \left\langle\chi^2\right\rangle + O\left(\left\langle\left|\chi^3\right|\right\rangle \left|\frac{K}{\sqrt{N}}\right|^3\right) \\ &= 1 - \frac{1}{2}\frac{K^2}{N} + O\left(\left|\frac{K}{\sqrt{N}}\right|^3\right).\end{aligned} \tag{A.96}$$

Here we applied the facts that $\langle\chi\rangle = 0$ and $\left\langle\chi^2\right\rangle = 1$, used the Taylor series remainder result

$$\left|e^{ix} - \left(1 + ix - \frac{x^2}{2}\right)\right| \leq \frac{\left|x^3\right|}{3!},$$

and assumed that $\left\langle\left|\chi^3\right|\right\rangle$ is finite. Inserting (A.96) into (A.95) and neglecting the last term yields

$$\hat{p}_Z(Z) = \left(1 - \frac{1}{2}\frac{K^2}{N}\right)^N.$$

In the limit $N \to \infty$, this becomes

$$\hat{p}_Z(Z) = e^{-K^2/2}. \tag{A.97}$$

The inverse Fourier transform of this is

$$p_Z(Z) = \frac{1}{\sqrt{2\pi}} e^{-Z^2/2}. \tag{A.98}$$

This result establishes the central limit theorem: $p_Z(Z)$ is a *Gaussian* or *normal* distribution with zero mean and unit variance. Similarly, when $N \gg 1$, the sum $Y$ is well approximated as

$$p_Y(Y) \sim \mathcal{N}(N\mu, N\sigma^2). \tag{A.99}$$

Central limit theorems have also been established under more general conditions than considered here; see for example Graham & Rawlings (2013) for further discussion.

Finally, consider a linear combination of independent Gaussian random variables:

$$Y = \sum_{j=1}^{N} \alpha_j X_j,$$

where $X_j \sim \mathcal{N}(\mu_j, \sigma_j^2)$. From (A.89) and (A.91), we have that

$$\begin{aligned}
\hat{p}_Y(Y) &= \prod_{j=1}^{N} \hat{p}_{X_j}\left(\alpha_j K\right) \\
&= \prod_{j=1}^{N} \exp\left(-i\alpha_j K \mu_j - \frac{1}{2}\alpha_j^2 K^2 \sigma_j^2\right) \\
&= \exp\left(-iK \sum_{j=1}^{N} \alpha_j \mu_j - \frac{1}{2} K^2 \sum_{j=1}^{N} \alpha_j^2 \sigma_j^2\right).
\end{aligned} \tag{A.100}$$

Letting

$$\mu_Y = \sum_{j=1}^{N} \alpha_j \mu_j, \tag{A.101}$$

$$\sigma_Y^2 = \sum_{j=1}^{N} \alpha_j^2 \sigma_j^2, \tag{A.102}$$

(A.100) becomes

$$\hat{p}_Y(Y) = \exp\left(-iK\mu_Y - \frac{1}{2}K^2\sigma_Y^2\right) \tag{A.103}$$

or

$$Y \sim \mathcal{N}(\mu_Y, \sigma_Y^2). \tag{A.104}$$

A linear combination of Gaussians is Gaussian.

### A.5.4 Time Correlation Functions and the Wiener–Khinchin Theorem

Consider a stochastic process $\boldsymbol{X}(t)$, which might be, for example, the solution to the Langevin equation encountered in Chapter 6. If this stochastic process is *stationary*, then all statistics involving $\boldsymbol{X}(t)$ are identical to those involving $\boldsymbol{X}(t')$:

$$\langle f(\boldsymbol{X}(t))\rangle = \langle f(\boldsymbol{X}(t'))\rangle \quad \forall t, t' \tag{A.105}$$

for arbitrary functions $f(\boldsymbol{X})$. Equivalently, if $\boldsymbol{X}(t)$ is stationary, then its probability distribution, called the *stationary distribution*, is time independent. Additionally, a stationary process $\boldsymbol{X}(t)$ is said to be *ergodic* if any particular realization of $\boldsymbol{X}(t)$ will sample the entire stationary probability distribution function if given enough time. All stochastic processes addressed in this book are ergodic.[1] For ergodic processes, ensemble averages are equivalent to time averages:

$$\langle f(\boldsymbol{X})\rangle = \lim_{T\to\infty} \frac{1}{2T}\int_{-T}^{T} f(\boldsymbol{X}(t))\, dt. \tag{A.106}$$

Now consider two scalar-valued stationary ergodic stochastic processes $u(t)$ and $v(t)$. The *time correlation function* $C_{uv}(\tau)$ is defined as

$$C_{uv}(\tau) = \langle u(t)v(t+\tau)\rangle \tag{A.107}$$

$$= \lim_{T\to\infty} \frac{1}{2T}\int_{-T}^{T} u(t)v(t+\tau)\, dt. \tag{A.108}$$

If $v = u$, then this is the *autocorrelation function* of $u$.

Fourier transforms provide a convenient tool for working with stochastic processes, so we will now prove an important result regarding the Fourier transform of $C_{uv}(\tau)$. First, we define the "clipped" process $u_T(t)$ as

$$u_T(t) = \begin{cases} u(t) & -T \le t \le T, \\ 0 & \text{otherwise} \end{cases} \tag{A.109}$$

and likewise for $v_T(t)$. Note that any experimental or computational realization of a stochastic process has finite duration and thus corresponds to a clipped process. More important for our present purposes is that, using the clipped process, we can write that

$$C_{uv}(\tau) = \lim_{T\to\infty} \frac{1}{2T}\int_{-\infty}^{\infty} u_T(t)v_T(t+\tau)\, dt. \tag{A.110}$$

The limits of $\pm\infty$ on the integration make this expression more amenable than (A.108) to treatment with Fourier transforms. Indeed, now we write $C_{uv}(\tau)$ in terms of the Fourier

[1] An example of a nonergodic stochastic process is diffusion of a particle subject to a potential with infinitely high energy barriers. In this case, a trajectory starting on one side of a barrier can never sample portions of state space on the other side.

transforms of $u_T(t)$ and $v_T(t)$:

$$\begin{aligned} C_{uv}(\tau) &= \lim_{T\to\infty} \frac{1}{2T}\frac{1}{(2\pi)^2}\int_{-\infty}^{\infty}\int_{-\infty}^{\infty}\int_{-\infty}^{\infty} \hat{u}_T(\omega)\hat{v}_T(\omega')e^{i\omega t}e^{i\omega'(t+\tau)}\,d\omega\,d\omega'\,dt \\ &= \lim_{T\to\infty} \frac{1}{2T}\frac{1}{(2\pi)^2}\int_{-\infty}^{\infty}\int_{-\infty}^{\infty}\int_{-\infty}^{\infty} \hat{u}_T(\omega)\hat{v}_T(\omega')e^{i(\omega+\omega')t}e^{i\omega'\tau}\,d\omega\,d\omega'\,dt. \end{aligned}$$

Applying the frequency-shift result

$$\int_{-\infty}^{\infty} e^{i(\omega+\omega')t}\,dt = 2\pi\delta(\omega+\omega')$$

yields that

$$\begin{aligned} C_{uv}(\tau) &= \lim_{T\to\infty} \frac{1}{2T}\frac{1}{2\pi}\int_{-\infty}^{\infty} \hat{u}_T(-\omega')\hat{v}_T(\omega')e^{i\omega'\tau}\,d\omega' \\ &= \lim_{T\to\infty} \frac{1}{2T}F^{-1}\{\hat{u}_T(-\omega')\hat{v}_T(\omega')\}. \end{aligned}$$

Finally, dropping the $'$ and taking the Fourier transform of both sides, we have that

$$F\{C_{uv}(\tau)\} = \hat{C}_{uv}(\omega) = \langle \hat{u}(-\omega)\hat{v}(\omega)\rangle. \tag{A.111}$$

This result is sometimes called the *cross-correlation theorem*. For the important special case $v = u$, this becomes

$$F\{C_{uu}(\tau)\} = \hat{C}_{uu}(\omega) = \langle \hat{u}(-\omega)\hat{u}(\omega)\rangle, \tag{A.112}$$

a result known as the *Wiener–Khinchin theorem* (Reichl 1998). The *power spectral density* or simply *power spectrum* of $u(t)$ is given by

$$S_u(\omega) = \left\langle |\hat{u}(\omega)|^2 \right\rangle. \tag{A.113}$$

If $u(t)$ is real-valued, then $\hat{u}(-\omega) = \bar{\hat{u}}(\omega)$, where $\bar{\ }$ denotes complex conjugate, so

$$|\hat{u}(\omega)|^2 = \hat{u}(\omega)\bar{\hat{u}}(\omega) = \hat{u}(\omega)\hat{u}(-\omega).$$

Therefore, we can write the Wiener–Khinchin theorem for a real-valued process as

$$\hat{C}_{uu}(\omega) = S_u(\omega). \tag{A.114}$$

# References


Alder, B. J. & Wainwright, T. E. (1970), "Decay of the Velocity Autocorrelation Function," *Phys. Rev. A* **1**(1), 18–21.

Allen, M. P. & Tildesley, D. J. (1987), *Computer Simulation of Liquids*, Oxford University Press, Oxford.

Alvarez, A. & Soto, R. (2005), "Dynamics of a Suspension Confined in a Thin Cell," *Phys. Fluids* **17**(9), 093103.

Ando, T., Chow, E., Saad, Y. & Skolnick, J. (2012), "Krylov Subspace Methods for Computing Hydrodynamic Interactions in Brownian Dynamics Simulations," *J. Chem. Phys.* **137**(6), 064106.

Anekal, S. & Bevan, M. (2005), "Interpretation of Conservative Forces from Stokesian Dynamic Simulations of Interfacial and Confined Colloids," *J. Chem. Phys.* **122**(3), 034903.

Aris, R. (1989), *Vectors, Tensors and the Basic Equations of Fluid Mechanics*, Dover, New York.

Balboa Usabiaga, F., Delmotte, B. & Donev, A. (2017), "Brownian Dynamics of Confined Suspensions of Active Microrollers," *J. Chem. Phys.* **146**(13), 134104.

Balducci, A., Mao, P., Han, J. & Doyle, P. S. (2006), "Double-Stranded DNA Diffusion in Slitlike Nanochannels," *Macromolecules* **39**(18), 6273–6281.

Ball, R. C. & Richmond, P. (1980), "Dynamics of Colloidal Dispersions," *Physics and Chemistry of Liquids* **9**(2), 99–116.

Banchio, A. & Brady, J. (2003), "Accelerated Stokesian Dynamics: Brownian Motion," *J. Chem. Phys.* **118**(22), 10323–10332.

Barthes-Biesel, D. & Rallison, J. M. (1981), 'The Time-Dependent Deformation of a Capsule Freely Suspended in a Linear Shear-Flow', *J. Fluid Mech.* **113**, 251–267.

Batchelor, G. K. (1967), *An Introduction to Fluid Dynamics*, Cambridge University Press, Cambridge.

Batchelor, G. K. (1970*a*), "Slender-Body Theory for Particles of Arbitrary Cross-Section in Stokes Flow," *J. Fluid Mech.* **44**, 419–440.

Batchelor, G. K. (1970*b*), "The Stress System in a Suspension of Force-Free Particles," *J. Fluid Mech.* **41**(3), 545–570.

Batchelor, G. K. & Green, J. T. (1972), "Determination of Bulk Stress in a Suspension of Spherical-Particles to Order C-2," *J. Fluid Mech.* **56**, 401–427.

Beams, N. N., Olson, L. N. & Freund, J. B. (2016), "A Finite Element Based $P^3M$ Method for $N$-Body Problems," *SIAM J. Sci. Comput.* **38**(3), A1538–A1560.

Berg, H. C. (1993), *Random Walks in Biology*, Princeton University Press, Princeton.

Bergeron, V., Bonn, D., Martin, J. & Vovelle, L. (2000), "Controlling Droplet Deposition with Polymer Additives," *Nature* **405**(6788), 772–775.

Berne, B. J. & Pecora, R. (1976), *Dynamic Light Scattering*, Wiley-Interscience, New York.

Bird, R. B., Armstrong, R. C. & Hassager, O. (1987), *Dynamics of Polymeric Liquids*, Vol. 1 of *Fluid Mechanics*, 2 edn, Wiley-Interscience, New York.

Bird, R. B., Curtiss, C. F., Armstrong, R. C. & Hassager, O. (1987), *Dynamics of Polymeric Liquids*, Vol. 2 of *Kinetic Theory*, 2 edn, Wiley-Interscience, New York.

Bird, R. B., Stewart, W. E. & Lightfoot, E. N. (2002), *Transport Phenomena*, 2nd edn, Wiley, New York.

Blake, J. R. (1971*a*), "Note on the Image System for a Stokeslet in a No-Slip Boundary," *Proc. Camb. Philos. S.-M* **70**, 303–310.

Blake, J. R. (1971*b*), "Spherical Envelope Approach to Ciliary Propulsion," *J. Fluid Mech.* **46**(01), 199–208.

Blawzdziewicz, J. (2007), "Boundary Integral Methods for Stokes Flows," in A. Prosperetti & G. Tryggvason, eds, *Computational Methods for Multiphase Flow*, Cambridge University Press, Cambridge.

Bocquet, L. (2004), "High Friction Limit of the Kramers Equation: The Multiple Time-Scale Approach," *Am. J. Phys.* **65**(2), 140–144.

Brady, J. F. & Bossis, G. (1988), "Stokesian Dynamics," *Annu. Rev. Fluid Mech.* **20**, 111–157.

Chaikin, P. M. & Lubensky, T. C. (1995), *Principles of Condensed Matter Physics*, Cambridge University Press, Cambridge.

Chan, P. & Leal, L. G. (1979), "The Motion of a Deformable Drop in a Second-Order Fluid," *J. Fluid Mech.* **92**, 131–170.

Chwang, A. T. & Wu, T. Y.-T. (1975), "Hydromechanics of Low-Reynolds-Number Flow. 2. Singularity Method for Stokes Flows," *J. Fluid Mech.* **67**, 787–815.

Cortez, R., Fauci, L. & Medovikov, A. (2005), "The Method of Regularized Stokeslets in Three Dimensions: Analysis, Validation, and Application to Helical Swimming," *Phys. Fluids* **17**(3), 031504.

Cui, B., Diamant, H., Lin, B. & Rice, S. (2004), "Anomalous Hydrodynamic Interaction in a Quasi-Two-Dimensional Suspension," *Phys. Rev. Lett.* **92**(25), 258301.

Darnton, N. C., Turner, L., Rojevsky, S. & Berg, H. C. (2007), "On Torque and Tumbling in Swimming Escherichia Coli," *J. Bact.* **189**(5), 1756–1764.

de Gennes, P. G. (1979), *Scaling Concepts in Polymer Physics*, Cornell University Press, Ithaca, NY.

Deen, W. M. (2012), *Analysis of Transport Phenomena*, 2nd edn, Oxford University Press, Oxford.

Dhont, J. K. G. & Briels, W. J. (2003), "Viscoelasticity of Suspensions of Long, Rigid Rods," *Colloid Surface A* **213**, 131–156.

Di Carlo, D. (2009), "Inertial Microfluidics," *Lab Chip* **9**(21), 3038.

Doi, M. & Edwards, S. F. (1986), *The Theory of Polymer Dynamics*, Oxford University Press, New York.

Doyle, P. S., Shaqfeh, E. & Gast, A. P. (1997), "Dynamic Simulation of Freely Draining Flexible Polymers in Steady Linear Flows," *J. Fluid Mech.* **334**, 251–291.

Drazin, P. G. & Reid, W. H. (1981), *Hydrodynamic Stability*, Cambridge Monographs on Mechanics and Applied Mathematics, Cambridge University Press, Cambridge.

Dufresne, E., Squires, T., Brenner, M. & Grier, D. (2000), "Hydrodynamic Coupling of Two Brownian Spheres to a Planar Surface," *Phys. Rev. Lett.* **85**(15), 3317–3320.

Durlofsky, L. & Brady, J. F. (1987), "Analysis of the Brinkman Equation as a Model for Flow in Porous Media," *Phys. Fluids* **30**(11), 3329–3341.

Durlofsky, L., Brady, J. F. & Bossis, G. (1987), "Dynamic Simulation of Hydrodynamically Interacting Particles," *J. Fluid Mech.* **180**, 21–49.

Einstein, A. (1998), *Einstein's Miraculous Year: Five Papers That Changed the Face of Physics*, Princeton University Press, Princeton.

Ermak, D. L. & McCammon, J. A. (1978), "Brownian Dynamics with Hydrodynamic Interactions," *J. Chem. Phys.* **69**(4), 1352–1360.

Fiore, A. M., Balboa Usabiaga, F., Donev, A. & Swan, J. W. (2017), "Rapid Sampling of Stochastic Displacements in Brownian Dynamics Simulations," *J. Chem. Phys.* **146**(12), 124116.

Fixman, M. (1978), "Simulation of Polymer Dynamics. 1. General Theory," *J. Chem. Phys.* **69**(4), 1527–1537.

Fixman, M. (1986), "Construction of Langevin Forces in the Simulation of Hydrodynamic Interaction," *Macromolecules* **19**(4), 1204–1207.

Fox, R. F. & Uhlenbeck, G. E. (1970), "Contributions to Non-Equilibrium Thermodynamics. I. Theory of Hydrodynamical Fluctuations," *Phys. Fluids* **13**, 1893–1901.

Franosch, T., Grimm, M., Belushkin, M., Mor, F. M., Foffi, G., Forró, L. & Jeney, S. (2011), "Resonances Arising from Hydrodynamic Memory in Brownian Motion," *Nature* **478**(7367), 85–88.

Fuller, G. G. & Leal, L. G. (1981), "The Effects of Conformation-Dependent Friction and Internal Viscosity on the Dynamics of the Nonlinear Dumbbell Model for a Dilute Polymer Solution," *J. Non-Newton. Fluid Mech.* **8**(3), 271–310.

Gardiner, C. W. (1985), *Handbook of Stochastic Methods*, Springer, Berlin.

Gasquet, C. & Witomski, P. (1998), *Fourier Analysis and Applications: Filtering, Numerical Computation, Wavelets*, Springer, New York.

Gimbutas, Z., Greengard, L. & Veerapaneni, S. (2015), "Simple and Efficient Representations for the Fundamental Solutions of Stokes Flow in a Half-Space," *J. Fluid Mech.* **776**, R1–10.

Goldstein, H. (1980), *Classical Mechanics*, 2nd edn, Addison Wesley, Reading, MA.

Gonzalez, O. & Stuart, A. M. (2008), *A First Course in Continuum Mechanics*, Cambridge University Press, Cambridge.

Götz, T. (2000), Interactions of Fibers and Flow: Asymptotics, Theory and Numerics, PhD thesis, Universität Kaiserslautern.

Graham, M. D. (2003), "Interfacial Hoop Stress and Instability of Viscoelastic Free Surface Flows," *Phys. Fluids* **15**(6), 1702–1710.

Graham, M. D. (2011), "Fluid Dynamics of Dissolved Polymer Molecules in Confined Geometries," *Annu. Rev. Fluid Mech.* **43**(1), 273–298.

Graham, M. D. (2014), "Drag Reduction and the Dynamics of Turbulence in Simple and Complex Fluids," *Phys. Fluids* **26**, 101301.

Graham, M. D. & Rawlings, J. B. (2013), *Modeling and Analysis Principles for Chemical and Biological Engineers*, Nob Hill Publishing, Madison.

Gray, J. & Hancock, G. J. (1955), "The Propulsion of Sea-Urchin Spermatozoa," *J. Experimental Biol.* **32**(4), 802–814.

Greenberg, M. D. (1978), *Foundations of Applied Mathematics*, Prentice Hall, Englewood Cliffs.

Guazzelli, E. & Morris, J. F. (2012), *A Physical Introduction to Suspension Dynamics*, Cambridge Texts in Applied Mathematics, Cambridge University Press, Cambridge.

Guyon, E., Hulin, J.-P., Petit, L. & Mitescu, C. D. (2001), *Physical Hydrodynamics*, Oxford University Press, Oxford.

Happel, J. & Brenner, H. (1965), *Low Reynolds Number Hydrodynamics*, Prentice Hall, Englewood Cliffs.

Harden, J. L. & Doi, M. (1992), "Diffusion of Macromolecules in Narrow Capillaries," *J. Phys. Chem.* **96**(10), 4046–4052.

Hasimoto, H. (1959), "On the Periodic Fundamental Solutions of the Stokes Equations and Their Application to Viscous Flow Past a Cubic Array of Spheres," *J. Fluid Mech.* **5**(2), 317–328.

Hernández-Ortiz, J. P., de Pablo, J. J. & Graham, M. D. (2007), "Fast Computation of Many-Particle Hydrodynamic and Electrostatic Interactions in a Confined Geometry," *Phys. Rev. Lett.* **98**(14), 140602.

Hernández-Ortiz, J. P., Ma, H., de Pablo, J. J. & Graham, M. D. (2006), "Cross-Stream-line Migration in Confined Flowing Polymer Solutions: Theory and Simulation," *Phys. Fluids* **18**(12), 123101.

Hernández-Ortiz, J. P., Ma, H., de Pablo, J. J. & Graham, M. D. (2008), "Concentration Distributions during Flow of Confined Flowing Polymer Solutions at Finite Concentration: Slit and Grooved Channel," *Korea-Aust. Rheol. J.* **20**(3), 143–152.

Hiemenz, P. C. & Rajagopalan, R. (1997), *Principles of Colloid and Surface Chemistry*, 3rd edn, Marcel Dekker, New York.

Hinch, E. J. (1975), "Application of the Langevin Equation to Fluid Suspensions," *J. Fluid Mech.* **72**(03), 499–511.

Hinch, E. J. (1977), "An Averaged-Equation Approach to Particle Interactions in a Fluid Suspension," *J. Fluid Mech.* **83**(04), 695–720.

Hinch, E. J. & Leal, L. G. (1972), "The Effect of Brownian Motion on the Rheological Properties of a Suspension of Non-Spherical Particles," *J. Fluid Mech.* **52**(04), 683–712.

Ho, B. P. & Leal, L. G. (1974), "Inertial Migration of Rigid Spheres in 2-Dimensional Unidirectional Flows," *J. Fluid Mech.* **65**, 365–400.

Howells, I. D. (2006), "Drag Due to the Motion of a Newtonian Fluid through a Sparse Random Array of Small Fixed Rigid Objects," *J. Fluid Mech.* **64**(03), 449.

Indei, T., Schieber, J. D., Córdoba, A. & Pilyugina, E. (2012), "Treating Inertia in Passive Microbead Rheology," *Phys. Rev. E* **85**(2), 021504.

Jeffery, G. B. (1922), "The Motion of Ellipsoidal Particles Immersed in a Viscous Fluid," *Proc. Roy. Soc. London A* **102**, 161–179.

Jendrejack, R., de Pablo, J. & Graham, M. D. (2002), "Stochastic Simulations of DNA in Flow: Dynamics and the Effects of Hydrodynamic Interactions," *J. Chem. Phys.* **116**(17), 7752–7759.

Jendrejack, R., Graham, M. D. & de Pablo, J. (2000), "Hydrodynamic Interactions in Long Chain Polymers: Application of the Chebyshev Polynomial Approximation in Stochastic Simulations," *J. Chem. Phys.* **113**(7), 2894–2900.

Jendrejack, R. M., Schwartz, D. C., de Pablo, J. J. & Graham, M. D. (2004), "Shear-Induced Migration in Flowing Polymer Solutions: Simulation of Long-Chain Deoxyribose Nucleic Acid in Microchannels," *J. Chem. Phys.* **120**(5), 2513–2529.

Jendrejack, R., Schwartz, D., Graham, M. D. & de Pablo, J. (2003), "Effect of Confinement on DNA Dynamics in Microfluidic Devices," *J. Chem. Phys.* **119**(2), 1165–1173.

Johnson, D. W. (1975), "A Fourier Series Method for Numerical Kramers–Kronig Analysis," *J. Phys. A-Math. Gen.* **8**(4), 490–495.

Johnson, R. E. (1980), "An Improved Slender-Body Theory for Stokes Flow," *J. Fluid Mech.* **99**(2), 411–431.

Joo, Y. L. & Shaqfeh, E. S. G. (1992), "A Purely Elastic Instability in Dean and Taylor–Dean Flow," *Phys. Fluids A* **4**(3), 524–543.

Kawaguchi, S., Imai, G., Suzuki, J., Miyahara, A. & Kitano, T. (1997), "Aqueous Solution Properties of Oligo- and Poly(ethylene Oxide) by Static Light Scattering and Intrinsic Viscosity," *Polymer* **38**(12), 2885–2891.

Kaye, A., Stepko, R. F. T., Work, W. J., Aleman, J. V. & Malkin, A. Y. (1998), "Definition of Terms Relating to the Non-Ultimate Mechanical Properties of Polymers," *Pure and Appl. Chem.* **70**(3), 701–754.

Keller, J. B. & Rubinow, S. I. (1976), "Slender-Body Theory for Slow Viscous-Flow," *J. Fluid Mech.* **75**, 705–714.

Kim, S. & Karrila, S. J. (1991), *Microhydrodynamics: Principles and Selected Applications*, Butterworth-Heinemann, Boston.

Kim, S., Ong, P. K., Yalcin, O., Intaglietta, M. & Johnson, P. C. (2009), "The Cell-Free Layer in Microvascular Blood Flow," *Biorheology* **46**(3), 181–189.

Kim, S. & Russel, W. B. (1985), "Modelling of Porous Media by Renormalization of the Stokes Equations," *J. Fluid Mech.* **154**, 269–286.

Kloeden, P. E. & Platen, E. (1992), *Numerical Solution of Stochastic Differential Equations*, Springer, Berlin.

Kubo, R. (1966), "The Fluctuation–Dissipation Theorem," *Reports on Progress in Physics* **29**(1), 255–284.

Kubo, R., Toda, M. & Hashitsume, N. (1991), *Statistical Physics II: Nonequilibrium Statistical Mechanics*, Vol. 31 of *Springer Series in Solid State Sciences*, 2nd edn, Springer-Verlag, Berlin.

Kumar, A. & Graham, M. D. (2012), "Accelerated Boundary Integral Method for Multiphase Flow in Non-Periodic Geometries,' *J. Comput. Phys.* **231**(20), 6682–6713.

Kumar, S. & Larson, R. G. (2001), "Brownian Dynamics Simulations of Flexible Polymers with Spring–Spring Repulsions," *J. Chem. Phys.* **114**(15), 6937–6941.

Ladyzhenskaya, O. A. (1969), *The Mathematical Theory of Viscous Incompressible Flow*, revised 2nd edn, Gordon and Breach, New York.

Lamb, H. (1932), *Hydrodynamics*, 6th edn, Cambridge University Press, Cambridge.

Landau, L. D. & Lifschitz, E. M. (1959), *Fluid Mechanics*, Pergamon, London.

Landau, L. D. & Lifschitz, E. M. (1980), *Statistical Physics*, Pergamon, New York.

Landau, L. D. & Lifschitz, E. M. (1984), *Fluid Mechanics*, Vol. 6 of *Course of Theoretical Physics*, 2nd edn, Pergamon, Oxford.

Larson, R. G. (1988), *Constitutive Equations for Polymer Melts and Solutions*, Butterworths, Boston.

Larson, R. G. (1999), *The Structure and Rheology of Complex Fluids*, Oxford University Press, New York.

Larson, R. G., Shaqfeh, E. S. G. & Muller, S. J. (1990), "A Purely Elastic Instability in Taylor–Couette Flow," *J. Fluid Mech.* **218**, 573–600.

Lasota, A. & Mackey, M. C. (1994), *Chaos, Fractals and Noise: Stochastic Aspects of Dynamics*, 2nd edn, Springer, New York.

Lauga, E. & Powers, T. R. (2009), "The Hydrodynamics of Swimming Microorganisms," *Reports on Progress in Physics* **72**(9), 096601.

Lax, M. (1966), "Classical Noise IV: Langevin Methods," *Rev. Mod. Phys.* **38**(3), 541–566.

Leal, L. G. (2007), *Advanced Transport Phenomena: Fluid Mechanics and Convective Transport Processes*, Cambridge University Press, Cambridge.

Lesnicki, D., Vuilleumier, R., Carof, A. & Rotenberg, B. (2016), "Molecular Hydrodynamics from Memory Kernels," *Phys. Rev. Lett.* **116**(14), 147804–5.

Lighthill, M. J. (1952), "On the Squirming Motion of Nearly Spherical Deformable Bodies through Liquids at Very Small Reynolds Numbers," *Comm. Pure Appl. Math.* **5**, 109–118.

Liron, N. & Mochon, S. (1976), "Stokes Flow for a Stokes-let between 2 Parallel Flat Plates," *J. Eng. Math.* **10**(4), 287–303.

Liu, B. & Dünweg, B. (2003), "Translational Diffusion of Polymer Chains with Excluded Volume and Hydrodynamic Interactions by Brownian Dynamics Simulation," *J. Chem. Phys.* **118**(17), 8061–8072.

Lovalenti, P. M. & Brady, J. F. (1993), "The Hydrodynamic Force on a Rigid Particle Undergoing Arbitrary Time-Dependent Motion at Small Reynolds Number," *J. Fluid Mech.* **256**, 561–605.

Ma, H. & Graham, M. D. (2005), "Theory of Shear-Induced Migration in Dilute Polymer Solutions Near Solid Boundaries," *Phys. Fluids* **17**(8), 083103.

Malvern, L. E. (1969), *Introduction to the Mechanics of a Continuous Medium*, Prentice Hall, Englewood Cliffs.

Marko, J. F. & Siggia, E. D. (1994), "Bending and Twisting Elasticity of DNA," *Macromolecules* **27**(4), 981–988.

Marko, J. F. & Siggia, E. D. (1995), "Stretching DNA," *Macromolecules* **28**(26), 8759–8770.

Mason, T. & Weitz, D. (1995), "Optical Measurements of Frequency-Dependent Linear Viscoelastic Moduli of Complex Fluids," *Phys. Rev. Lett.* **74**(7), 1250–1253.

McQuarrie, D. A. (2000), *Statistical Mechanics*, University Science Books, Sausalito.

Metsi, E. (2000), Large Scale Simulations of Bidisperse Emulsions and Foams, PhD thesis, University of Illinois at Urbana–Champaign.

Misiunas, K., Pagliara, S., Lauga, E., Lister, J. R. & Keyser, U. F. (2015), "Nondecaying Hydrodynamic Interactions along Narrow Channels," *Phys. Rev. Lett.* **115**(3), 038301.

Morozov, A. N. & Spagnolie, S. E. (2015), "Introduction to Complex Fluids," *in* S. E. Spagnolie, ed., *Complex Fluids in Biological Systems*, Springer, New York.

Mucha, P., Tee, S., Weitz, D., Shraiman, B. & Brenner, M. (2004), "A Model for Velocity Fluctuations in Sedimentation," *J. Fluid Mech.* **501**, 71–104.

Nguyen, H. N. & Cortez, R. (2014), "Reduction of the Regularization Error of the Method of Regularized Stokeslets for a Rigid Object Immersed in a Three-Dimensional Stokes Flow," *Commun. Comput. Phys.* **15**(1), 126–152.

Oldroyd, J. G. (1950), "On the Formulation of Rheological Equations of State," *Proc. Roy. Soc. A* **200**(1063), 523–541.

Onishi, Y. & Jeffrey, D. J. (1984), "Calculation of the Resistance and Mobility Functions for 2 Unequal Rigid Spheres in Low-Reynolds-Number Flow," *J. Fluid Mech.* **139**, 261–290.

Öttinger, H. C. (1988), "A Note on Rigid Dumbbell Solutions at High Shear Rates," *J. Rheol.* **32**(2), 135–143.

Öttinger, H. C. (1996), *Stochastic Processes in Polymeric Fluids*, Springer, Berlin.

Perrin, J. (1916), *Atoms*, 4th edn., Constable & Company Ltd., London.

Phillips, R. J., Brady, J. F. & Bossis, G. (1988), "Hydrodynamic Transport Properties of Hard-Sphere Dispersions. I. Suspensions of Freely Mobile Particles," *Phys. Fluids* **31**(12), 3462.

Phillips, R., Kondev, J. & Theriot, J. A. (2009), *Physical Biology of the Cell*, Garland Science Publishers, New York.

Poole, R. J. (2012), "The Deborah and Weissenberg Numbers," *British Society of Rheology, Rheology Bulletin*, **53**(2), 32–39.

Pozrikidis, C. (1992), *Boundary Integral and Singularity Methods for Linearized Viscous Flow*, Cambridge University Press, Cambridge.

Pozrikidis, C. (1997), *Theoretical and Computational Fluid Dynamics*, Oxford University Press, Oxford.

Prosperetti, A. & Tryggvason, G., eds (2007), *Computational Methods for Multphase Flow*, Cambridge University Press, Cambridge.

Proudman, I. & Pearson, J. (1957), "Expansions at Small Reynolds Numbers for the Flow Past a Sphere and a Circular Cylinder," *J. Fluid Mech.* **2**, 237–262.

Purcell, E. M. (1977), "Life at Low Reynolds Number," *Am. J. Phys.* **45**(1), 3–11.

Reichl, L. E. (1998), *A Modern Course in Statistical Physics*, 2nd edn, Wiley-Interscience, New York.

Robertson, H. S. (1993), *Statistical Thermophysics*, PTR Prentice Hall, Englewood Cliffs.

Rotne, J. & Prager, S. (1969), "Variational Treatment of Hydrodynamic Interaction in Polymers," *J. Chem. Phys.* **50**(11), 4831–4837.

Rubinstein, M. & Colby, R. H. (2003), *Polymer Physics*, Oxford University Press, Oxford.

Russel, W. B., Saville, D. A. & Schowalter, W. R. (1989), *Colloidal Dispersions*, Cambridge University Press, Cambridge.

Saintillan, D., Darve, E. & Shaqfeh, E. S. G. (2005), "A Smooth Particle-Mesh Ewald Algorithm for Stokes Suspension Simulations: The Sedimentation of Fibers," *Phys. Fluids* **17**(3), 033301.

Schieber, J. D., Córdoba, A. & Indei, T. (2013), "The Analytic Solution of Stokes for Time-Dependent Creeping Flow around a Sphere: Application to Linear Viscoelasticity as an Ingredient for the Generalized Stokes–Einstein Relation and Microrheology Analysis," *J. Non-Newton. Fluid Mech.* **200**(C), 3–8.

Schmidt, J. & Skinner, J. (2003), "Hydrodynamic Boundary Conditions, the Stokes–Einstein Law, and Long-Time Tails in the Brownian Limit," *J. Chem. Phys.* **119**(15), 8062–8068.

Schmidt, J. & Skinner, J. (2004), "Brownian Motion of a Rough Sphere and the Stokes–Einstein Law," *J. Phys. Chem. B* **108**(21), 6767–6771.

Schuss, Z. (2010), *Theory and Applications of Stochastic Processes: An Analytical Approach*, Vol. 170 of *Applied Mathematical Sciences*, Springer, New York.

Segel, L. A. (1987), *Mathematics Applied to Continuum Mechanics*, Dover, New York.

Segre, G. & Silberberg, A. (1962), "Behaviour of Macroscopic Rigid Spheres in Poiseuille Flow Part 2. Experimental Results and Interpretation," *J. Fluid Mech.* **14**(1), 136–157.

Shaqfeh, E. S. G. (1996), "Purely Elastic Instabilities in Viscometric Flows," *Annu. Rev. Fluid Mech.* **28**, 129–185.

Shaqfeh, E. S. G. (2005), "The Dynamics of Single-Molecule DNA in Flow," *J. Non-Newton. Fluid Mech.* **130**, 1–28.

Sierou, A. & Brady, J. F. (2001), "Accelerated Stokesian Dynamics Simulations," *J. Fluid Mech.* **448**, 115–146.

Smart, J. R. & Leighton, D. T. (1991), "Measurement of the Drift of a Droplet Due to the Presence of a Plane," *Phys Fluids A* **3**(1), 21–28.

Smith, D. E., Perkins, T. T. & Chu, S. (1996), "Dynamical Scaling of DNA Diffusion Coefficients," *Macromolecules* **29**(4), 1372–1373.

Smith, S. B., Cui, Y. J. & Bustamante, C. (1996), "Overstretching B-DNA: The Elastic Response of Individual Double-Stranded and Single-Stranded DNA Molecules," *Science* **271**(5250), 795–799.

Snook, I. (2007), *The Langevin and Generalised Langevin Approach to the Dynamics of Atomic, Polymeric and Colloidal Systems*, Elsevier Science, Amsterdam.

Squires, T. M. & Mason, T. G. (2010), "Fluid Mechanics of Microrheology," *Annu. Rev. Fluid Mech.* **42**(1), 413–438.

Stakgold, I. (1998), *Green's Functions and Boundary Value Problems*, 2nd edn, Wiley-Interscience, New York.

Stoltz, C., de Pablo, J. & Graham, M. D. (2006), "Concentration Dependence of Shear and Extensional Rheology of Polymer Solutions: Brownian Dynamics Simulations," *J. Rheol.* **50**(2), 137–167.

Stone, H. A. & Samuel, A. (1996), "Propulsion of Microorganisms by Surface Distortions," *Phys. Rev. Lett.* **77**(19), 4102–4104.

Strobl, G. (1996), *The Physics of Polymers*, Springer, Berlin.

Taylor, G. I. (1922), "Diffusion by Continuous Movements," *Proceedings of the London Mathematical Society* **s2-20**(1), 196–212.

Tlusty, T. (2006), "Screening by Symmetry of Long-Range Hydrodynamic Interactions of Polymers Confined in Sheets," *Macromolecules* **39**(11), 3927–3930.

Tornberg, A. & Shelley, M. (2004), "Simulating the Dynamics and Interactions of Flexible Fibers in Stokes Flows," *J. Comput. Phys.* **196**(1), 8–40.

von Kármán, T. (1934), "Some Aspects of the Turbulence Problem," *in Collected Works of Theodore von Kármán*, Vol. III (1955), Butterworths Scientific Publications, London, pp. 120–155.

White, C. M. & Mungal, M. G. (2008), "Mechanics and Prediction of Turbulent Drag Reduction with Polymer Additives," *Annu. Rev. Fluid Mech.* **40**, 235–256.

Wiener, N. (1976), *Norbert Wiener: Collected Works with Commentaries*, Vol. 1 of *Mathematicians of Our Time*, MIT Press, Cambridge.

Xu, K., Forest, M. & Klapper, I. (2007), "On the Correspondence between Creeping Flows of Viscous and Viscoelastic Fluids," *J. Non-Newton. Fluid Mech.* **145**(2-3), 150–172.

Yamakawa, H. (1970), "Transport Properties of Polymer Chains in Dilute Solution: Hydrodynamic Interaction," *J. Chem. Phys.* **53**(1), 436–443.

Zhang, Y., de Pablo, J. J. & Graham, M. D. (2012), "An Immersed Boundary Method for Brownian Dynamics Simulation of Polymers in Complex Geometries: Application to DNA Flowing through a Nanoslit with Embedded Nanopits," *J. Chem. Phys.* **136**(1), 014901–014901.

Zhu, L., Rorai, C., Mitra, D. & Brandt, L. (2014), "A Microfluidic Device to Sort Capsules by Deformability: A Numerical Study," *Soft Matter* **10**(39), 7705–7711.

Zwanzig, R. (2001), *Nonequilibrium Statistical Mechanics*, Oxford University Press, Oxford.

Zwanzig, R. & Bixon, M. (1970), "Hydrodynamic Theory of the Velocity Correlation Function," *Phys. Rev. A* **2**(5), 2005–2012.

# Index

Printed in the United States
by Baker & Taylor Publisher Services